International Review of

# Cytology

A Survey of

# Cell Biology

## VOLUME 165

# International Review of Cytology

# A Survey of Cell Biology

Edited by

**Kwang W. Jeon**
Department of Zoology
University of Tennessee
Knoxville, Tennessee

**VOLUME 165**

**ACADEMIC PRESS**
San Diego New York Boston London Sydney Tokyo Toronto

This book is printed on acid-free paper. 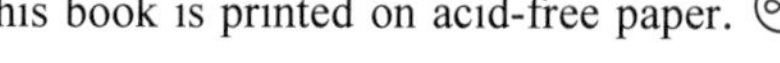



Academic Press, Inc.
A Division of Harcourt Brace & Company
525 B Street, Suite 1900, San Diego, California 92101-4495

*United Kingdom Edition published by*
Academic Press Limited
24-28 Oval Road, London NW1 7DX

International Standard Serial Number: 0074-7696

International Standard Book Number: 0-12-364569-7

PRINTED IN THE UNITED STATES OF AMERICA
96 97 98 99 00 01 EB 9 8 7 6 5 4 3 2 1

# CONTENTS

## Ependymins: Meningeal-Derived Extracellular Matrix Proteins at the Blood–Brain Barrier

Werner Hoffmann and Heinz Schwarz



## Cadherin Cell Adhesion Molecules in Differentiation and Embryogenesis

James A. Marrs and W. James Nelson



## Escape and Migration of Nucleic Acids between Chloroplasts, Mitochondria, and the Nucleus

Peter E. Thorsness and Eric R. Weber

## Cytoplasmic Mechanisms of Axonal and Dendritic Growth in Neurons

Steven R. Heidemann

# CONTRIBUTORS

Numbers in parentheses indicate the pages on which the authors' contributions begin.

Steven R. Heidemann (235), *Department of Physiology and Microbiology, Michigan State University, East Lansing, Michigan 48824*

Werner Hoffmann (121), *Institut für Molekularbiologie und Medizinische Chemie, Otto-von-Guericke Universität, D-39127 Magdeburg, Germany*

James A. Marrs (159), *Departments of Medicine, Physiology, and Biophysics, Indiana University Medical Center, Indianapolis, Indiana 46202*

W. James Nelson (159), *Department of Molecular and Cellular Physiology, Stanford University School of Medicine, Beckman Center for Molecular and Genetic Medicine, Stanford, California 94305*

Heinz Schwarz (121), *Max-Planck-Institut für Entwicklungsbiologie, D-72076 Tübingen, Germany*

Ramón Serrano (1), *Instituto de Biologia Molecular y Celular de Plantas, Universidad Politécnica de Valencia-C.S.I.C., 46022 Valencia, Spain*

Peter E. Thorsness (207), *Department of Molecular Biology, University of Wyoming, Laramie, Wyoming 82071*

Eric R. Weber (207), *Department of Molecular Biology, University of Wyoming, Laramie, Wyoming 82071*

Eugenia V. Zybina (53), *Laboratory of Cell Morphology, Institute of Cytology of the Russian Academy of Sciences, 194064 St. Petersburg, Russia*

Tatiana G. Zybina (53), *Laboratory of Cell Pathology, Institute of Cytology of the Russian Academy of Sciences, 194064 St. Petersburg, Russia*

# Salt Tolerance in Plants and Microorganisms: Toxicity Targets and Defense Responses

Ramón Serrano
Instituto de Biologia Molecular y Celular de Plantas, Universidad Politécnica de Valencia–CSIC, 46022 Valencia, Spain

Salt tolerance of crops could be improved by genetic engineering if basic questions on mechanisms of salt toxicity and defense responses could be solved at the molecular level. Mutant plants accumulating proline and transgenic plants engineered to accumulate mannitol or fructans exhibit improved salt tolerance. A target of salt toxicity has been identified in *Saccharomyces cerevisiae:* it is a sodium-sensitive nucleotidase involved in sulfate activation and encoded by the *HAL2* gene. The major sodium-extrusion system of *S. cerevisiae* is a P-ATPase encoded by the *ENA1* gene. The regulatory system of *ENA1* expression includes the protein phosphatase calcineurin and the product of the *HAL3* gene. In *Escherichia coli,* the $Na^+$–$H^+$ antiporter encoded by the *nhaA* gene is essential for salt tolerance. No sodium transport system has been identified at the molecular level in plants. Ion transport at the vacuole is of crucial importance for salt accumulation in this compartment, a conspicuous feature of halophytic plants. The primary sensors of osmotic stress have been identified only in *E. coli.* In *S. cerevisiae,* a protein kinase cascade (the HOG pathway) mediates the osmotic induction of many, but not all, stress-responsive genes. In plants, the hormone abscisic acid mediates many stress responses and both a protein phosphatase and a transcription factor (encoded by the *ABI1* and *ABI3* genes, respectively) participate in its action.

**KEY WORDS:** Osmolytes, Sodium transport, Potassium transport, Stress, Signal transduction, *HAL* genes.

## I. Introduction

The progressive salinization of irrigated land compromises the future of agriculture in the most productive areas of our planet (Ashraf, 1994; Serrano and Gaxiola, 1994). Historical records demonstrate that salinity is not a new problem; a clear example is the decline of Sumerian civilization in Mesopotamia (Jacobson and Adams, 1958). Arid regions offer optimal light and temperature for most crops, but insufficient precipitation causes extensive reliance on irrigation. Water supplies always contain some dissolved salt, which on evaporation becomes more and more concentrated in the soil. This cumulative effect over decades or centuries (depending on salt concentration, rain, and soil permeability) results in deterioration of soil structure and incompatibility with plant life.

Classic genetic improvement by breeding requires no basic knowledge of the mechanisms of the phenomena under consideration. It only needs close relatives of crop plants containing the desired traits, which are then transferred by crosses. However, in the case of modern genetic engineering a basic understanding of the phenomena is essential for the manipulation. Relevant genes need to be selected for expression in transgenic plants, and this is the beauty of modern biotechnology—that it is inexorably linked to the molecular basis of the phenomena, to basic science. The reason crop plants are so sensitive to salinity is the key issue we need to address before genetic engineering can be applied to the salt tolerance problem.

With the exception of a few halophytic species, which tolerate seawater and which probably correspond to secondary adaptations, most plants cannot tolerate salinity higher than about 10–20% seawater (about 50–100 m*M* NaCl) (Downton, 1984). This is remarkable, because most animal cells live with (and even require) an external salt concentration of approximately 150 m*M* (physiological saline). The reason for such differences can be traced back to the different chemiosmotic mechanisms operating in plant and animal plasma membranes (Serrano, 1990). In plant (and fungal) cells the $H^+$ gradient generated by the plasma membrane $H^+$-ATPase is the driving force for secondary active transports mediated by $H^+$ cotransports. In animal cells, the $Na^+$ gradient generated by the $Na^+$,$K^+$-ATPase drives $Na^+$-cotransport systems. There is a correlation among plants in the existence of a primary $H^+$ pump, tolerance to acidity, and sensitivity to salinity. Also, the existence of a primary $Na^+$ pump in animal cells correlates with tolerance to salinity and sensitivity to low pH. In evolutionary terms, typical animal cells likely evolved in seawater while typical plant cells evolved in freshwater. As discussed by Wilson and Lin (1980), primitive cells, devoid of rigid cell walls, probably relied on active sodium efflux for volume regulation. In this way, the active extrusion of salt counteracted colloidal osmotic swelling of the cells. According to this view, proton pumps evolved from sodium pumps when the development of a rigid cell wall provided volume regulation. The primordial sodium pump could be found on present marine Protozoa and algae, but there is little information about the plasma membrane ATPases of these organisms.

Salinity has two possible toxic effects on cells: the nonspecific osmotic effect and the specific toxicity of the ions on defined cellular systems. Confronted with this threat, organisms develop defenses based on osmotic adjustment, ion transport systems, and stress-defense proteins (Serrano, 1994). The major problem for the understanding of salt toxicity is the paucity of information about molecular mechanisms and about the physiological relevance of different targets and defense responses. Most of the data collected in salt tolerance research (as in the case of other stresses) are correlative and cause-and-effect relationships have rarely been addressed (Nguyen and Joshi, 1994). This type of research has been described as "phenomenological" and a more "functional" genetic approach, which addresses the crucial reactions of salt tolerance in a more direct way, has been advocated (Serrano and Gaxiola, 1994).

The genetic analysis of complex phenomena such as salt tolerance requires the isolation of mutants that have either decreased or increased salt tolerance. Genes responsible for the mutations need to be isolated, while the encoded proteins identify crucial reactions in salt tolerance. This mutational analysis is possible only in a few model experimental systems where mutants can be easily selected and the affected genes isolated. Crop plants do not belong to these convenient systems. The molecular basis of salt tolerance needs to be investigated in model organisms and information needs to be transferred to crop plants. The correlative information collected with plants at the physiological and molecular levels can then be combined with results of mutational analyses to obtain a complete picture and to design biotechnological approaches to improve salt tolerance in transgenic plants.

This kind of analysis has been started with the yeast *Saccharomyces cerevisiae,* one of the most powerful model systems for the molecular biology of eukaryotic cells. The suitability of yeast as a model for plants was suggested because fungi and plants have similar ion transport systems at their plasma membranes (Serrano and Gaxiola, 1994). The classic bacterium *Escherichia coli* has also been investigated extensively in terms of salt stress (Csonka and Hanson, 1991). Although its prokaryotic nature distances *E. coli* from plants, this model system may also illuminate aspects of salt tolerance, as is discussed. The small plant *Arabidopsis thaliana* is reaching the status of a model system in plant genetics (Meyerowitz, 1989), although isolation of mutated genes still offers considerable difficulties. Some salt-tolerant *Arabidopsis* mutants are discussed. Finally, plant tissue culture, because of "somaclonal variation," is an important source of mutants and some lines with increased salt tolerance have already been obtained (Dix, 1993).

The developing picture of salt tolerance is far from complete, but some important guidelines are already apparent. This article attempts to combine results from microorganisms and plants in order to elucidate general principles of salt tolerance. Concomitant with a description of the biology of salt stress at the physiological and molecular levels, the insights provided by mutants are highlighted, and the most promising biotechnological approaches are indicated.

## II. Toxicity and Responses at the Cellular Level

### A. Osmotic versus Specific Ion Toxicities

In the closely related bacteria *Salmonella typhimurium* and *E. coli* the major component of salt toxicity is the osmotic effect of NaCl. This is demonstrated by the fact that (1) isoosmotic concentrations of sucrose and NaCl have similar inhibitory effects (Csonka, 1981); (2) salt tolerance increases when osmotic adaptation is improved by either uptake of external proline or increased metabolic production of this osmolyte in mutant strains (Csonka, 1981; Csonka and Hanson, 1991); and (3) mutant strains with increased sodium transport capability exhibit no improvement in salt tolerance (Carmel *et al.,* 1994). Therefore, under normal conditions osmotic adjustment limits salt tolerance in *E. coli.* This does not mean that sodium toxicity is not relevant, because mutants with decreased sodium extrusion capability are salt sensitive (Schuldiner and Padan, 1992). Rather, in wild-type *E. coli* the capability for osmotic adjustment is relatively less potent than the capability for sodium extrusion. An active $Na^+/H^+$-antiporter system maintains the intracellular sodium level at a sufficiently low level to avoid toxic effects on intracellular systems (Schuldiner and Padan, 1992).

In the yeast *S. cerevisiae* growing on glucose media, NaCl is much more toxic than equivalent osmotic concentrations of KCl or sorbitol (Gaxiola *et al.,* 1992; Gläser *et al.,* 1993). Therefore, sodium toxicity is the major problem raised by salinity. The situation is different, however, on media with other carbon sources, such as galactose. In this case NaCl, KCl, and sorbitol have similar toxicities (G. Rios, unpublished observations). Apparently, with glucose as the carbon source osmotic adjustment is more active than sodium extrusion whereas in nonglucose media the opposite is true. The high rate of glucose metabolism in *S. cerevisiae* results in optimal synthesis of the osmolyte glycerol. Other yeast and fungi, however, have a much slower rate of glucose metabolism (Gancedo and Serrano, 1989), so that osmotic adjustment by conversion of glucose to glycerol limits salt tolerance even in glucose media (Beever and Laracy, 1986; R. Ali, M. J. Garcia, and R. Serrano, unpublished observations).

The situation with plants is far from clear. Some studies (Munns and Termaat, 1986) indicate that specific NaCl toxicity—not osmotic stress—is responsible for the inhibitory effects of salt on wheat seedlings. These results were based on a comparison of 0.1 *M* NaCl with an equivalent osmotic concentration (−0.48 MPa) obtained by using a six-fold concentrated nutrient solution. The first, transient stoppage of shoot growth after salt stress is clearly due to reduced water supply to the shoot, caused by the nonspecific osmotic effect of the salt (Yeo *et al.,* 1991). In this case equivalent osmotic concentrations of NaCl and mannitol had similar inhibitory effects. After several hours osmotic adjustment occurs, water supply is restored, and shoot growth resumes but at a reduced rate. The longer-term (days and weeks) effects of salt are not due to water deficit because turgor of salt-stressed plants is similar or higher than that of controls (Munns, 1993). During the recovery from the first osmotic phase mannitol is more toxic than NaCl because it cannot be absorbed by the plant and does not contribute to the osmotic adjustment. The second, nonosmotic phase of shoot growth inhibition can be explained by hormonal signals from the root to the shoot (Munns, 1993). The root senses the soil environment, and water stress causes a decrease in cytokinin and an increase in abscisic acid transported from the root to the shoot (Lerner and Amzallag, 1994). Therefore, it is difficult to evaluate the role of ion toxicities in whole plants. An important observation made on the basis of X-ray microanalysis is that, as predicted by Oertli some time ago (Oertli, 1968), the NaCl concentration in the leaf apoplast is much higher than in the soil solution (Flowers *et al.,* 1991). Whether the eventual cause of salt toxicity in leaves is osmotic dehydration or cytoplasmic ion toxicities was not clarified.

The primary mechanism of salt toxicity in plants remains obscure. In some plants, such as *Citrus sinensis* (Bañuls and Primo-Millo, 1992) and soybeans (Abel, 1969), chloride is the toxic component of the salt because NaCl and KCl have similar toxicities and $NaNO_3$ is less toxic. In tomato (Rush and Epstein, 1981), wheat (Gorham *et al.,* 1990), and *A. thaliana* (Sheahan *et al.,* 1993) NaCl is more toxic than KCl, pointing to sodium toxicity. Studies on tobacco cells in suspension culture indicate that equivalent osmotic concentrations of sorbitol and mannitol are less inhibitory than NaCl. However, KCl is even more toxic than NaCl (LaRosa *et al.,* 1985). The same is observed in many halophytic plants (Rush and Epstein, 1981), and the toxicity of KCl may be due to greater chloride uptake from KCl than from NaCl.

A reasonable conclusion that can be obtained from this analysis is that the capabilities for osmotic adjustment and ion transport in different organisms are closely balanced, although one or another of these processes may be more active in specific organisms under certain conditions. Therefore, in order to improve salt tolerance of crop plants, both osmotic adjustment and ion transport will probably need to be manipulated simultaneously.

An important difference between microorganisms and plants is that in the former chloride has relatively little effect, whereas in many plants it is toxic.

Other mechanisms of salt toxicity need to be considered, such as disturbed mineral nutrition (Laüchli *et al.,* 1994). Salt stress induces potassium and calcium deficiencies and this effect is partially corrected by high calcium concentrations in the medium (10 m*M*). In plants, high external calcium restores not only calcium, but potassium as well, and also reduces sodium uptake. These effects are mediated by the complex interactions of these cations with their transport systems (Epstein *et al.,* 1963; Epstein and Rains, 1965; Rains and Epstein, 1967). Calcium also reduces chloride uptake in *Citrus* plants (Bañuls and Primo-Millo, 1992).

## B. Osmolyte Synthesis and Transport

A striking example of convergent evolution is provided by the existence in most organisms of *osmolytes,* or organic solutes, which accumulate during osmotic stress and which are compatible with protein and membrane structure (Yancey *et al.,* 1982; Brown, 1990). The accumulation of osmolytes allows osmotic adjustment without the accumulation of high salt concentrations toxic to proteins and membranes (see Section C). The major osmolytes in microorganisms and plants are polyols (glycerol, sorbitol, and mannitol), nonreducing sugars (sucrose and trehalose), and amino acids (glutamate, proline, and betaine). Some animals utilize urea and trimethylamine-*N*-oxide.

These compounds are not only compatible with protein and membrane structure but they also stabilize these structures when challenged with adverse conditions such as dehydration, high temperatures, or denaturing chemicals. Therefore, osmolyte synthesis may be triggered not only by osmotic stress but also by other stresses such as heat shock. Protein stabilization by osmolytes is based mostly on their exclusion from the protein surface (Timasheff and Arakawa, 1989). As the native, folded state of proteins has less exposed surface than the unfolded form, this exclusion stabilizes the native state. Many osmolytes are excluded from the surface layer of proteins because they interact more with water molecules. In technical terms, they increase the cohesive force of water and, hence, its surface tension and are excluded from the water–protein interface. Glycerol has a special mechanism (the so-called solvophobic effect) because it has affinity for polar regions of proteins but interacts unfavorably with hydrophobic regions. Therefore, glycerol favors native states, in which hydrophobic regions are buried.

Membrane stabilization by osmolytes is partially due to protection of membrane proteins but, in addition, the phospholipid bilayer is also stabilized (Crowe *et al.,* 1992). Dehydration of phospholipid bilayers during osmotic stress generates a rigid gel state of the lipid phase. Phase transition on rehydration produces transmembrane leakage. Osmolytes interact with the polar head groups of the phospholipids and maintain their liquid crystalline state during dehydration.

Osmolytes are important both in metabolism and stress response, and the synthesis and degradation of these compounds are subject to multiple regulation, reflecting the dual role of these compounds as carbon and nitrogen stores and as compatible solutes. The metabolic synthesis of these compounds is usually accompanied by their active transport into the cells. This allows the uptake of osmolytes present in the external medium or leaked out from the cells.

## 1. *Escherichia coli*

The bacterial model *E. coli* synthesizes two osmolytes in response to osmotic stress: glutamate and trehalose (Csonka and Hanson, 1991). Glutamate accumulation makes only a small contribution to osmotic adjustment and its physiological role has not been demonstrated. It seems to be synthesized when glutamate dehydrogenase is activated by the high intracellular ionic strength occurring during osmotic stress (Csonka and Hanson, 1991). However, mutants in the trehalose biosynthetic pathway are osmotically sensitive in minimal medium, pointing to the crucial role of this osmolyte in salt tolerance (Giaever *et al.,* 1988). The trehalose pathway diverges from glycolysis at the level of hexose phosphates (Fig. 1). There is evidence for a futile

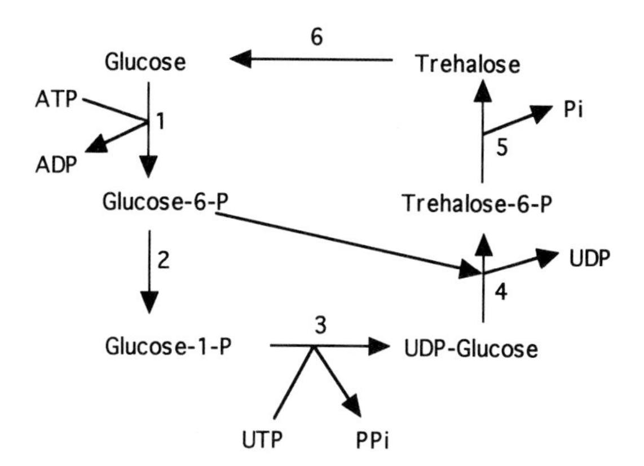


FIG. 1 The trehalose metabolic cycle. The enzymatic steps are as follows: 1, eukaryotic hexokinase (EC 2.7.1.1) or bacterial phosphoenolpyruvate-dependent phosphotransferase systems; 2, phosphoglucomutase (EC 5.4.2.2); 3, glucose-1-P uridylyltransferase (EC 2.7.7.9); 4, trehalose-6-P synthase (UDP forming, EC 2.4.1.15); 5, trehalose-6-P phosphatase (EC 3.1.3.12); 6, trehalase (EC 3.2.1.28).

cycle involving production and degradation of trehalose in osmotically stressed *E. coli* (Styrvold and Ström, 1991). Synthesis of one trehalose molecule from two glucoses consumes three ATP high-energy bonds and, therefore, recycling into glucose is an expensive pathway. This cycle probably exists because it allows a fine tuning of trehalose levels during osmotic adaptation. The genes encoding trehalose-6-phosphate (trehalose-6-P) synthase (*otsA*) and trehalose-6-P phosphatase (*otsB*) constitute an operon induced by osmotic stress [the *ots* operon (osmoregulated trehalose synthesis)] (Kaasen *et al.,* 1991). This operon is also induced during stationary phase and contributes to the thermotolerance developed by *E. coli* under these conditions (Hengge-Aronis *et al.,* 1991).

Osmolytes that can be taken from the external medium, instead of internally synthesized, have been called *osmoprotectants* (LeRedulier *et al.,* 1984). In *E. coli* betaine and proline fulfill this role and the active accumulation of these compounds is induced by osmotic stress. The *proU* operon encodes a high-affinity transporter for betaine with some affinity for proline. The operon consists of three structural genes, *proV, proW,* and *proX,* corresponding to the subunits of an ATP-driven, periplasmic binding protein-dependent trnasporter of the ABC family (Higgins, 1992). Increased *proU* gene dosage results in enhanced osmotolerance in media supplemented with proline or betaine (Gowrishankar *et al.,* 1986). This transport system, therefore, is limiting for the improved osmotic adjustment obtained by accumulation of external osmolytes. External choline can be utilized as a source of betaine by a metabolic pathway encoded by the osmoregulated *bet* operon (Andresen *et al.,* 1988). It contains three genes, *betT, betA,* and *betB,* corresponding to the high-affinity choline transporter, membrane-bound choline oxidase, and soluble $NAD^+$-dependent betaine-aldehyde dehydrogenase (EC 1.2.1.8).

Clearly, the capability of *E. coli* for osmoinduced synthesis of osmolytes is limited because uptake of external proline and betaine improve salt tolerance. Although proline biosynthesis is not stimulated by osmotic stress in this organism, a mutant resistant to the analog azetidine-2-carboxylate overproduces proline and exhibits enhanced salt tolerance (Csonka, 1981). This beneficial metabolic disturbance results from deregulation of the proline biosynthetic pathway. The first committed step of this pathway is catalyzed by glutamate 5-kinase (Fig. 2) and this enzyme is subject to feedback inhibition by proline. Different point mutations in the *proB* gene encoding this enzyme reduce its allosteric inhibition (Rushlow *et al.,* 1984; Smith, 1985; Csonka *et al.,* 1988) and cause proline overproduction.

### 2. *Saccharomyces cerevisiae*

Baker's yeast accumulates glycerol as the major osmolyte during osmotic stress, although trehalose is also involved in osmotic adjustment and stress

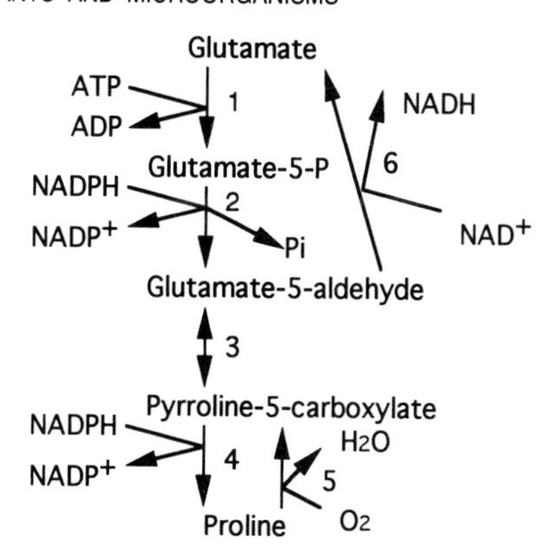


FIG. 2 The proline metabolic cycle. The enzymatic steps are as follows: 1, glutamate 5-kinase (EC 2.7.2.11); 2, glutamate-5-semialdehyde dehydrogenase (EC 1.2.1.41); 3, spontaneous cycle formation; 4, 1-pyrroline-5-carboxylate reductase (EC 1.5.1.2); 5, proline oxidase (probably EC 1.5.99.8); 6, 1-pyrroline-5-carboxylate dehydrogenase (EC 1.5.1.12).

protection (Blomberg and Adler, 1992). The metabolic cycle of trehalose synthesis and hydrolysis in yeast is similar to that in *E. coli* (Fig. 1). In actuality, the yeast genes encoding trehalose-6-P synthase (*TPS1*) (Bell *et al.*, 1992; Vuorio *et al.*, 1993) and trehalose-6-P phosphatase (*TPS2*) (De Virgilio *et al.*, 1993; Vuorio *et al.*, 1993) are highly homologous to the bacterial *otsA* and *otsB* genes, respectively (Kaasen *et al.*, 1994). The *TPS1* gene, as happens with many yeast genes, is fortuitously expressed in *E. coli* from its own promoter (Vuorio *et al.*, 1993) and it restores the osmotolerance and trehalose accumulation of an *E. coli otsA* mutant (McDougall *et al.*, 1993). In yeast the products of the *TPS1* and *TPS2* genes (polypeptides of 56 and 100 kDa, respectively) form a complex with another polypeptide of 123 kDa encoded by the *TSL1* gene (Vuorio *et al.*, 1993). A functional *TPS1* gene is required for expression of phosphatase activity encoded by the *TPS2* gene (Vuorio *et al.*, 1993), suggesting that complex formation is essential for trehalose-6-P phosphatase activity. However, a functional *TPS2* gene is not needed for expression of trehalose-6-P synthase activity by the *TPS1* gene (De Virgilio *et al.*, 1993) and, as previously indicated, the *TPS1* gene alone expresses synthase activity in *E. coli.*

In yeast trehalose synthesis is induced by heat stress, which increases the expression of the *TPS2* gene (De Virgilio *et al.*, 1993) and probably also of the *TPS1* gene. Disruption of either the *TPS1* or *TPS2* gene caused an inability to accumulate trehalose on mild heat shock or on initiation of stationary phase and significantly reduced the levels of heat-induced and stationary phase-induced thermotolerance (De Virgilio *et al.*, 1994). This constitutes genetic evidence indicating a role for trehalose as a thermoprotectant. No evidence suggesting trehalose plays a role in osmotic protection of yeast has been described.

The intracellular concentrations of trehalose (as most 0.4 *M*) are much lower than those of glycerol, which can reach more than 2 *M*. Therefore, glycerol accumulation is responsible for most of the osmotic adjustment of *S. cerevisiae* (Blomberg and Adler, 1992). This yeast continuously produces glycerol from glucose to dispose of reducing equivalents under fermentative conditions (Gancedo and Serrano, 1989). Synthesis of yeast biomass from sugars reduces $NAD^+$ to NADH and glycerol production regenerates $NAD^+$ under fermentative conditions. Most of this growth-associated glycerol leaks to the external medium. Increased glycerol accumulation during osmotic stress is due to both enhanced production and retention inside cells (Blomberg and Adler, 1989). Glycerol is also a carbon source for yeast and the metabolic cycle of synthesis and degradation is depicted in Fig. 3. *Sacchacomyces cerevisiae* seems to have only passive glycerol transport (Gancedo and Serrano, 1989), but other yeasts such as *Debaryomyces hansenii* (Lucas *et al.*, 1990), *Zygosaccharomyces rouxii* (van Zyl *et al.*, 1990), and *Pichia sorbitophila* (Lages and Lucas, 1995) have active glycerol transport, probably mediated by $H^+$ symport. A glycerol symport with either sodium or potassium has been postulated in *D. hansenii* (Lucas *et al.*, 1990).

The crucial enzyme in osmoregulated glycerol synthesis is glycerol-3-P dehydrogenase. The corresponding gene, *GPD1*, is induced by osmotic stress and inactivating mutations cause osmosensitivity (Larsson *et al.*, 1993; Albertyn *et al.*, 1994a,b).

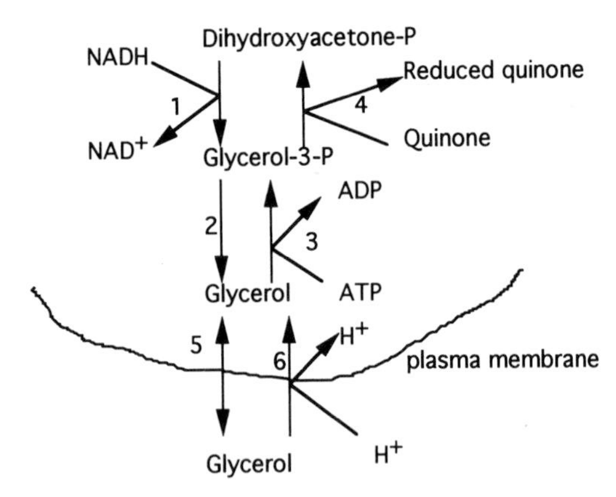


FIG. 3 The glycerol metabolic cycle. The enzymatic steps are as follows: 1, cytoplasmic glycerol-3-P dehydrogenase ($NAD^+$) (EC 1.1.1.8); 2, glycerol-3-P phosphatase (EC 3.1.3.21); 3, glycerol kinase (EC 2.7.1.30); 4, mitochondrial glycerol-3-P dehydrogenase (probably EC 1.1.99.5); 5, passive transport; 6, active transport.

Although glycerol is the predominant osmoresponsive polyol in fungi, other polyols such as erythrytol, arabitol, mannitol, and sorbitol may contribute to osmotic adjustment and stress protection in some fungi (Blomerg and Adler, 1992).

### 3. Higher Plants

The major osmolytes characterized in higher plants are sucrose, proline, and betaine. Although these compounds may be present in cytosol and vacuoles, cytoplasmic concentrations of proline and betaine are much higher than in the vacuole because organic and inorganic ions make the major contribution to osmotic concentrations inside the vacuole (Leigh *et al.*, 1981; McCue and Hanson, 1990). Sucrose may play an essential role in the tolerance to desiccation of pollen (Hoekstra and van Roekel, 1988) and in the osmotic adjustment of tobacco (Binzel *et al.*, 1987; Pilon-Smits *et al.*, 1995). A more general response is observed with proline, which appears to be the most widely distributed osmolyte accumulated under osmotic stress by plants (Delauney and Verma, 1993; Larher *et al.*, 1993). Proline in plants can be synthesized either from glutamate, by the same pyrroline-5-carboxylate pathway operating in bacteria (Fig. 2), or from ornithine. However, only the pyrroline-5-carboxylate pathway is activated by osmotic stress (Delauney and Verma, 1993). The stress hormone abscisic acid (see Section II,F,3) also induces proline accumulation (Finkelstein and Somerville, 1990). The first two steps in osmoregulated proline biosynthesis are catalyzed by a bifunctional enzyme, a polypeptide of 73 kDa that contains the activities of both glutamate 5-kinase and glutamate-5-semialdehyde dehydrogenase, corresponding to the products of the *proB* and *proA* genes of *E. coli* (Hu *et al.*, 1992). The unstable nature of the glutamate-5-P intermediate may favor a close interaction between the first two enzymes of proline biosynthesis. This bifunctional enzyme is called 1-pyrroline-5-carboxylate synthase and its expression is moderately induced by osmotic stress (Hu *et al.*, 1992). Another factor contributing to osmoinduced proline accumulation may be the release of feedback inhibition of glutamate 5-kinase by proline in stressed tissues (Boggess *et al.*, 1976). Although 1-pyrroline-5-carboxylate reductase is also induced by osmotic stress (Williamson and Slocum, 1992), proline concentration was not significantly increased in transgenic plants exhibiting high levels of this enzyme, indicating that reduction of pyrroline-5-carboxylate is not rate limiting in proline production (Szoke *et al.*, 1992).

Betaine is also accumulated by some plants under osmotic stress. Its biosynthesis from choline involves two enzymatic steps catalyzed by chloroplast choline monooxigenase and betaine-aldehyde dehydrogenase (EC

1.2.1.8) (McCue and Hanson, 1990). The latter is induced at the level of transcription by osmotic stress (Weretilnyk and Hanson, 1990).

Proline and betaine may also act as exogenous osmoprotectants in some plant tissues (Handa *et al.,* 1986; Lone *et al.,* 1987). Uptake may be stimulated by osmotic stress, with each organic solute inhibiting the endogenous synthesis of the other (Lone *et al.,* 1987). The evolution of osmolyte diversity within the family Plumbaginaceae has suggested that $\beta$-alanine betaine may be superior to glycine betaine under hypoxic saline conditions and proline betaine may be the most effective nitrogen-containing osmoprotectant (Hanson *et al.,* 1994).

Mannitol and sorbitol are present in some plant species as phloem-translocated carbohydrates together with sucrose (Loescher, 1987). Reduction of glucose 6-P by aldose-6-P reductase (NADPH) (EC 1.1.1.200) produces sorbitol 6-P, which is hydrolyzed by sorbitol-6-phosphatase (EC 3.1.3.50). In the same way, reduction of mannose 6-P by mannose-6-P 6-reductase (NADPH) (EC 1.1.1.224) produces mannitol 1-P, which is hydrolyzed by mannitol-1-phosphatase (EC 3.1.3.22). The utilization of these polyols involves sorbitol dehydrogenase ($NAD^+$) (EC 1.1.1.14) or mannitol 2-dehydrogenase ($NAD^+$) (EC 1.1.1.67), either of which causes fructose to enter the glycolytic pathway. Evidence has been presented for increased activity of mannose-6-P 6-reductase and increased mannitol accumulation during salinity stress of celery (*Apium graveolens;* Everard *et al.,* 1994). In the halophyte *Mesembryanthenum crystallinum* the mono-methylated cyclitol pinitol accumulates to high levels during osmotic stress and the *S*-adenosylmethionine-dependent methyltransferase involved in its biosynthesis is induced at the level of transcription (Vernon and Bohnert, 1992). In this respect, a gene induced during acquired desiccation tolerance of barley embryos encodes a protein with homology to aldose reductases (Bartels *et al.,* 1991), although the precise reaction catalyzed under physiological conditions has not been determined.

The physiological role of osmolyte synthesis in plants has been demonstrated by the improved osmostress tolerance of mutant plants with enhanced synthesis of osmolytes. These mutants were obtained by either chemical mutagenesis, somaclonal variation, or genetic engineering. Barley mutants selected for resistance to hydroxyproline accumulated proline and its growth was less inhibited by salt stress (Kueh and Bright, 1982). In an alternative approach, salt-resistant somaclones of *Nicotiana plumbaginifolia* (Sumaryati *et al.,* 1992) and *Brassica juncea* (Jain *et al.,* 1991; Kirti *et al.,* 1991) accumulated more proline than parental plants. The genetic nature of the mutation was demonstrated by regeneration of the somaclones and transmission of the salt tolerance to the next generation. In the case of the *Nicotiana* mutant, genetic analysis demonstrated a single nuclear dominant gene (Sumaryati *et al.,* 1992). It would be interesting to investigate if these mutations correspond to the first rate-limiting enzyme of proline biosynthesis, glutamate 5-kinase (see above).

Two different tobacco mutants generated by genetic engineering also established a correlation between osmolyte synthesis and salt tolerance. In one case, the *E. coli* gene encoding mannitol-1-P 5-dehydrogenase ($NAD^+$) (EC 1.1.1.17), which produces fructose 6-P in the bacteria during mannitol utilization, was expressed in tobacco and the plants accumulated moderate levels of mannitol (Tarczynski *et al.,* 1992). This implies that the enzyme was working inside the plant cells in the direction of mannitol 1-P synthesis and that some undefined phosphatase was generating mannitol. Although the levels of mannitol were relatively low (on the same order as sucrose, 1–4 $\mu$mol/g fresh weight), these transgenic plants were protected from salt stress as compared with controls (Tarczynski *et al.,* 1993). Even more surprising is the effect of the *SacB* gene of *Bacillus subtilis,* encoding levansucrase (sucrose:2,6-$\beta$-D-fructan 6-$\beta$-D-fructosyltransferase; EC 2.4.1.10) on tobacco plants. The transgenic plants accumulated low levels of fructan during osmotic stress (up to 0.3 mg/g fresh weight versus 0.9 mg/g fresh weight for sucrose) and both root and shoot growth were significantly protected during polyethylene glycol (PEG) 10,000-mediated drought stress (Pilon-Smits *et al.,* 1995). The significance for osmotic adjustment and protein and membrane protection of the small amounts of osmolytes generated in these transgenic plants is difficult to explain. Either mannitol and fructan accumulate in particularly sensitive tissues or these osmolytes have regulatory effects independent of their role as compatible solutes. It seems that transgenic plants producing trehalose are the next target for various plant biotechnology companies (Kidd and Devorak, 1994).

### C. Salt-Sensitive Cellular Functions

Low salt concentrations (50–150 m*M*) have positive effects on many enzymes because of general electrostatic effects on protein structure. This corresponds to the ionic strength of the cytoplasm of most organisms and it is logical that cytoplasmic enzymes have become adapted for optimal function under these conditions, where protein hydration is maximal and the electrical layer around the protein surface is suppressed (Wyn Jones and Pollard, 1983).

High salt concentrations (greater than 0.3–0.5 *M* NaCl) inhibit most enzymes because of the perturbation of the hydrophobic–electrostatic balance between the forces maintaining protein structure (Wyn Jones and Pollard, 1983). High ionic strength increases hydrophobic forces (salting-out effect) and decreases electrostatic forces (screening effect). High concentrations of some anions (relative effectiveness: $SCN^- > I^- > ClO_4^- >$

$NO_3^- > Cl^- > SO_4^{2-}$) and of some cations (relative effectiveness: $Ca^{2+} > Mg^{2+} > Li^+ > Na^+ > K^+ > NH_4^+$) break down water structure and decrease hydrophobic interactions (salting-in, lyotropic, or chaotropic effect) (Wyn Jones and Pollard, 1983). In salts such as $(NH_4)_2SO_4$ and NaCl, salting out due to ionic strength is the predominant effect.

Individual enzymes vary widely in their sensitivities to these salt effects, but it seems that many metabolic reactions would be strongly inhibited at salt concentrations greater than 0.3–0.5 *M* (Yancey *et al.*, 1982; Gimmler *et al.*, 1984). In addition, membrane functions are also affected by high salt concentrations, probably because of structurai changes in membrane proteins. Salt tolerance must involve salt exclusion from the cytoplasm, with osmotic adjustment effected by compatible organic solutes or "osmolytes" (see above).

A different situation is found among the halophytic archaebacteria (e.g., members of the Halobacteriaceae, such as *Halobacterium halobium*), where massive amino acid substitutions of their unique proteins result in optimal function at high salt concentrations. The major difference between halobacterial and "normal" proteins is the presence in the former of a high proportion of charged amino acids and a small proportion of large hydrophobic amino acids (Yancey *et al.*, 1982; Brown, 1990). In this way, the optimal balance between hydrophobic and electrostatic interactions in these halobacterial proteins occurs at high salt concentrations. This strategy has not been successful in the course of evolution, probably because it lacks the flexibility offered by the combination of "normal" proteins and osmolyte synthesis during osmotic stress. Halophytic archaebacteria are restricted to rare ecological niches with high salt concentrations and cannot survive under more normal conditions.

In addition to these relatively nonspecific salt effects occurring at high concentrations, some enzyme systems may be especially sensitive to inhibition by either $Na^+$ or $Cl^-$ at much lower concentrations, reflecting specific interactions of the ions with inhibitory binding sites. If these salt-sensitive enzymes catalyzed important metabolic reactions and their activity levels did not greatly exceed the required metabolic fluxes, they could qualify as primary salt toxicity targets determining the sensitivity of the organisms to salt stress. The identification of these targets is of crucial importance for the understanding of salt damage and for the manipulation of salt tolerance. Protein synthesis has been considered as a possible primary target of salt toxicity because *in vitro* protein synthesis systems are dependent on physiological potassium (0.1–0.15 *M*) and are inhibited by sodium and chloride concentrations greater than 0.1 *M* (Wyn Jones and Pollard, 1983). Sodium must interfere with different potassium sites on proteins and RNA, whereas chloride must interfere with anionic sites involved in RNA binding.

Phosphoenolpyruvate carboxylase, glutathione reductase, and ribulose-bisphosphate carboxylase are enzymes with anionic substrates especially sensitive to chloride inhibition (half-maximal inhibition at 0.1–0.2 *M*) (Gimmler *et al.,* 1984). Sodium inhibition may be expected in the case of potassium-activated enzymes such as pyruvate kinase, phosphofructokinase, and ADPG-starch synthetase (Wyn Jones and Pollard, 1983), but there is no information on the salt sensitivity of these enzymes.

The major sodium toxicity target of yeast cells has been identified as a 3′,5′-bisphosphate nucleotidase involved in sulfate activation (Gläser *et al.,* 1993; Murguía *et al.,* 1995). This finding resulted from a general search for genes that, on overexpression, improve salt tolerance (Gaxiola *et al.,* 1992; Serrano, 1994; Serrano and Gaxiola, 1994). These genes could identify crucial ion transport and metabolic reactions in salt tolerance. Yeast cells are especially suited for this search because genomic libraries in multicopy plasmids are readily available. After transformation, a popupation of transgenic yeasts with increased gene dosage (10- to 20-fold normal) of different genomic regions may be selected for salt tolerance. At present three yeast salt tolerance genes (*HAL* genes) have been identified and named *HAL1* (Gaxiola *et al.,* 1992), *HAL2* (Gläser *et al.,* 1993), and *HAL3* (Ferrando *et al.,* 1995). *HAL1* and *HAL3* influence intracellular sodium and potassium concentrations (see Section II,F,2) and therefore are part of the regulatory machinery for ion homeostasis. *HAL2* has a different mechanism. It does not affect intracellular ion concentrations but is required for methionine biosynthesis, being allelic to the *MET22* gene defined by methionine auxotrophic mutants (Gläser *et al.,* 1993). Interestingly, the enzyme encoded by *HAL2* does not correspond to any of the metabolic enzymes involved in methionine biosynthesis but to a side reaction frequently overlooked although essential for this metabolic pathway (Metzler, 1977).

Before the inorganic sulfate group can be incorporated into organic molecules, it is activated by reaction with ATP (catalyzed by ATP-sulfurylase) to produce adenosine-5′-phosphosulfate (APS) and, with the help of pyrophosphatase, inorganic phosphate ($P_i$):

$$\mathrm{ATP} + \mathrm{SO_4}^{-2} = \mathrm{APS} + 2\mathrm{P_i}$$

This reaction is thermodynamically unfavorable ($\Delta G = +12$ kJ/mol) and therefore the activated sulfate represented by APS does not accumulate inside cells in significant amounts. One solution to this problem is specifically found in plant chloroplasts: APS reacts with reduced glutathione (GSH, reaction catalyzed by thiol sulfotransferase (EC 2.8.2.16)) to produce higher concentrations of another activated form of sulfate, glutathione thiosulfonate:

$$\mathrm{APS} + \mathrm{GSH} = \mathrm{AMP} + \mathrm{GS\text{-}SO_3}^-$$

This is reduced by ferredoxin-coupled thiosulfonate reductase to sulfide, which is incorporated into cysteine and methionine (Anderson and Beardall, 1991).

Another more general pathway for accumulation of activated sulfate is found among heterotrophic organisms such as yeast and humans, and also in chloroplasts. It is based on the phosphorylation of APS by ATP catalyzed by APS kinase and producing 3′-phosphoadenosine-5′-phosphosulfate (PAPS):

$$APS + ATP = PAPS + ADP$$

This extra phosphorylation of the adenosine of APS drives the accumulation of activated sulfate in the form of PAPS. However, when this activated sulfate is utilized by either reduction to sulfite or transfer to other molecules, an unusual nucleotide is formed, 3′-phosphoadenosine-5′-phosphate (PAP), which needs to be hydrolyzed to AMP to recycle adenosine. The specific nucleotidase catalyzing this reaction (Metzler, 1977) is encoded by the yeast *HAL2* gene and is sensitive to inhibition by lithium ($K_i = 0.1$ m$M$) and sodium ($K_i = 20$ m$M$), whereas potassium at high concentrations ($>0.1$ $M$) activates the enzyme and counteracts the inhibition produced by lithium and sodium (Murguía *et al.*, 1995). During salt stress the enzyme is inhibited *in vivo* and PAP accumulates inside the cells. This results in product inhibition of the enzymatic reactions generating PAP from PAPS. In yeast PAPS reduction to sulfite during methionine biosynthesis seems the most important affected reaction because growth under salt stress is ameliorated by methionine supplementation (Gläser *et al.*, 1993). PAPS-dependent sulfate transfer reactions are significant in animal cells (sulfoprotein and sulfopolysaccharide formation) and plant cells (sulfolipid formation).

The *HAL2*-encoded nucleotidase has significant homology with the enzyme family of animal inositol phosphatases (Gläser *et al.*, 1993). Some important differences are that inositol phosphatases are inhibited by lithium but not by sodium and that the *HAL2*-encoded nucleotidase does not hydrolyze inositol phosphates. HAL2 homologs exist in animal and plant cells and the tomato nucleotidase is sensitive to lithium and sodium (Murguía *et al.*, 1995). Therefore, it is likely that one of the toxic effects of salt in plants, as occurs in yeast, is the inhibition of the PAP nucleotidase. Given the peculiarities of sulfate metabolism in plants, where sulfate assimilation occurs by reaction with glutathione (see above), it could be predicted that lithium and sodium inhibition of plant PAP nucleotidase will interfere mostly with sulfolipid synthesis and not with cysteine and methionine biosynthesis. This, however, could be equally toxic to the plant because of the probable essential role of these lipids in chloroplast functions.

There is a precedent for the participation of the methionine biosynthetic pathway in stress tolerance (Neidhardt *et al.,* 1990). In the bacterium *E. coli* the enzyme homoserine succinyltransferase (encoded by the *metA* gene) is sensitive to temperatures higher than 40°C and in the range of 40 to 45°C the growth rate is limited by the availability of methionine in the medium. At higher temperatures other factors become rate limiting.

## D. Ion Transport at the Plasma Membrane and Salt Tolerance

### 1. Potassium and Sodium Uptake

Potassium uptake in *E. coli* occurs by three different mechanisms (Bakker, 1993a,b; Epstein *et al.,* 1993). The major transporter under normal conditions is the Trk system, constitutively expressed and of moderate affinity ($K_m$ for $K^+$ about $10^{-3}$ *M*). This is a multisubunit transporter ATPase of the ABC family (Higgins, 1992). The ATP-binding site is in the *trkE*-encoded cytoplasmic peripheral subunit. Another similar subunit is encoded by the *trkA* gene, although in this case it binds $NAD^+$ instead of ATP. *trkG* and *trkH* encode homologous transmembrane proteins and either one suffices. This system has a dual requirement for both ATP and a proton gradient and therefore it is not a simple ion-pumping ATPase. The much simpler transporter encoded by the *kup* gene is probably an $H^+$–$K^+$ symport. The affinity of this system is similar to that of the Trk system, but much less active. The Kdp system is inducible by growth at low $K^+$ and it corresponds to a P-type ATPase of great complexity. The subunit encoded by the *kdpB* gene is homologous to eukaryotic P-ATPases and contains the sites for ATP binding and formation of the acyl-phosphate intermediate. The subunits encoded by the *kdpA* and *kdpC* genes have no counterpart in eukaryotic P-ATPases; they are, however, essential for activity and the *kdpA* subunit determines $K^+$ affinity. The most remarkable property of this system is the $K_m$ for $K^+$ uptake of about $10^{-6}$ *M*.

Cation transport systems with homology to the bacterial systems previously described have not been described in fungi and plants. Ion transport in the yeast *S. cerevisiae* has been amenable to a fruitful genetic analysis, which has identified the major transport systems operating at the plasma membrane of this model organism (Serrano, 1991; Gaber, 1992). Basically, a plasma membrane $H^+$-ATPase encoded by the *PMA1* gene generates an electrochemical proton gradient that regulates pH and drives the secondary active transport systems catalyzing nutrient uptake. Two homologous genes, *TRK1* and *TRK2,* seem responsible for most of the high-affinity $K^+$ uptake, which allows optimal growth at 0.2 m*M* $K^+$. In *trk1 trk2* double mutants, a nonidentified pathway for low-affinity $K^+$ uptake still permits growth in media with 100 m*M* $K^+$ (Ko and Gaber, 1991). *TRK2* is functionally homologous to *TRK1* but it is less expressed under normal conditions (Ramos *et al.*, 1994). Therefore, *TRK1* is the major high-affinity $K^+$ uptake system of *S. cerevisiae* (Ko and Gaber, 1991). The TRK proteins have no homologies in databanks and it has never been demonstrated that they are actual transporters. It has been suggested that they are components of a $K^+$ uptake system composed of several subunits and that the role of the TRK proteins is to modulate the affinity of the transporter in response to intracellular $K^+$ depletion and salt stress (Ramos *et al.*, 1994). In media with either low $K^+$ or high NaCl, the TRK system experiences an increase in affinity for $K^+$ and $Rb^+$ but not for $Na^+$ or $Li^+$. The mechanism of this kinetic change is unknown but it dramatically increases the discrimination between $K^+$ ($Rb^+$) and $Na^+$ ($Li^+$).

The defective growth of *trk1* mutants in low-$K^+$ media has allowed complementation cloning of plant $K^+$ transport systems by expression in yeast (Anderson *et al.*, 1992; Sentenac *et al.*, 1992; Schachtman and Schroeder, 1994). Expression in yeast of the *Arabidopsis AKT1* (Sentenac *et al.*, 1992) and *KAT1* (Anderson *et al.*, 1992) genes improves $K^+$ uptake by the *trk1* mutant. These two genes encode similar proteins with homology to the animal $K^+$ efflux channels activated by depolarization. However, these plant channels function both in yeast and in oocytes (Schachtman *et al.*, 1992) as $K^+$ influx channels activated by hyperpolarization. The structural features that determine the different functions of plant and animal $K^+$ channels have not been identified.

The *AKT1*- (Sentenac *et al.*, 1992) and *KAT1*- (Kochian *et al.*, 1993) encoded channels exhibit complicated kinetics for $K^+$ transport. In addition to a saturable component of moderate affinity ($K_m$ in the $10^{-3}$ to $10^{-2}$ *M* range), a diffusion component without apparent saturation operates in the $10^{-2}$ to $10^{-1}$ *M* range (Sentenac *et al.*, 1992). Therefore, the properties of these channels may explain the multiphase $K^+$ uptake kinetics observed in plants (Nissen, 1991). The high-affinity component of $K^+$ uptake identified in plant roots ($K_m$ in the $10^{-5}$ *M* range; Epstein *et al.*, 1963) seems to correspond to a different system encoded by the *HKT1* gene of wheat (Schachtman and Schroeder, 1994). This gene was also cloned by complementation of the yeast *trk1* mutant but it encodes a membrane protein without homology to animal $K^+$ channels. Actually, $K^+$ transport by the product of the *HKT1* gene expressed in oocytes ($K_m$ about $10^{-5}$ *M*) seems to occur by $H^+$ symport, a mechanism originally suggested in *Neurospora crassa* (Rodriguez-Navarro *et al.*, 1986) and not expected from a channel. Electrophysiological evidence for an $H^+$–$K^+$ symport of high affinity in *Arabidopsis* roots has also been reported (Maathuis and Sanders, 1994). As *HKT1* is highly expressed in roots (Schachtman and Schroeder, 1994), this type of system may account for high-affinity $K^+$ uptake in plants. The channels, however, may account for the low-affinity $K^+$ uptake (Schroeder and Fang, 1991). Another mechanism for $K^+$ transport described in *Chara australis* is an $Na^+$–$K^+$ symport (Smith and Walker, 1989; McCulloch *et al.*, 1990) and could be related to the $H^+$–$K^+$ symport previously described.

The plant $K^+$ channels expressed in yeast maintain accumulation ratios ($K_i^+/K_o^+$ slightly greater than $10^5$, which corresponded to an equilibrium electrical potential difference of at least 300 mV (Sentenac *et al.*, 1992). This is much higher than estimated in *S. cerevisiae* by different methods (Serrano, 1991). However, determinations of yeast membrane potential based on the accumulation of cytosine mediated by the reversible $H^+$–cytosine symport (Eddy *et al.*, 1994) indicate that such high values may be possible and therefore that no active $K^+$ transport needs to be invoked.

The discrimination between $Na^+$ and $K^+$ seems much better in the case of the low-affinity plant $K^+$ channel KAT1 ($Na^+$ conductance/$K^+$ conductance = 0.07; Schachtman *et al.*, 1992) than for the putative $H^+$–$K^+$ symporter of high-affinity HKT1 ($Na^+$ conductance/$K^+$ conductance = 0.3; Schachtman and Schroeder, 1994). Therefore, the relative activity of these two types of $K^+$ transport systems may influence salt tolerance. Sodium ion uptake in barley roots occurs by both the high-affinity system ($K_m$ about $10^{-4}$ *M*) and by the low-affinity component ($K_m$ about $10^{-2}$ *M* and linear diffusion). However, in these experiments the discrimination between $Na^+$ and $K^+$ was much better at low cation concentrations (Rains and Epstein, 1967). This is the opposite of what is expected from the properties of the KAT1 and HKT1 systems. Cation uptake at high concentrations may occur by unknown mechanisms of low $Na^+/K^+$ discrimination in addition to the highly discriminative KAT1 channels.

Mutations in the *ERG6* (*LIS1*) gene of *S. cerevisiae,* encoding a methyltransferase of the ergosterol pathway, result in altered sterol composition and hypersensitivity to $Li^+$ and $Na^+$ (Welihuida *et al.*, 1994). This phenotype is caused by increased uptake of the toxic cations without noticeable effects on their efflux. A similar mechanism could account for the observed differences in sodium permeability detected in maize cultivars (Schubert and Laüchli, 1990).

The complementation of the $K^+$ uptake defect of the yeast *trk1* mutant has provided some unexpected results. For example, mutant alleles of the yeast sugar transport genes *HXT1* and *HXT3* restore $K^+$ uptake in cells deleted for *TRK1* and *TRK2* (Ko *et al.*, 1993). This implies an unexpected flexibility of membrane transporters for mutational alterations of substrate specificity. When complementation was attempted with mouse cDNAs, a putative transcriptional regulator (HBP1; Lesage *et al.*, 1994), a membrane fusion ATPase (SKD2; Perier *et al.*, 1994), and a related ATPase (SKD1; Perier *et al.*, 1994) were found. Although the mechanisms of these complementations were not investigated, putative transcriptional regulators like HBP1 may increase the expression of cryptic $K^+$ uptake systems and putative intracellular membrane fusion proteins like SKD2 may mitigate defects in $K^+$-dependent membrane fusion responsible for slow growth under $K^+$ deprivation.

In addition to hyperpolarization-activated $K^+$ uptake systems, both yeast (Bertl *et al.,* 1993) and plants (Cao *et al.,* 1992) contain depolarization-activated $K^+$ efflux systems identified by electrophysiological studies. These transport systems may participate in balancing charge movements during secondary active transport, mostly when the proton pump is not operative. These channels also transport $Na^+$ and their activity is reduced during adaptation of tobacco cells to salinity (Murata *et al.,* 1994). As high $Na^+$ concentrations induce depolarization, this kind of channel may mediate both $K^+$ efflux and $Na^+$ entry.

Mechanosensitive or stretch-activated channels have been identified in yeast (Gustin *et al.,* 1988) and plants (Cosgrove and Hedrich, 1991). They exert low-level discrimination between anions and cations and may contribute to osmotic adjustment during the initial stages of osmotic stress.

The role of cation uptake systems in salt tolerance has been demonstrated by mutant studies. In the yeast *S. cerevisiae,* $Trk1^-$ cells have less $K^+/Na^+$ discrimination and are more salt sensitive than $Trk1^+$ cells (Haro *et al.,* 1993). Apparently, the TRK system has a greater specificity for $K^+$ versus $Na^+$ (especially after the regulatory change triggered by salinity; see above) compared to the unidentified cation uptake system of low efficiency operating in $Trk^-$ cells. Two yeast genes, *HAL1* (Gaxiola *et al.,* 1992) and *HAL3* (Ferrando *et al.,* 1995), have been identified that by overexpression improve $K^+/Na^+$ discrimination and salt tolerance. They encode soluble proteins without homologies in any databank and may define a novel regulatory pathway for ion homeostasis. Although their effects are partially mediated through the sodium efflux pump (see Section II,D,2), an important determinant of the salt tolerance conferred by *HAL1* and *HAL3* is the stimulation of $K^+$ uptake under salt stress. The significance of enhanced $K^+$ uptake in salt tolerance is corroborated by physiological studies with plant variants differing in salt tolerance. The clearest case is posed by the *stl1* and *stl2* mutations of the fern *Ceratopteris richardii* (Vogelien *et al.,* 1993). Both mutations confer salt tolerance to gametophytes and reduced $Na^+$ accumulation. In addition, the stronger *stl2* mutation increased $K^+$ accumulation. NaCl-adapted cells of tobacco (Watad *et al.,* 1991), *Poncirus trifoliata* (Beloualy and Bouharmont, 1992), and *Citrus aurantium* (Ben-Hayyim *et al.,* 1985) exhibit higher $K^+$ accumulation than control cells. Finally, wild halophytic tomato (*Lycopersicum cheesmanii*) is more efficient in $K^+$ absorption than cultivated tomato (*Lycopersicum esculentum*) (Rush and Epstein, 1981). Recessive mutants of *A. thaliana* capable of germination under saline conditions have been isolated (Saleki *et al.,* 1993). Although these mutants are primarily osmotolerant (they are also tolerant to mannitol), seeds exposed to salinity absorbed more $K^+$ than wild-type controls, again pointing to an important role of enhanced cation uptake in salt tolerance. However, cesium-resistant mutants of *A. thaliana* had reduced rubidium and phosphate uptake (Sheahan *et al.,* 1993). Increased $K^+$ uptake is beneficial for salt tolerance because $K^+$ counteracts the inhibitory effects of $Na^+$ on enzymatic systems (see above).

### 2. Sodium Efflux

*Escherichia coli* contains two homologous $Na^+$–$H^+$ antiporters encoded by the *nhaA* and *nhaB* genes (Schuldiner and Padan, 1993). These systems catalyze electrogenic exchange of 2 $H^+$ per 1 $Na^+$ and have $K_m$ values for $Na^+$ in the $10^{-3}$ *M* range. They are essential for salt tolerance. *nhaA* is also involved in pH homeostasis because its activity has a steep pH dependence and it is inactive at pH values below the intracellular homeostatic pH (7.6–7.8). Expression of *nhaA* is specifically induced by intracellular $Na^+$ and $Li^+$, which are probably "sensed" by the DNA-binding protein encoded by the *nhaR* gene (Carmel *et al.,* 1994). Induction by high pH has been shown to result from increased intracellular $Na^+$ under these conditions. In the gram-positive bacterium *Enterococcus hirae* (previously called *Streptococcus faecalis*) two systems for sodium efflux have been described: an $Na^+$-pumping F-ATPase of the vacuolar type (Takase *et al.,* 1994; Solioz and Davies, 1994) and an $Na^+$-$H^+$ antiporter with no homology to the one in *E. coli* (Waser *et al.,* 1992). The $Na^+$-ATPase contains a hydrophobic subunit (NtpJ) with homology to the TrkG/H subunits of the *E. coli* Trk $K^+$ uptake system. It also has homology to the *S. cerevisiae TRK1* and *TRK2* genes. This may suggest that the yeast Trk system is an ATPase of either the ABC or F-vacuolar type and that the ATP-binding catalytic subunits have not yet been isolated.

Two different genetic approaches have led to the identification of $Na^+$-efflux systems in yeast. In *S. cerevisiae,* the $Na^+$ and $Li^+$ hypersensitivity of a natural strain was complemented by a genomic plasmid library from a strain of normal tolerance to these cations (Haro *et al.,* 1991). A clone containing the previously described P-ATPase gene *PRM2* (Rudolph *et al.,* 1989) was shown to confer tolerance to $Li^+$ and $Na^+$ and enhanced efflux of these cations. In addition, disruption of this gene in other strains resulted in reduced efflux of $Li^+$ and $Na^+$ and hypersensitivity to growth inhibition by these cations. The product of the *PMR2* gene was originally proposed to be a $Ca^+$-pumping ATPase on the basis of homology to animal plasma membrane $Ca^+$-ATPases (Rudolph *et al.,* 1989). Although no biochemical studies have been reported, it probably corresponds to an $Na^+$-pumping

ATPase and the gene has been renamed as *ENA1* (for efflux of natrium). *ENA1* (*PMR2*) is the first of a tandem of four highly homologous genes. The function of the other three genes is unknown. *ENA1* is the most highly expressed and seems responsible for most of the salt tolerance of yeast (Garciadeblas *et al.,* 1993).

In a different approach, nitroguanidine-generated mutants of *Schizosaccharomyces pombe* with enhanced tolerance to $Li^+$ have been isolated. These mutants exhibited enhanced $Li^+$ and $Na^+$ efflux and contained an amplified locus (*sod2*) encoding a putative $H^+$–$Na^+$ antiporter, with homology to the animal and bacterial antiporters (Jia *et al.,* 1992). Disruption of *sod2* resulted in salt sensitivity. It is not clear why two related yeasts have completely different mechanisms for sodium efflux. More information is needed about the distribution of $Na^+$-ATPases and $H^+$–$Na^+$ antiporters in different fungi.

In *S. cerevisiae ENA1* is an important determinant of salt tolerance (Haro *et al.,* 1993). The expression of this gene is induced by osmotic stress, starvation, and high pH (Garciadeblas *et al.,* 1993). Although the salt induction mechanism seems common among other stress-induced genes (see Section II,F,2), the general level of expression of *ENA1* under both basal and induced conditions is specifically dependent on the protein phosphatase calcineurin (Mendoza *et al.,* 1994). This explains the observation that disruption of the calcineurin gene or addition of calcineurin inhibitors (immunosuppressive drugs such as FK506) reduces sodium efflux and salt tolerance (Nakamura *et al.,* 1993). The product of the *HAL3* gene has also been shown to influence *ENA1* expression (Ferrando *et al.,* 1995). Disruption of *HAL3* decreases both salt tolerance and *ENA1* expression and overexpression of *HAL3* increases both salt tolerance and *ENA1* expression. *HAL3* encodes a cytoplasmic protein of unknown function. The important conclusion from these studies is that, in addition to the ion transporters themselves, regulatory proteins affecting the degree of expression of the transporter genes are of crucial importance in salt tolerance. This suggests that regulatory pathways controlling the expression of crucial stress defense genes such as *ENA1* may be of great biotechnological value.

The mechanism of sodium efflux in plant cells has not been identified. Physiological studies have shown that $Na^+$ efflux from plant tissues is stimulated at low external pH (Colombo *et al.,* 1979; Jacoby and Teomy, 1988). This suggested the existence of an $Na^+$–$H^+$ antiport in plant plasma membranes. However, electrophysiological measurements indicated that $Na^+$ efflux is electrogenic and probably mediated by an $Na^+$-pumping ATPase (Cheeseman, 1982). Biochemical evidence is scarce but an $Na^+$–$H^+$ antiporter has been identified in plasma membrane vesicles from the halophytic plant *Atriplex nummularia* (Braun *et al.,* 1988) and from the marine algae *Dunaliella salina* (Pick, 1992) and *Platymonas viridis* (Popova and Balnokin,

1992). Furthermore, an $Na^+$-activated ATPase has been identified in plasma membrane preparations from the marine algae *Heterosigma akashiwo* (Wada *et al.,* 1989). The formation of an $Na^+$-dependent acyl-phosphate intermediate (Wada *et al.,* 1989) and the cross-reaction with antiserum against animal $Na^+$,$K^+$-ATPase (Wada *et al.,* 1992) suggests that the algal $Na^+$-ATPase is related to the well-characterized animal P-ATPase involved in $K^+$ uptake and $Na^+$ efflux (Stein, 1990). A similar enzyme has been proposed to exist in *Dunaliella* (Pick, 1992). In higher plants there is no biochemical evidence for an $Na^+$-ATPase. However, putative $Ca^{2+}$-ATPase genes from tomato (Wimmers *et al.,* 1992) and tobacco (Perez-Prat *et al.,* 1992) are induced by NaCl. This resembles the situation previously described for the *ENA1/PMR2* gene of *S. cerevisiae,* which has some homology to the genes encoding animal $Ca^{2+}$-ATPases but is induced by NaCl and participates in $Na^+$ efflux. Biochemical studies performed in a suitable expression system are needed to clarify the cation(s) pumped by these putative $Ca^{2+}$-ATPases from higher plants.

## 3. Chloride Transport

Bacteria and yeast cells do not require chloride for growth and exhibit low permeability to this anion (Rothstein, 1964). However, plants have a requirement for $Cl^-$ as a micronutrient essential for photosynthesis and also utilize this anion for osmotic adjustment within the vacuole (Leigh and Wyn Jones, 1985). Accordingly, active $Cl^-$ accumulation at the plant plasma membrane occurs by $Cl^-$–$H^+$ symport with stoichiometry of more than one $H^+$ per $Cl^-$ (Sanders, 1980; Sanders and Hansen, 1981; Felle, 1994). Different anion channels regulated by either voltage or calcium have been identified at the electrophysiological level (Hedrich, 1994). These channels may participate in volume and turgor regulation by allowing downhill $Cl^-$ efflux.

Unfortunately, the molecular bases for all these $Cl^-$ transporters are unknown and it is not yet possible to evaluate if manipulation of $Cl^-$ transport can modify salt tolerance. In plants especially sensitive to $Cl^-$ toxicity, such as *Citrus sinensis* (Bañuls and Primo-Millo, 1992), genetic engineering of $Cl^-$ efflux channels may be beneficial. In the absence of isolated genes for plant $Cl^-$ channels, cloned animal $Cl^-$ channels such as ClC-2 (Gründer *et al.,* 1992) could be tested in transgenic plants.

## 4. Role of Plasma Membrane $H^+$-ATPase in Salt Tolerance

The major plasma membrane ATPases of fungi and plants are homologous enzymes that operate as electrogenic proton pumps. The electrochemical proton gradient generated by these ATPases drives the secondary active transport of different molecules by proton cotransport (Serrano, 1989). In the yeast *S. cerevisiae,* salt tolerance of mutants with reduced expression of $H^+$-ATPase is similar to that of wild-type cells (Vallejo and Serrano, 1989). In addition, the level of expression of the $H^+$-ATPase is not increased by salt stress (R. Serrano, unpublished observations). Therefore, the ATPase does not seem to be among the factors limiting salt tolerance in this organism. Actually, mutants of *S. cerevisiae* and *S. pombe* with reduce $H^+$-ATPase activity have increased tolerance to very high NaCl concentrations (2.5 *M*), high ethanol concentrations (12.5%), and heat shock (25–48°C shock). Therefore, high activity of the enzyme may even be detrimental for stress tolerance (Panaretou and Piper, 1990). The situation is different in the yeast *Zygosaccharomyces rouxii,* where NaCl increases the expression of the plasma membrane $H^+$-ATPase (Watanabe *et al.,* 1993). It could be speculated that the $H^+$-ATPase is induced by NaCl when $Na^+$ extrusion occurs by $H^+$–$Na^+$ antiport but not when an $Na^+$-APase is operative. In the latter case, NaCl induces the $Na^+$-ATPase gene and the $H^+$-ATPase is of little relevance for salt tolerance.

In normal plants, such as sunflower, salt stress does not increase the expression of the plasma membrane $H^+$-ATPase (Roldan *et al.,* 1991) and in tomato, salinity stress even reduces the activity of the enzyme (Gronwald *et al.,* 1990). The situation is different in the halophytic plant *Atriplex nummularia,* where salinity during growth increases the activity of the plasma membrane $H^+$-ATPase (Braun *et al.,* 1986) and the level of its mRNA (Niu *et al.,* 1993). This effect correlates with the presence of an $Na^+$–$H^+$ antiporter in the plasma membrane of this plant (see above).

### E. Role of the Vacuole in Salt Tolerance

One obvious role of the vacuole in salt tolerance is in the accumulation of toxic $Na^+$ and $Cl^-$, to the benefit of sensitive cytoplasmic and organellar enzymes (Leigh and Wyn Jones, 1985). For example, in tobacco cells adapted to 0.43 *M* NaCl, the concentrations of $Na^+$ and $Cl^-$ in different compartments (measured by X-ray microanalysis) were as follows: cytoplasm, 0.1 *M;* vacuole, 0.6–0.8 *M* (Binzel *et al.,* 1988). In *Dunaliella parva* cells grown in medium with 0.4 *M* NaCl, the values were as follows: cytoplasm, 0.03 *M;* vacuole, 0.3 *M.* However, the concentration of $K^+$ was 0.1 *M* in the cytoplasm and 0.03 *M* in the vacuole (Hajibagheri and Flowers, 1993). The activity of vacuolar $H^+$-ATPase is increased by NaCl treatment in barley roots (Matsumoto and Chung, 1988) and in tobacco cells (Reuveni *et al.,* 1990). This increase in activity corresponds both to enhanced expression of vacuolar ATPase genes (Narasimhan *et al.,* 1991) and to enhanced intrinsic activity of the enzyme (Reuveni *et al.,* 1990). The change made in response to salt stress is completed by the salt activation of the vacuolar $Na^+$–$H^+$ antiporter observed in vacuolar vesicles from barley roots (DuPont, 1992). This activation is rapid and independent of protein synthesis and therefore must correspond to a modification of the antiporter protein. In any case, both the vacuolar $H^+$ gradient driving $Na^+$ accumulation and the activity of the $Na^+$–$H^+$ antiporter are increased during salt stress, pointing to the important role of vacuolar sodium accumulation in salt tolerance. This is also suggested by the observation that the antiporter activity of vacuolar vesicles is much higher in the salt-tolerant *Plantago maritima* than in the salt-sensitive *Plantago media* (Staal *et al.*, 1991).

No information exists about salt modulation of $Cl^-$ transport systems in plant vacuoles. Electrophysiological studies have demonstrated the existence of anion channels in plant vacuoles (Hedrich, 1994) and these channels could mediate $Cl^-$ uptake driven by the positive-inside vacuolar membrane potential.

One unexpected role of the vacuole in osmotic tolerance has been uncovered in the yeast *S. cerevisiae.* Mutants selected for defects in vacuolar biogenesis (secreting vacuolar proteins to the periplasm) were osmosensitive and mutants selected for inability to grow at high osmotic concentrations were defective in vacuolar biogenesis and secreted vacuolar proteins to the periplasm (Latterich and Watson, 1991). One interpretation of these results is that certain functions required for vacuolar biogenesis are also required for osmoregulation. Cytoskeletal elements and proteins mediating vesicle budding and fusion are likely candidates and osmosensitivity has been reported in some actin mutants (Novick and Botstein, 1985). Alternatively, the vacuole itself may be involved in osmoregulation as a water source during osmotic stress of the cytoplasm. Many vacuolar mutants have either no visible vacuoles or a highly reduced vacuolar compartment and they lose viability instantly on osmotic challenge (Latterich and Watson, 1993). The flow of water from vacuole to cytoplasm may play an important role in the initial response to osmotic stress (Mager and Varela, 1993).

## F. Signal Transduction during Salt Stress and Interaction with Other Abiotic Stresses

### 1. *Escherichia coli*

Stress responses in *E. coli* involve several well-established mechanisms that could serve as models for eukaryotic systems (Neidhardt *et al.*, 1990). Induction of about 12 genes by oxidative stress operates by oxidative inactivation of the repressor protein encoded by the *oxyR* gene. DNA damage induces about 20 repair genes (SOS response) by activating the protease activity of the RecA protein, which then inactivates the *lexA* repressor. These are examples of negative control, where stress counteracts gene repression. A different situation, positive control activated by stress, is posed by heat shock, which activates transcription and translation of the *htpR* gene encoding the RNA polymerase $\sigma^{32}$ factor. Polymerase programmed by $\sigma^{32}$ preferentially transcribes promoters (about 20) recognized by this factor.

An interesting example of physical control of gene expression is provided by the osmotic induction of the *proU* operon, encoding a betaine transport system (Higgins *et al.*, 1988; Hulton *et al.*, 1990). An increase in extracellular osmolarity increases *in vivo* DNA supercoiling and this determines *proU* expression. Selection for trans-acting mutations that affect *proU* expression has yielded only mutations that alter DNA supercoiling, either in the *topA* gene (encoding topoisomerase) or in the *osmZ* gene (encoding histone-like protein H1). In *E. coli* osmotic stress increases intracellular $K^+$ (Epstein, 1986) and therefore it has been proposed that the changes in supercoiling determining expression of *proU* are mediated by an increase in ionic strength (Higgins *et al.*, 1987). There is some evidence for this kind of mechanism from *in vitro* transcription studies (Prince and Villarejo, 1990).

A different mechanism based on two characteristic protein kinases (Neidhardt *et al.*, 1990; Stock *et al.*, 1990) mediates the osmotic regulation of the *kdpABC* operon and the porin genes *ompF* and *ompC*. This "two-component" signal transduction pathway consists of a "sensor protein kinase" autophosphorylated at histidine in response to some signal and a "response regulator" protein phosphorylated at aspartate by transfer from the histidine phosphate. The histidine phosphate functions as a catalytic intermediate of the activated kinase. The regulator may be a transcription factor activated by phosphorylation but it may also interact with other proteins to modulate its activity. The *kdpABC* operon, encoding a high-affinity $K^+$-uptake ATPase (see above), is regulated by a two-component system encoded by the *kdpD* and *kdpE* genes (Walderhaug *et al.*, 1992). The KdpD protein is the sensor protein kinase and it is a membrane protein with four transmembrane stretches. KdpE is a transcription factor binding to the promoter of the *kdpABC* operon (Sugiura *et al.*, 1992) and phosphorylated by KdpD in response to osmotic stress (Nakashima *et al.*, 1993). The original model for KdpD regulation (Epstein, 1986) postulated that this membrane protein interacts with the bacterial cell wall across the periplasmic space. Changes in turgor would alter the width of the periplasmic space and consequently the conformation of KdpD. However, more recent evidence indicates that turgor loss is not the signal for transcriptional control of the *kdpABC* operon (Asha and Gowrishankar, 1993; Sugiura *et al.*, 1994). The operon is not induced by nonionic osmotica such as sucrose and the most likely induction signal seems to be intracellular potassium deprivation.

This primary signal may be modulated by other inducing conditions such as high NaCl concentrations or acidic pH, which may inhibit $K^+$ uptake and are counteracted by potassium supplementation. Interestingly, mutant forms of KdkD, with altered amino acids in the transmembrane region, become insensitive to the $K^+$ signal but respond to osmotic stress and to perturbation of membrane structure by ethanol or local anesthetics (Sugiura *et al.*, 1994). This illustrates the complexities of sensor proteins in terms of exciting stresses.

The regulation of the porin genes *ompF* and *ompC* also involves a "two-component" system (Neidhardt *et al.*, 1990; Csonka and Hanson, 1991). The sensor protein kinase is also a membrane protein, encoded by the *envZ* gene, that phosphorylates a transcription factor encoded by the *ompR* gene. The phosphorylated OmpR protein activates transcription of *ompC* and inhibits transcription of *ompF*.

### 2. *Saccharomyces cerevisiae*

The best characterized stress response at the molecular level is the classic heat shock response of eukaryotic cells (Sorger, 1991; Nover, 1994). The transcription of genes containing at their promoters the so-called heat shock element (HSE) is increased by high temperatures as a consequence of the activation of the heat shock transcription factor (HSF). Heat shock elements are formed by contiguous arrays of three 5-bp recognition units (XGAAX) arranged in alternating orientation (Sorger, 1991). "X" denotes less strongly conserved nucleotides that may also be important, and the first "X" has been proposed by Fernandes *et al.* (1994) to have a strong tendency to be an "A." The HSF is a trimer and it recognizes HSEs by cooperative binding of the three monomers to the three alternating 5-bp units (Perisic *et al.*, 1989; Fernandes *et al.*, 1994).

There are two mechanisms for activation of the HSF on heat shock: one general and another specific to animal cells (Sorger, 1991). In both animal and yeast cells the HSF is phosphorylated on heat shock and this modification may enhance its transcriptional activity, although this point has not been demonstrated. The nature of the heat-activated protein kinase acting on the HSF and its mechanism of activation are unknown. In yeast the trimeric HSF is bound to responsive promoters under basal conditions and contributes to the basal expression of many genes. In animal cells, however, monomeric HSF is complexed with the HSP70 chaperone and only on heat shock dissociates from it, trimerizes, and binds to the HSEs. Chaperones are part of the protein-folding machinery of the cells and many of them are induced by heat shock to cope with the increase in denatured proteins at high temperature (Nover, 1994). Signal transduction on heat shock in animal cells may simply involve recruitment of the HSP70 chaperone by denatured proteins and concomitant release of the HSF to activate transcription of HSE-containing genes. Ulterior phosphorylation of HSF may reinforce its activity. This release mechanism may apply to other regulatory systems because chaperones account for more than 5% of the cell protein and have been shown to form complexes *in vivo* with many signaling proteins (Rutherford and Zuker, 1994). Stresses that result in protein denaturation would dissociate chaperones from the complexes and release active signaling proteins as demonstrated for the HSF. The hypothetical protein kinase activated by heat shock and acting on the HSF could be regulated in this way. Heat shock also induces proton uptake and intracellular acidification in yeast (Weitzel *et al.*, 1987) and this pH decrease may also be part of the signal transduction (Piper, 1993).

Despite the widespread nature of the classic heat shock response, mutational analyses in yeast call into question its physiological significance for heat tolerance (Piper, 1993; Nover, 1994). Gene disruption of most heat shock-inducible genes does not compromise heat tolerance. This may be explained by the redundancy of many of these genes in the yeast genome. A more definitive argument is provided by a mutation in the yeast HSF (*hsf1-m3*) that prevents HSF activation on heat shock (Smith and Yaffe, 1991). This mutation results in a general block in heat shock induction but does not affect the acquisition of thermotolerance. The HSF, however, is essential for yeast growth under normal conditions because it contributes to the basal expression of many essential genes. The product of the *HSP104* gene is the only heat shock protein required for optimal thermotolerance (Winkler *et al.*, 1991; Sanchez *et al.*, 1992). However, on prolongation of the preconditioning heat treatment, trehalose (and probably also glycerol; see Trollmo *et al.*, 1988) accumulates and results in optimal thermotolerance independently of *HSP104* (De Virgilio *et al.*, 1991).

In the yeast *S. cerevisiae*, the HSE is exclusively induced by heat shock stress, being unresponsive to osmotic stress, oxidative stress, DNA damage, and drugs that inhibit protein synthesis and cause translational errors. In addition, this promoter element is insensitive to glucose repression, and maximal heat induction is observed during exponential phase in glucose media (Kirk and Piper, 1991). Another promoter element responsive to heat shock but also to other stresses and subjected to glucose repression has been described in yeast. The DNA damage-inducible gene *DDR2* of unknown function (Kobayashi and McEntee, 1990) and the cytosolic catalase gene *CTT1* induced by oxidative stress (Wieser *et al.*, 1991) are also induced by heat shock by a mechanism independent of HSEs. A novel element designated STRE (stress response element; Marchler *et al.*, 1993) or TRS (thermal stress response sequence; Kobayashi and McEntee, 1993), with core consensus sequence CCCCT, operates as a positive promoter element mediating the induction of these genes by different stresses such as heat shock, osmotic stress, nitrogen starvation, and oxidative stress ($H_2O_2$). In addition, it mediates negative regulation of transcription by the cAMP–protein kinase A pathway (Belazzi *et al.,* 1991; Marchler *et al.,* 1993). This novel element is not responsive to DNA damage by ultraviolet (UV) light or alkylating agents (Kobayashi and McEntee, 1993), which utilize a different promoter element (Friedberg *et al.,* 1991). Blocking of DNA replication can provide specific signals (Zhou and Elledge, 1992) transduced by phosphorylation and activation of the protein kinase encoded by the *DUN1* gene (Zhou and Elledge, 1993). Activation of transcription of DNA damage-inducible genes occurs, at least in part, by counteracting the action of the general repressor of transcription encoded by the *SSN6* and *TUP1* genes (Keleher *et al.,* 1992). The Ssn6–Tup1 repressor is recruited to target promoters by sequence-specific DNA-binding proteins.

In addition to *CTT1* and *DDR2,* other stress-inducible genes containing putative STREs at their promoters are *HSP104, GAC1, UB14, HSP12, HSP26* (Mager and Varela, 1993; Schüller *et al.,* 1994), *TPS1* (Vuorio *et al.,* 1993), *TPS2* (Vuorio *et al.,* 1993; Gounalaki and Thireos, 1994), *HAL1* (Gaxiola *et al.,* 1992), *GPD1* (Larsson *et al.,* 1993), and *ALD2, DDR48, PAI3,* and *SIP18* (Miralles and Serrano, 1995). The functionality of these putative STREs, however, needs to be demonstrated by promoter analysis because the CCCCT sequence is only a core consensus not necessarily functional in every sequence context. For example, the putative STREs in the *ALD2* (Miralles and Serrano, 1995) and *HAL1* (J. A. Marquez and R. Serrano, unpublished) genes are not functional.

Yeast proteins specifically binding to the STRE sequence have been identified by gel retardation experiments (Kobayashi and McEntee, 1993; Marchler *et al.,* 1993), but there are no clues about the nature of these putative transcription factors. More progress has been made on the upstream components of this regulatory pathway. A search for mutants defective in glycerol synthesis and osmotic tolerance identified a mitogen-activated protein (MAP) kinase pathway (Brewster *et al.,* 1993) involved in the osmotic induction mediated by STREs (Schüller *et al.,* 1994). The MAP kinase pathways are intracellular signaling modules made up of three sequentially acting characteristic protein kinases (Nishida and Gotoh, 1993; Cooper, 1994; Levin and Errede, 1995). The three types of kinases are designated MAPKs or ERKs (mitogen-activated protein kinases or extracellular regulated protein kinases), MEKs (MAPK/ERK kinases), and MEKKs (MEK kinases) and they act in the sequence MEKK → MEK → MAPK. MEKKs are indirectly activated by extracellular signals, such as mitogens in animal cells, via specific receptors, G proteins, and different protein kinases. MEKs are double-specificity kinases, phosphorylating both threonine and tyrosine in MAPKs, which become activated and may phosphorylate and activate transcription factors such as the protooncogene product c-Jun of animal cells. The so-called HOG (high osmolarity glycerol response) pathway in yeast is defined by the kinases encoded by the genes *SSK2* (Levin and Errede, 1995), *PBS2* (Boguslawski, 1992), and *HOG1* (Brewster *et al.,* 1993), corresponding to the MEKK, MEK, and MAPK, respectively. The HOG1 kinase is phosphorylated at tyrosine residues in response to mild osmotic stress (maximum response at 0.2–0.3 *M* NaCl) in a rapid and *PBS2*-dependent reaction required for optimal induction of glycerol synthesis (Brewster *et al.,* 1993). This response is mediated by the HOG-dependent induction of the *GPD1* gene (encoding glycerol-3-phosphate dehydrogenase; Albertyn *et al.,* 1994b), probably through STREs. Interestingly, although STREs mediate gene induction by several stresses such as heat shock, nitrogen starvation, oxidative stress, acidic conditions, and osmotic stress, only the latter two stresses (acid and osmotic) are dependent on the HOG pathway (Schüller *et al.,* 1994). In addition, the *ALD2* gene is induced by osmotic stress in an HOG-dependent way but does not contain a functional STRE (Miralles and Serrano, 1995). Also, the *DDR48* (Miralles and Serrano, 1995) and *HAL1* (J. A. Marquez and R. Serrano, unpublished) genes are induced by osmotic stress but in an HOG-independent way. Therefore, the complexity of signal transduction during osmostress responses in yeast may include several promoter elements in addition to an STRE and several signaling pathways in addition to the HOG–MAP kinase pathway.

Our knowledge of the HOG pathway is rather limited because the upstream and downstream components of the three-kinase modules are not defined. A two-component system similar to the bacterial sensor systems previously described has been found to modulate the HOG pathway negatively (Maeda *et al.,* 1994). The product of the *SLN1* gene is a transmembrane kinase with two potential bacterial-like autophosphorylation sites, one histidine and one aspartate. The product of the *SSk1* gene contains an aspartate that may be phosphorylated by activation of the SLN1 protein. A null mutation of *SLN1* is lethal and this phenotype is suppressed by mutations in either *SSK1* or in the HOG pathway. Apparently, *SLN1* is a negative regulator of the HOG pathway, which, when unchecked, becomes lethal. Overexpression of *SSK1* has been shown to increase tyrosine phosphorylation of the HOG1 protein. Therefore, it is plausible that the SLN1 protein is a negative regulator of the SSK1 protein, which is a positive regulator of HOG. Accordingly, overexpression of different protein phosphatases such as the products of the *PTC1, PTC3,* and *PTP2* genes also suppress the lethality caused by lack of *SLN1* function. Although the couple *SLN1–SSK1* is clearly involved in activation of HOG, the proposal that the SLN1 protein is an osmosensor has not been demonstrated because *ssk1* mutants are not osmosensitive. Therefore, in the absence of the *SLN1–*

*SSK1* couple, the osmotic activation of HOG, essential for osmotic tolerance, still occurs (Maeda *et al.,* 1994).

A different type of abiotic stress is caused by drugs that interfere with basic cellular functions such as ribosomal protein synthesis (cycloheximide, neomycin, and chloramphenicol), oxidative phosphorylation (oligomycin, antimycin, valinomycin, and uncouplers), and DNA structure (acriflavine, 4-nitroquinoline *N*-oxide, *N*-methyl-*N'*-nitro-*N*-nitrosoguanidine). Heavy metals such as cadmium may also be included as chemical stressors affecting general protein function. A mutational analysis involving either increased resistance by overexpression or decreased resistance by mutation has identified a complex cellular pathway involved in multidrug resistance (Balzi and Goffeau, 1994). At one end, multidrug-resistance ATPases of the ABC family (Higgins, 1992) export drugs and heavy metals from the cells. These are the products of the *PDR5, SNQ2,* and *YDR1* (Hirata *et al.,* 1994) genes for drug export and the product of the *YCF1* gene for cadmium export. At the other end of the pathway are transcription factors of either the zinc finger (products of the *PDR1* and *PDR3* genes) or leucine zipper (products of the *YAP1* and *YAP2* genes) type, which determine the expression of the efflux pumps. The mechanism of regulation of these pathways is unknown but it must involve sensors of the drugs or of their cellular effects and signal transduction to activate the transcription factors.

Although the YAP1 transcription factor binds to a promoter element very different from the STRE, it is also involved (by some indirect mechanism) in the induction of STRE-containing promoters by osmotic, thermal (Gounalaki and Thireos, 1994), and oxidative stress (Kuge and Jones, 1994). These studies also demonstrated the unexpected crucial role of some *YAP1*-regulated genes in stress tolerance. The thioredoxin encoded by the *TRX2* gene is important for tolerance to oxidative stress, probably by catalyzing the regeneration of enzymes damaged by oxidation of critical cysteine residues (Kuge and Jones, 1994). The trehalose-P phosphatase encoded by the *TPS2* gene is unexpectedly required for drug (cycloheximide) resistance conferred by overexpression of *YAP1* and, accordingly, drug stress induces trehalose synthesis (Gounalaki and Thireos, 1994). In this respect, it must be pointed out that the *PBS2* gene encoding the MEK of the HOG pathway was initially identified as a gene required for polymyxin B resistance [polymyxin B sensitivity (PBS)] and also involved in thermal tolerance (Boguslawski, 1992).

There is a general correlation between conditions favoring fast growth and stress sensitivity (Mager and Varela, 1993; Piper, 1993). In *S. cerevisiae* fast growth is concomitant with glucose-rich medium, and this carbon source has a negative effect on the expression of stress defense genes. Glucose effects in yeast are mediated by two major pathways modulated by protein kinases: the Snf1 protein kinase and protein kinase A (Thevelein, 1994).

Fast glucose phosphorylation mediated by the *HXK2* hexokinase gene is required for triggering both pathways. The role of this particular hexokinase in glucose-induced signal transduction is explained by the fact that *HXK2* is highly expressed in glucose media while the other hexokinase genes (*HXK1* and *GLK1*) are glucose repressed (Herrero *et al.,* 1995). Fast glucose phosphorylation somehow inactivates the Snf1 protein kinase and activates protein kinase A. The mechanisms of these effects are unknown and may involve "fine sensing" of hexose phosphate levels.

Protein kinase A activity is essential for yeast growth and the enzyme is encoded by three homologous genes, *TPK1, TPK2,* and *TPK3* (Thevelein, 1994). It is modulated by cAMP, which binds and inactivates the regulatory (inhibitory) subunit encoded by the *BCY1* gene. The levels of cAMP are determined by the concerted operation of adenylate cyclase (encoded by the *CYR1* or *CDC35* gene) and phosphodiesterase (encoded by the *PDE1* and *PDE2* genes). Yeast cyclase is activated by the GTP-bound form of the Ras protein (encoded by the *RAS1* and *RAS2* genes). The GTP and GDP forms of Ras depend on the activities of the Ira protein (encoded by the *IRA1* and *IRA2* genes) and of the CDC25 protein. Ira stimulates GTP hydrolysis by Ras, producing the inactive GDP form. Cdc25 stimulates exchange of GDP by GTP, resulting in the active GTP form. This pathway is activated by glucose phosphorylation at the level of Cdc25, although the mechanism is unknown. This favors the GTP form of Ras, which activates the cyclase and increases cAMP levels. Elevated cAMP dissociates the inhibitory Bcy subunit from the catalytic Tpk subunits of protein kinase A. When glucose is exhausted from the medium, the levels of cAMP decrease (François *et al.,* 1987) and the decreased activity of protein kinase A may contribute to the increase in tolerance to several types of stress imposed on stationary cultures (Thevelein, 1994). Accordingly, mutants with reduced protein kinase A activity are more stress tolerant and mutants with hyperactive protein kinase A are stress sensitive. These effects are mediated, at least in part, by the negative effect of protein kinase A on the STRE-dependent expression of many stress defense genes (Marchler *et al.,* 1993; Schüller *et al.,* 1994). The YAP1 transcription factor is involved in the expression of many stress genes and it has been shown to be epistatic to protein kinase A, that is, a null mutation in *YAP1* prevents changes in STRE-dependent expression in mutants of the protein kinase A pathway (Gounalaki and Thireos, 1994). Therefore, this kinase may be a negative modulator of Yap1. Yap1 influences the expression of promoters that do not contain binding sites for this transcription factor (Gounalaki and Thireos, 1994). Therefore, additional components of this complicated network are missing. A paradigm is offered by the mechanism of glucose repression of the *ADH2* gene, where a specific transcription factor (encoded by the *ADR1* gene) is inactivated on phosphorylation by protein kinase A (Cherry *et al.,* 1989). The putative transcription factor binding to the STRE may experience similar regulation. To fit Yap1 within this model requires one to postulate an unidentified Yap1-dependent gene required for STRE-dependent expression. Protein kinase A and the HOG protein kinases define parallel pathways converging with opposite effects on STRE-dependent expression (Schüller *et al.,* 1994). Accordingly, overexpression of the HOG kinase *PBS2* enhances the stress tolerance of mutants compromised by hyperactivation of protein kinase A (Boguslawski, 1992). The precise interaction between these two pathways is unknown.

The other pathway of glucose signaling in yeast, the *SNF1* pathway, is also related to stress tolerance. The Snf1 protein kinase phosphorylates and inactivates the Ssn6–Tup1 general repressor (Keleher *et al.,* 1992). Glucose repression of many genes is mediated by the inhibitory effect on transcription exerted by the combination of Ssn6–Tup1 with the DNA-binding protein Mig1. Derepression in the absence of glucose occurs by reactivation of the Snf1 protein kinase, which inhibits Ssn6–Tup1 action on the sensitive genes (Gancedo, 1992). A *snf1* null mutant is sensitive to heat stress and starvation (Thompson-Jaeger *et al.,* 1991), pointing to an important role of this signaling pathway in stress tolerance. The stress genes controlled by the Snf1 pathway, however, have not been identified. The stress phenotypes of inactive *snf1* mutants are similar to those of hyperactive protein kinase A mutants and mutations that decrease the activity of protein kinase A moderate the stress sensitivity of *snf1* mutants (Thompson-Jaeger *et al.,* 1991). This may explain the cAMP-independent effects of glucose exhaustion during yeast growth (Cameron *et al.,* 1988). Mutations on the *SNF1* and *MIG1* genes have no significant effect on the activity of yeast glycerol-phosphate dehydrogenase (Albertyn *et al.,* 1994a), as if the Snf1 pathway were not involved in the expression of the HOG-dependent gene *GPD1* (Albertyn *et al.,* 1994b). Other stress-signaling pathways could be modulated by the Snf1 pathway. Invertase derepression is fully dependent on the Snf1 pathway but mutants partially defective in this pathway can be suppressed by overexpression of *PDE2,* a cAMP phosphodiesterase gene (Hubbard *et al.,* 1992). This suggests cross-talk between the two pathways, as if protein kinase A inhibited Snf1.

The regulatory pathways previously discussed (HOG, protein kinase A, and Snf1) are general and affect many stress-inducible genes. We have previously indicated that the *ENA1* gene, an important determinant of salt tolerance, is specifically regulated by the calcium-activated protein phosphatase calcineurin (Mendoza *et al.,* 1994) and by the HAL3 protein (Ferrando *et al.,* 1995). It would be interesting to determine how the *ENA1*-specific calcium–HAL3 pathway interacts with the other general pathways. In this respect, other regulatory proteins such as casein kinase I (encoded by the *YCK1* and *YCK2* genes) have been shown to increase salt tolerance on overexpression (Robinson *et al.*, 1992). *YCK1* was isolated as a multicopy suppressor of *snf1* mutants and therefore these two protein kinases may regulate the same components of the salt stress defense machinery.

### 3. Higher Plants

Stress responses in higher plants may be mediated either by stress perception at responsive cells or by local production of stress hormones such as abscisic acid (ABA), ethylene, and jasmonic acid. These hormones travel within the plant and act on sensitive cells anticipating the direct sensing of the stresses (Nover, 1994). In isolated plant tissues and cells, many genes are similarly induced either by direct water stress or by application of ABA (Bray, 1993; Chandler and Robertson, 1994), suggesting that stress hormones may utilize similar signaling pathways other than physical stress. Actually, salinity stress increases cytoplasmic calcium in isolated cells (Lynch *et al.*, 1989) and the same intracellular signal is generated by application of ABA (Schroeder and Hagiwara, 1990; Giraudat, 1995). There are, however, a number of water deficit-induced genes that do not respond to ABA application (Bray, 1993). This may suggest that there is a signaling pathway specific to osmotic stress and different from the calcium-mediated pathway triggered by either ABA or osmotic stress. The two pathways have been traced back to different promoter elements responsive to ABA [the abscisic acid responsive element (ABRE)] and to some unknown signal generated by water deficit and different from the ABA signal [the drought response element (DRE)]. A final complication is that some genes induced by osmotic stress in normal plants fail to do so in mutants blocked in the ABA biosynthesis pathway. This suggests that in whole plants only ABA, not physical stress directly, generates the necessary signals to trigger expression of these genes (Bray, 1993).

One ABRE with consensus sequence CACGTGGC has been identified in many ABA-responsive genes (Bray, 1993; Chandler and Robertson, 1994) and a DNA-binding protein of the leucine zipper family (EmBP-1 from wheat) interacts specifically with this sequence (Guiltinan *et al.*, 1990). It remains to be shown through functional analysis that EmBP-1 acts as a transcriptional activator. This demonstration has been provided for another ABRE-binding protein, the tobacco nuclear factor TAF-1: transient expression of TAF-1 increased $\beta$-glucuronidase activity of plant cells containing an ABRE-*GUS* reporter (Chandler and Robertson, 1994). The mechanism of activation by ABA is not known. Concerning the DRE, a consensus sequence TACCGACAT has been demonstrated to mediate ABA-independent osmotic induction of *GUS* reporter gene in transgenic *Arabidopsis* and tobacco plants (Yamaguchi-Shinozaki and Shinozaki, 1994).

The power of mutant analysis in *A. thaliana* has provided considerable insight into the mechanisms of ABA action (Rock and Quatrano, 1994; Giraudat, 1995). Several ABA-insensitive mutants have been isolated and epistatic interactions and phenotypes have defined two different pathways. The *ABI1* and *ABI2* genes define a pathway mostly affecting vegetative tissues (including stomatal regulation) while the *ABI3* and *ABI5* genes constitute a pathway mostly involved in desiccation tolerance during embryogenesis. Two insensitivity genes have been cloned by map-based positional cloning: *ABI3* is a transcription factor homologous to the seed-specific ABA-sensitivity gene *VP1* from maize and *ABI1* encodes a calcineurin-like protein phosphatase regulated by calcium. This confirms at the molecular level the role of calcium in ABA signal transduction. Calcineurin inhibitors, such as cyclophilin–cyclosporin A and FKBP protein–FK506 complexes, block the calcium-induced inactivation of $K^+$ uptake channels (Luan *et al.*, 1993), again pointing to the participation of a calcineurin-like protein phosphatase in the ABA signaling pathway.

The activation of ABA synthesis by osmotic stress requires induction of gene expression and occurs at the cleavage step converting 9′-*cis*-neoxanthin to xanthoxin (Bray, 1993). The primary water stress signal transduction pathway independent of ABA (see above) may mediate this crucial phenomenon but the mechanism is not known.

## III. Salt Stress at the Whole-Plant Level

One crucial aspect in salt and drought tolerance at the whole-plant level concerns the regulation of stomatal opening to reduce water loss from the plant. Saline and osmotic stresses cause a decrease in growth-promoting cytokinin and an increase in growth-inhibiting abscisic acid (ABA) transported from the root to the shoot. An increase in 1-aminocyclopropane-1-carboxylate (ACC), the precursor of the inhibitory hormone ethylene, is also associated with stress (Lerner and Amzallag, 1994). This hormonal change causes decreased growth and increased stress tolerance. It seems that growth inhibition during salt stress is more the result of an adaptative hormonal response than a consequence of salt or osmotic toxicity. The growth-inhibitory hormones ABA and ethylene are involved in the signal transduction for expression of stress defense genes. The opposing effects of plant hormones on growth and stress tolerance parallel the opposing effects of the cAMP pathway on growth and stress tolerance in yeast (see above). Apparently, optimal conditions for stress tolerance are incompatible with growth and different signal transduction pathways coordinate this antagonism.

Soil drying below a water potential of around −0.3 MPa results in a substantial increase in the ABA content of roots, which is subsequently transported to the shoots (Zhang and Davies, 1989). Abscisic acid induces the closing of stomata, an effect mediated by changes in the activity of different ion transport systems of guard cells. Basically, in addition to inhibiting the electrogenic $H^+$-ATPase, ABA inactivates inward-rectifying $K^+$ channels and activates outward-rectifying $K^+$ and $Cl^-$ channels (Blatt and Thiel, 1993). This results in depolarization and KCl efflux, with concomitant turgor loss and stomatal closure. These effects are probably mediated by an increase in cytoplasmic $Ca^{2+}$ triggered by ABA (see above). Closing of stomata saves water for the plant but reduces photosynthesis and inhibits growth. Therefore, the ideal solution is crassulacean acid metabolism (CAM; Lüttge, 1993). This is a photosynthetic accommodation to water stress, in which stomata open at night and nocturnal $CO_2$ fixation saves loss of water by transpiration. In the absence of light, this fixation occurs in the form of the so-called crassulacean acids, mostly malic acid. During the day these acids release $CO_2$ to allow for photosynthesis with closed stomata. Some annual plants such as *Mesenbryanthemum crystallinum* (ice plant) develop CAM metabolism only during water stress. However, perennial cacti such as *Cereus validus* have constitutive CAM. Development of CAM in *M. crystallinum* correlates with induction of multiple genes at the transcriptional and translational levels (Bohnert *et al.*, 1994). Light is the major activator of stomatal opening in normal plants, a phenomenon mediated by light-induced activation of the electrogenic $H^+$-ATPase (Kearns and Assmann, 1993). It would be interesting to determine the molecular basis for the altered light response of the $H^+$-ATPase in CAM plants.

In addition to stomatal closure, ABA participates in other physiological responses to water stress. For example, increased ABA is required for enhanced proline synthesis during osmotic stress (Ober and Sharp, 1994) and exogenous ABA accelerates adaptation of cultured tobacco cells (LaRosa *et al.*, 1985) and *Sorghum* seedlings (Lerner and Amzallag, 1994) to salt stress. The targets for all these physiological effects of ABA are unknown.

Another crucial aspect for salt tolerance at the whole-plant level is the balance between the root supply of salt to the shoots and the capability for salt uptake of the leaves. Sufficient ions for osmotic adjustment must be transferred to the leaves without exceeding the capacity of leaf cells to accumulate them at the vacuole. Failure in either respect has lethal consequences (Flowers and Yeo, 1988). When the capacity for vacuolar salt accumulation is high, as seems to be the case with halophytic plants, the root supply of salt has a positive effect on growth under conditions of salt stress and there is a correlation between salt tolerance and salt inclusion in the shoot (Glenn *et al.*, 1994). Salt at the vacuole serves as "cheap osmoticum" to fill most of the cellular space, with "expensive" osmolytes filling the cytoplasm. This correlation also has been found with maize cultivars, suggesting that in this crop plant vacuolar compartmentation is adequate (Aberico and Cramer, 1993). However, when the vacuolar accumulation of salt is limited, plants have much reduced salt tolerance and this trait then correlates with salt exclusion from the shoot (Laüchli, 1984). In nonhalophytic plants organic osmolytes constitute the major component of osmotic adjustment, even at the vacuole. The capability of plants for osmolyte synthesis is limited, usually less than 0.4 *M* osmotic concentration (1-MPa water potential).

In some crop plants such as soybean (Abel, 1969) and wheat (Dvorak *et al.*, 1994), genetic analysis indicates that the differences in salt exclusion between related cultivars and species is due to a single gene. The *Knal* gene of *Triticum aestivum* has been located on chromosome 4D (Dvorak *et al.*, 1994). Salt exclusion from the shoot is a property of the root xylem parenchymal cells, which become salt sinks, accumulate salt from the xylem sap, and "sacrifice" for the benefit of the leaves (Laüchli, 1984). A similar sacrifice is observed in the case of the salt glands or hairs developed by some halophytic plants. These epidermal cells, specialized for salt accumulation, die and release their contents on the leave surface. In the halophytic wild rice *Porteresia coarctata,* the salt secreted by the hairs is an important factor in the salt balance of the leaves and the genes responsible for these structures could be useful to improve salt tolerance of cultivated rice (Flowers *et al.*, 1990). Initial salt accumulation in these xylem and epidermal cells occurs at the vacuole and, therefore, one of their conspicuous features must be an exacerbated capability for $Na^+$ and $Cl^-$ transport at the vacuolar membrane.

## IV. Perspectives

Our understanding of living organisms proceeds by formulating hypotheses that need to be tested by experiments. True understanding requires manipulation of model systems, and this provides the link between basic and applied science. An example concerning oxidative stress has been provided by experiments with the model fly *Drosophila melanogaster,* in which simultaneous overexpression in transgenic flies of both superoxide dismutase and catalase resulted in substantial extension of life and less protein oxidative damage (Orr and Sohal, 1994). These results not only provide support for the free radical hypothesis of aging but also have enormous practical implications. For example, in crop plants environmental adversity often leads to oxidative stress and transgenic tobacco plants with elevated levels of superoxide dismutase (manganese isoform) exhibit increased tolerance of damage mediated by oxygen radicals (Bowler *et al.,* 1991). In the same manner, a correlation between life span and general stress resistance (starvation, heat shock) has been demonstrated in yeast (Kennedy *et al.,* 1995). The generation of transgenic tobacco plants with heterologous acyltransferases and altered chilling sensitivity (Hayashi *et al.,* 1994) has confirmed hypotheses linking membrane fluidity, temperature, and fatty acid unsaturation and, at the same time, has produced a biotechnologically useful crop. The manipulation of stress resistance by genetic engineering of crucial components of either the defense machinery or the cellular stress targets could serve as an "acid test" for our understanding of cellular mechanisms and in this way produce novel knowledge and generate useful applications.

As demonstrated by the stress genes identified in *S. cerevisiae,* mutants defective in stress tolerance and transformed cells overexpressing gene libraries have been of great value in characterizing crucial components of the stress defense machinery and weak points in the stress sensitivity of organisms. In the future, the application of these approaches to model plants such as *A. thaliana* may expand our knowledge from the cellular level (modeled with microorganisms) to the complexities of higher plants. The usefulness of model organisms needs to be emphasized. Most studies on plant stress, even at the molecular level, are simple correlations. A review (Chandler and Robertson, 1994) indicates that ". . . we are still unable to detail any single example demonstrating that an ABA-regulated protein plays a role in stress tolerance." Therefore, the postulated roles of many ABA-induced proteins (Bray, 1993) need to be substantiated by experiments. Only a molecular genetic analysis, generating transgenic organisms with gain and loss of function, can demonstrate cause–effect relationships as in the physical sciences. Although hundreds of stress-inducible genes have been described, very few genes, some of constitutive expression, have proved to be crucial for tolerance to particular stresses.

Biological stress responses appear to be conserved throughout evolution, and therefore information obtained with model systems such as microorganisms may be useful in our efforts to manipulate higher plants. The time has come to demonstrate that molecular biologists understand the principles of plant responses to salinity and drought stresses, to the extent of engineering stress-tolerant crops. The tools are ready and the first round of transgenic plants with genes for osmolyte synthesis (Tarczynski *et al.,* 1993; Pilon-Smits *et al.,* 1995) has satisfied most expectations. In addition to osmolyte synthesis, sodium transport systems such as those encoded by the yeast *ENA1* and *SOD2* genes or the bacterial *nhaA* and *nhaB* genes are crucial components of salt stress defense that could be tested in higher plants. Of course, the elusive plant sodium transporter could provide a superior tool for engineering salt tolerance. Chloride transport still needs to be manipulated, but no genes are yet available.

It is expected that salt tolerance could be improved by successive steps, in which the removal of one limiting factor sets the tolerance to the next limiting factor. In this respect, the manipulation of signal transduction pathways may result in pleiotropic effects on several crucial stress defenses. The salt tolerance resulting from overexpression of the yeast *YCK1* (casein kinase I), *YAP1* (transcription factor), and *HAL3* (signaling component of a calcium–calcineurin pathway) genes illustrates the value of this approach. Changes in signal transduction components may explain the stress adaptation phenomena observed by many physiologists and interpreted as the awakening of "dormant" stress defense genes. Interestingly, this adaptation may be transmitted to the next generation (Lerner and Amzallag, 1994) as a kind of "imprinting."

Stress regulatory pathways impinge on basic homeostatic mechanisms such as those involved in ion homeostasis, protein folding, growth control, and differentiation and therefore stress research is producing valuable information for our understanding of basic biological mechanisms. Chaperone-assisted protein folding is an offshoot of heat shock research and salinity research is also expected to contribute novel biological principles. One emerging generalization is the connection between the different stress responses. Organisms challenged by one single stress probably experience other, unexpected stresses because of metabolic alterations that compromise performance. For example, drought induces oxidative stress in plants (Gosset *et al.,* 1994a; Moran *et al.,* 1994). Consequently, stressed organisms trigger a general response to defend themselves against all kind of stresses. Osmotic stress induces antioxidant enzymes such as catalase, superoxide, dismutase, ascorbate peroxidase, peroxidase, and glutathione reductase (Gosset *et al.,* 1994b; Lopez *et al.,* 1994). Also, heat shock induces many of the genes induced by water stress (Borkird *et al.,* 1991). This may result in cross-protection, with enormous practical consequences. For example, heat shock improves storage of apple fruit (Lurie *et al.,* 1994) and salt resistance of cotton plants (Kuznetsov *et al.,* 1993).

One of the lessons to be learned from the great scientist Efraim Racker is that, as soon as one tries to do applied research, the need for fundamental knowledge arises, with the result that basic science is done at the same time. This mixture of basic and applied science is at the root of modern biological science as exemplified by the work of Louis Pasteur. An effort has been made in this article to convey this message in the case of salinity stress; it is hoped that this will help define an exciting area of research oriented to the need for pressing biotechnological solutions, but with important biological principles remaining to be discovered.

## Acknowledgments

This work was supported by grants from the European Union (AIR3 PL92 1508) and Project of Technological Priority, BIOTECH Programme, 1993–1996.

## References

Abel, G. H. (1969). Inheritance of the capacity for chloride inclusion and chloride exclusion by soybeans. *Crop Sci.* **9,** 697–699.

Aberico, G. J., and Cramer, G. R. (1993). Is the salt tolerance of maize related to sodium exclusion? *J. Plant Nutr.* **16,** 2289–2303.

Albertyn, J., Hohmann, S., and Prior, B. A. (1994a). Characterization of the osmotic-stress response in *Saccharomyces cerevisiae:* Osmotic stress and glucose repression regulate glycerol-3-phosphate dehydrogenase independently. *Curr. Genet.* **25,** 12–18.

Albertyn, J., Hohmann, S., Thevelein, J. M., and Prior, B. A. (1994b). *GPD1,* which encodes glycerol-3-phosphate dehydrogenase, is essential for growth under osmotic stress in *Saccharomyces cerevisiae,* and its expression is regulated by the high-osmolarity glycerol response pathway. *Mol. Cell. Biol.* **14,** 4135–4144.

Anderson, J. A., Huprikar, S. S., Kochian, L. V., Lucas, W. J., and Gaber, R. F. (1992). Functional expression of a probable *Arabidopsis thaliana* potassium channel in *Saccharomyces cerevisiae. Proc. Natl. Acad. Sci. U.S.A.* **89,** 3736–3740.

Anderson, J. W., and Beardall, J. (1991). "Molecular Activities of Plant Cells. An Introduction to Plant Biochemistry," pp. 230–232. Blackwell, Oxford.

Andresen, P. A., Kaasen, I., Styrvold, O. B., Boulnois, G., and Ström, A. R. (1988). Molecular cloning, physical mapping and expression of the *bet* genes governing the osmoregulatory choline-glycine betaine pathway of *Escherichia coli. J. Gen. Microbiol.* **134,** 1737–1746.

Asha, H., and Gowrishankar, J. (1993). Regulation of *kdp* operon expression in *Escherichia coli:* Evidence against turgor as signal for transcriptional control. *J. Bacteriol.* **175,** 4528–4537.

Ashraf, M. (1994). Breeding for salinity tolerance in plants. *Crit. Rev. Plant Sci.* **13,** 17–42.

Bakker, E. P. (1993a). Cell $K^+$ and $K^+$ transport systems in prokaryotes. *In* "Alkali Cation Transport Systems in Prokaryotes" (E. P. Bakker, ed.), pp. 205–224. CRC Press, Boca Raton, FL.

Bakker, E. P. (1993b). Low-affinity $K^+$ uptake systems *In* "Alkali Cation Transport Systems in Prokaryotes" (E. P. Bakker, ed.), pp. 253–276. CRC Press, Boca Raton, FL.

Balzi, E., and Goffeau, A. (1994). Genetics and biochemistry of yeast multidrug resistance. *Biochim. Biophys. Acta* **1187,** 152–162.

Bañuls, J., and Primo-Millo, E. (1992). Effects of chloride and sodium on gas exchange parameters and water relations of *Citrus* plants. *Physiol. Plant.* **86,** 115–123.

Bartels, D., Engelhardt, K., Roncarati, R., Schneider, K., Rotter, M., and Salamini, F. (1991). An ABA and GA modulated gene expressed in the barley embryo encodes an aldose reductase related protein. *EMBO J.* **10,** 1037–1043.

Beever, R. E., and Laracy, E. P. (1986). Osmotic adjustment in the filamentous fungus *Aspergillus nidulans. J. Bacteriol.* **168,** 1358–1365.

Belazzi, T., Wagner, A., Wieser, R., Schanz, M., Aam, G., Hartig, A., and Ruis, H. (1991). Negative regulation of transcription of the *Saccharomyces cerevisiae* catalase T (*CTT1*) gene by cAMP is mediated by a positive control element. *EMBO J.* **10,** 585–592.

Bell, W., Klaassen, P., Ohnacker, M., Boller, T., Herweijer, M., Schoppink, P., Van der Zee, P., and Wiemken, A. (1992). Characterization of the 56-kDa subunit of yeast trehalose-6-phosphate synthase and cloning of its gene reveal its identity with the product of *CIF1,* a regulator of carbon catabolite inactivation. *Eur. J. Biochem.* **209,** 951–959.

Beloualy, N., and Bouharmont, J. (1992). NaCl-tolerant plants of *Poncirus trifoliata* regenerated from tolerant cell lines. *Theor. Appl. Genet.* **83,** 509–514.

Ben-Hayyim, G., Speigel-Roy, P., and Neumann, H. (1985). Relation between ion accumulation of salt-sensitive and isolated stable salt-tolerant cell lines of *Citrus aurantium. Plant Physiol.* **78,** 144–148.

Bertl, A., Slayman, C. L., and Gradmann, D. (1993). Gating and conductance in an outward-rectifying $K^+$ channel from the plasma membrane of *Saccharomyces cerevisiae*. *J. Membr. Biol.* **132,** 183–199.

Binzel, M. L., Hasegawa, P. M., Rhodes, D., Handa, S., Handa, A. K., and Bressan, R. (1987). Solute accumulation in tobacco cells adapted to NaCl. *Plant Physiol.* **84,** 1408–1415.

Binzel, M. L., Hess, F. D., Bressan, R. A., and Hasegawa, P. M. (1988). Intracellular compartmentation of ions in salt adapted tobacco cells. *Plant Physiol.* **86,** 607–614.

Blatt, M. R., and Thiel, G. (1993). Hormonal control of ion channel gating. *Annu. Rev. Plant Physiol. Plant Mol. Biol.* **44,** 543–567.

Blomberg, A., and Adler, L. (1989). Roles of glycerol and glycerol-3-phosphate dehydrogenase ($NAD^+$) in acquired osmotolerance of *Saccharomyces cerevisiae*. *J. Bacteriol.* **171,** 1087–1092.

Blomberg, A., and Adler, L. (1992). Physiology of osmotolerance in fungi. *Adv. Microb. Physiol.* **33,** 145–212.

Bogges, S. F., Aspinal, D., and Paleg, L. G. (1976). Stress metabolism. IX. The significance of end-product inhibition of proline biosynthesis and of compartmentation in relation to stress-induced proline accumulation. *Aust. J. Plant Physiol.* **3,** 513–525.

Boguslawski, G. (1992). *PBS2,* a yeast gene encoding a putative protein kinase, interacts with the *RAS2* pathway and affects osmotic sensitivity of *Saccharomyces cerevisiae*. *J. Gen. Microbiol.* **138,** 2425–2432.

Bohnert, H. J., Thomas, J. C., DeRocher, E. J., Michalowski, C. B., Breiteneder, H., Vernon, D. M., Deng, W. D., Yamada, S., and Jensen, R. G. (1994). Responses to salt stress in the halophyte *Mesenbryanthemum crystallinum*. *In* "Biochemical and Cellular Mechanisms of Stress Tolerance in Plants" (J. H. Cherry, ed.), pp. 415–428. Springer-Verlag, Berlin.

Borkird, C., Claes, B., Caplan, A., Simoens, C., and Van Montagu, M. (1991). Differential expression of water-stress associated genes in tissues of rice plants. *J. Plant Physiol.* **138,** 591–595.

Bowler, C., Slooten, L., Vandenbranden, S., De Rycke, R., Botterman, J., Sybesma, C., Van Montagu, M., and Inze, D. (1991). Manganese superoxide dismutase can reduce cellular damage mediated by oxygen radicals in transgenic plants. *EMBO J.* **10,** 1723–1732.

Braun, Y., Hassidim, M., Lerner, H. R., and Reinhold, L. (1986). Studies on $H^+$-translocating ATPases in plants of varying resistance to salinity. I. Salinity during growth modulates the proton pump in the halophyte *Atriplex nummularia*. *Plant Physiol.* **81,** 1050–1056.

Braun, Y., Hassidim, M., Lerner, H. R., and Reinhold, L. (1988). Evidence for a $Na^+/H^+$ antiporter in membrane vesicles isolated from roots of the halophyte *Atriplex nummularia*. *Plant Physiol.* **87,** 104–108.

Bray, E. A. (1993). Molecular responses to water deficit. *Plant Physiol.* **103,** 1035–1040.

Brewster, J. L., de Valoir, T., Dwyer, N. D., Winter, E., and Gustin, M. C. (1993). An osmosensing signal transduction pathway in yeast. *Science* **259,** 1760–1763.

Brown, A. D. (1990). "Microbial Water Stress Physiology. Principles and Perspectives." Wiley, New York.

Cameron, S., Levin, L., Zoller, M., and Wigler, M. (1988). cAMP-independent control of sporulation, glycogen metabolism, and heat shock resistance in *S. cerevisiae*. *Cell* (*Cambridge, Mass.*) **53,** 555–566.

Cao, Y., Anderova, M., Crawford, N. M., and Schroeder, J. I. (1992). Expression of an outward-rectifying potassium channel from maize mRNA and complementary RNA in *Xenopus* oocytes. *Plant Cell* **4,** 961–969.

Carmel, O., Dover, N., Rahav-Manor, O., Dibrov, P., Kirsch, D., Karpel, R., Schuldiner, S., and Padan, E. (1994). A single amino acid substitution (Glu134 → Ala) in NhaR1 increases the inducibility by $Na^+$ of the product of *nhaA,* a $Na^+/H^+$ antiporter gene in *Escherichia coli*. *EMBO J.* **13,** 1981–1989.

Chandler, P. M., and Robertson, M. (1994). Gene expression regulated by abscisic acid and its relation to stress tolerance. *Annu. Rev. Plant Physiol. Plant Mol. Biol.* **45,** 113–141.

Cheeseman, J. M. (1982). Pump-leak sodium fluxes in low salt corn roots. *J. Membr. Biol.* **70,** 157–165.

Cherry, J. R., Johnson, T. R., Dollard, C., Schuster, J. R., and Denis, C. L. (1989). Cyclic AMP-dependent protein kinase phosphorylates and inactivates the yeast transcriptional activator ADR1. *Cell (Cambridge, Mass.)* **56,** 409–419.

Colombo, R., Bonetti, A., and Lado, P. (1979). Promoting effect of fusicoccin on $Na^+$ efflux in barley roots: Evidence for a $Na^+$–$H^+$ antiport. *Plant Cell Environ.* **2,** 281–285.

Cooper, J. A. (1994). MAP kinase pathways. Straight and narrow or tortuous and intersecting? *Curr. Biol.* **4,** 1118–1121.

Cosgrove, D. J., and Hedrich, R. (1991). Stretch-activated chloride, potassium and calcium channels coexisting in plasma membranes of guard cells of *Vicia faba* L. *Planta* **186,** 143–153.

Crowe, J. H., Hoekstra, F. A., and Crowe, L. M. (1992). Anhydrobiosis. *Annu. Rev. Physiol.* **54,** 579–599.

Csonka, L. N. (1981). Proline over-production results in enhanced osmotolerance in *Salmonella typhimurium. Mol. Gen. Genet.* **182,** 82–86.

Csonka, L. N., and Hanson, L. N. (1991). Prokaryotic osmoregulation: Genetics and physiology. *Annu. Rev. Microbiol.* **45,** 569–606.

Csonka, L. N., Gelvin, S. B., Goodner, B. W., Orser, C. S., Siemieniak, D., and Slightom, J. L. (1988). Nucleotide sequence of a mutation in the *proB* gene of *Escherichia coli* that confers proline overproduction and enhanced tolerance to osmotic stress. *Gene* **64,** 199–205.

Delauney, A. J., and Verma, D. P. S. (1993). Proline biosynthesis and osmoregulation in plants. *Plant J.* **4,** 215–223.

De Virgilio, C., Piper, P., Boller, T., and Wiemken, A. (1991). Acquisition of thermotolerance in *Saccharomyces cerevisiae* without heat shock protein hsp104 and in the absence of protein synthesis. *FEBS Lett.* **288,** 86–90.

De Virgilio, C., Bürckert, N., Bell, W., Jenö, P., Boller, T., and Wiemken, A. (1993). Disruption of *TPS2,* the gene encoding the 100-kDa subunit of the trehalose-6-phosphate synthase/phosphatase complex in *Saccharomyces cerevisiae,* causes accumulation of trehalose-6-phosphate and loss of trehalose-6-phosphate phosphatase activity. *Eur. J. Biochem.* **212,** 315–323.

De Virgilio, C., Hottiger, T., Dominguez, J., Boller, T., and Wiemken, A. (1994). The role of trehalose synthesis for the acquisition of thermotolerance in yeast. I. Genetic evidence that trehalose is a thermoprotectant. *Eur. J. Biochem.* **219,** 179–186.

Dix, P. J. (1993). The role of mutant cell lines in studies on environmental stress tolerance: An assessment. *Plant J.* **3,** 309–313.

Downton, W. J. S. (1984). Salt tolerance of food crops: Perspectives for improvements. *Crit. Rev. Plant Sci.* **1,** 183–201.

DuPont, F. M. (1992). Salt-induced changes in ion transport: Regulation of primary pumps and secondary transporters. *In* "Transport and Receptor Proteins of Plant Membranes. Molecular Structure and Function" (D. T. Cooke and D. T. Clarkson, eds.), pp. 91–100. Plenum, New York.

Dvorak, J., Noaman, M. M., Goyal, S., and Gorham, J. (1994). Enhancement of salt tolerance of *Triticum turgidum* L. by the *Kna1* locus transferred from the *Triticum aestivum* L. chromosome 4D by homologous recombination. *Theor. Appl. Genet.* **87,** 872–877.

Eddy, A. A., Hopkins, P., and Shaw, R. (1994). Proton and charge circulation through substrate symport in *Saccharomyces cerevisiae:* Non-classical behaviour of the cytosine symport. *Symp. Soc. Exp. Biol.* **48,** 123–139.

Epstein, E., and Rains, D. W. (1965). Carrier-mediated cation transport in barley roots: Kinetic evidence for a spectrum of active sites. *Proc. Natl. Acad. Sci. U.S.A.* **53,** 1320–1324.

Epstein, E., Rains, D. W., and Elzam, O. E. (1963). Resolution of dual mechanisms of potassium absorption by barley roots. *Proc. Natl. Acad. Sci. U.S.A.* **49,** 684–692.

Epstein, W. (1986). Osmoregulation by potassium transport in *Escherichia coli. FEMS Microbiol. Rev.* **39,** 73–78.

Epstein, W., Buurman, E., McLaggan, D., and Naprstek, J. (1993). Multiple mechanisms, roles and controls of $K^+$ transport in *Escherichia coli. Biochem. Soc. Trans.* **21,** 1006–1010.

Everard, J. D., Gucci, R., Kann, S. C., Flore, J. A., and Loescher, W. H. (1994). Gas exchange and carbon partitioning in the leaves of celery (*Apium graveolens* L.) at various levels of root zone salinity. *Plant Physiol.* **106,** 281–292.

Felle, H. H. (1994). The $H^+/Cl^-$ symporter in root-hair cells of *Sinapis alba.* An electrophysiological study using ion-selective microelectrodes. *Plant Physiol.* **106,** 1131–1136.

Fernandes, M., Xiao, H., and Lis, J. T. (1994). Fine structure analyses of the *Drosophila* and *Saccharomyces* heat shock factor-heat shock element interactions. *Nucleic Acids Res.* **22,** 167–173.

Ferrando, A., Kron, S. J., Rios, G., Fink, G. R., and Serrano, R. (1995). Regulation of cation transport in *Saccharomyces cerevisiae* by the salt tolerance gene *HAL3. Mol. Cell. Biol.* **15,** 5470–5481.

Finkelstein, R. R., and Somerville, C. R. (1990). Three classes of abscisic acid (ABA)-insensitive mutations of *Arabidopsis* define genes that control overlapping subsets of ABA responses. *Plant Physiol.* **94,** 1172–1179.

Flowers, T. J., and Yeo, A. R. (1988). Ion relations of salt tolerance. *In* "Solute Transport in Plant Cells and Tissues" (D. A. Baker and J. L. Halls, eds.), pp. 392–416. Longman, London.

Flowers, T. J., Flowers, S. A., Hajibagheri, M. A., and Yeo, A. R. (1990). Salt tolerance in the halophytic wild rice, *Porteresia coarctata* Tateoka. *New Phytol.* **114,** 675–684.

Flowers, T. J., Hajibagheri, M. A., and Yeo, A. R. (1991). Ion accumulation in the cell walls of rice plants growing under saline conditions: Evidence for the Oertli hypothesis. *Plant, Cell Environ.* **14,** 319–325.

François, J., Eraso, P., and Gancedo, C. (1987). Changes in the concentration of cAMP, fructose 2,6-bisphosphate and related metabolites and enzymes in *Saccharomyces cerevisiae* during growth on glucose. *Eur. J. Biochem.* **164,** 369–373.

Friedberg, E. C., Siede, W., and Cooper, A. J. (1991). Cellular responses to DNA damage in yeast. *In* "The Molecular and Cellular Biology of the Yeast *Saccharomyces:* Genome Dynamics, Protein Synthesis and Energetics" ( J. R. Broach, J. R. Pringle, and E. W. Jones, eds.), pp. 147–192. Cold Spring Harbor Lab. Press, Cold Spring Harbor, NY.

Gaber, R. F. (1992). Molecular genetics of yeast ion transport. *Int. Rev. Cytol.* **137A,** 299–353.

Gancedo, C., and Serrano, R. (1989). Energy yielding metabolism. *In* "The Yeast" (J. S. Harrison and A. H. Rose, eds.), 2nd ed., Vol. 3, pp. 205–259. Academic Press, San Diego, CA.

Gancedo, J. M. (1992). Carbon catabolite repression in yeast. *Eur. J. Biochem.* **206,** 297–313.

Garciadeblas, B., Rubio, F., Quintero, F. J., Bañuelos, M. A., Haro, R., and Rodriquez-Navarro, A. (1993). Differential expression of two genes encoding isoforms of the ATPase involved in sodium efflux in *Saccharomyces cerevisiae. Mol. Gen. Genet.* **236,** 363–368.

Gaxiola, R., Larrinoa, I. F., Villalba, J. M., and Serrano, R. (1992). A novel and conserved salt-induced protein is an important determinant of salt tolerance in yeast. *EMBO J.* **11,** 3157–3164.

Giaever, H. M., Styrvold, O. B., Kaasen, I., and Ström, A. R. (1988). Biochemical and genetic characterization of osmoregulatory trehalose synthesis in *Escherichia coli. J. Bacteriol.* **170,** 2841–2849.

Gimmler, H., Kaaden, R., Kirchner, U., and Weyand, A. (1984). The chloride sensitivity of *Dunaliella parva* enzymes. *Z. Pflanzenphysiol.* **114,** 131–150.

Giraudat, J. (1995). Abscisic acid signaling. *Curr. Opin. Cell Biol.* **7,** 232–238.

Gläser, H. U., Thomas, D., Gaxiola, R., Montrichard, F., Surdin-Kerjan, Y., and Serrano, R. (1993). Salt tolerance and methionine biosynthesis in *Saccharomyces cerevisiae* involve a putative phosphatase gene. *EMBO J.* **12,** 3105–3110.

Glenn, E. P., Olsen, M., Frye, R., Moore, D., and Miyamoto, S. (1994). How much sodium accumulation is necessary for salt tolerance in subspecies of the halophyte *Atriplex canescens? Plant Cell Environ.* **17,** 711–719.

Gorham, J., Wyn Jones, R. G., and Bristol, A. (1990). Partial characterization of the trait for enhanced $K^+$–$Na^+$ discrimination in the D genome of wheat. *Planta* **180,** 590–597.

Gosset, D. R., Millhollon, E. P., and Lucas, M. C. (1994a). Antioxidant response to NaCl stress in salt-tolerant and salt-sensitive cultivars of cotton. *Crop Sci.* **34,** 706–714.

Gosset, D. R., Millhollon, E. P., Lucas, M. C., Banks, S. W., and Marney, M. M. (1994b). The effect of NaCl on antioxidant enzyme activities in callus tissue of salt tolerant and salt sensitive cotton cultivars. *Plant Cell Rep.* **13,** 498–503.

Gounalaki, N., and Thireos, G. (1994). Yap1p, a yeast transcriptional activator that mediates multidrug resistance, regulates the metabolic stress response. *EMBO J.* **13,** 4036–4041.

Gowrishankar, J., Jayashree, P., and Rajkumari, K. (1986). Molecular cloning of an osmoregulatory locus in *Escherichia coli:* Increased *proU* gene dosage results in enhanced osmotolerance. *J. Bacteriol.* **168,** 1197–1204.

Gronwald, J. W., Suhayda, C. G., Tal, M., and Shannon, M. C. (1990). Reduction in plasma membrane ATPase activity of tomato roots by salt stress. *Plant Sci.* **66,** 145–153.

Gründer, S., Thiemann, A., Pusch, M., and Jentsch, T. J. (1992). Regions involved in the opening of ClC-2 chloride channel by voltage and cell volume. *Nature* (*London*) **360,** 759–762.

Guiltinan, M. J., Marcotte, W. R., and Quatrano, R. S. (1990). A plant leucine zipper protein that recognizes an abscisic acid responsive element. *Science* **250,** 267–271.

Gustin, M. C., Zhou, X.-L., Martinac, B., and Kung, C. (1988). A mechanosensitive ion channel in the yeast plasma membrane. *Science* **242,** 762–765.

Hajibagheri, M. A., and Flowers, T. J. (1993). Use of freeze-substitution and molecular distillation drying in the preparation of *Dunaliella parva* for ion localization studies by x-ray microanalysis. *Microsc. Res. Tech.* **24,** 395–399.

Handa, S., Handa, A. K., Hasegawa, P. M., and Bressan, R. A. (1986). Proline accumulation and the adaptation of cultured plant cells to water stress. *Plant Physiol.* **80,** 938–945.

Hanson, A. D., Rathinasabapathi, B., Rivoal, J., Burnet, M., Dillon, M. O., and Gage, D. A. (1994). Osmoprotective compounds in the Plumbaginaceae: A natural experiment in metabolic engineering of stress tolerance. *Proc. Natl. Acad. Sci. U.S.A.* **91,** 306–310.

Haro, R., Garciadeblas, B., and Rodriguez-Navarro, A. (1991). A novel P-type ATPase from yeast involved in sodium transport. *FEBS Lett.* **291,** 189–191.

Haro, R., Bañuelos, M. A., Quintero, F. J., Rubio, F., and Rodriguez-Navarro, A. (1993). Genetic basis of sodium exclusion and sodium tolerance in yeast. A model for plants. *Physiol. Plant.* **89,** 868–874.

Hayashi, H., Nishida, I., Ishizaki-Nishizawa, O., Nishiyama, Y., and Murata, N. (1994). Genetically engineered modification of plant chilling sensitivity and characterization of cyanobacterial heat shock proteins. *In* "Biochemical and Cellular Mechanisms of Stress Tolerance in Plants" (J. H. Cherry, ed.), pp. 543–555. Springer-Verlag, Berlin.

Hedrich, R. (1994). Voltage-dependent chloride channels in plant cells: Identification, characterization and regulation of a guard cell anion channel. *Curr. Top. Membr.* **42,** 1–33.

Hengge-Aronis, R., Klein, W., Lange, R., Rimmele, M., and Boos, W. (1991). Trehalose synthesis genes are controlled by the putative sigma factor encoded by *rpoS* and are involved in stationary-phase thermotolerance in *Escherichia coli. J. Bacteriol.* **173,** 7918–7924.

Herrero, P., Galindez, J., Ruiz, N., Martinez-Campa, C., and Moreno, F. (1995). Transcriptional regulation of the *Saccharomyces cerevisiae HXK1, HXK2* and *GLK1* genes. *Yeast* **11,** 137–144.

Higgins, C. F. (1992). ABC transporters: From microorganisms to man. *Annu. Rev. Cell Biol.* **8,** 67–113.

Higgins, C. F., Cairney, J. C., Stirling, D. A., Sutherland, L., and Booth, I. R. (1987). Osmotic regulation of gene expression: Ionic strength as an intracellular signal? *Trends Biochem. Sci.* **12,** 339–344.

Higgins, C. F., Dorman, C. J., Stirling, D. A., Waddell, L., Booth, I. R., May, G., and Bremer, E. (1988). A physiological role for DNA supercoiling in the osmotic regulation of gene expression in *S. typhimurium* and *E. coli. Cell* (*Cambridge, Mass.*) **52,** 569–584.

Hirata, D., Yano, K., Miyahara, K., and Miyakawa, T. (1994). *Saccharomyces cerevisiae* Ydr1, which encodes a member of the ATP-binding cassette (ABC) superfamily, is required for multidrug resistance. *Curr. Genet.* **26,** 285–294.

Hoekstra, F. A., and van Roekel, T. (1988). Desiccation tolerance of *Papaver dubium* L. pollen during its development in the anther. Possible role of phospholipid composition and sucrose content. *Plant Physiol.* **88,** 626–632.

Hu, C. A., Delauney, A. J., and Verma, D. P. S. (1992). A bifunctional enzyme ($\Delta^1$-pyrroline-5-carboxylate synthetase) catalyzes the first two steps in proline biosynthesis in plants. *Proc. Natl. Acad. Sci. U.S.A.* **89,** 9354–9358.

Hubbard, E. J. A., Yang, X., and Carlson, M. (1992). Relationship of the cAMP-dependent protein kinase pathway to the *SNF1* protein kinase and invertase expression in *Saccharomyces cerevisiae. Genetics* **130,** 71–80.

Hulton, C. S. J., Seirafi, A., Hinton, J. C. D., Sidebotham, J. M., Waddell, L., Pavitt, J. M., Owen-Hughes, T., Spassky, A., Buc, H., and Higgins, C. F. (1990). Histone-like protein H1 (H-NS), DNA supercoiling, and gene expression in bacteria. *Cell* (*Cambridge, Mass.*) **63,** 631–642.

Jacobson, T., and Adams, R. M. (1958). Salt and silt in ancient Mesopotamian agriculture. *Science* **128,** 1251–1258.

Jacoby, B., and Teomy, S. (1988). Assessment of a $Na^+/H^+$ antiport in ATP-depleted red beet slices and barley roots. *Plant Sci.* **55,** 103–106.

Jain, S., Nainawatee, H. S., Jain, R. K., and Chowdhury, J. B. (1991). Proline status of genetically stable salt-tolerant *Brassica juncea* L. somaclones and their parent cv. Prakash. *Plant Cell Rep.* **9,** 684–687.

Jia, Z.-P., McCullough, N., Martel, R., Hemmingsen, S., and Young, P. G. (1992). Gene amplification at a locus encoding a putative $Na^+/H^+$ antiporter confers sodium and lithium tolerance in fission yeast. *EMBO J.* **11,** 1631–1640.

Kaasen, I., Falkenberg, P., Styrvold, O. B., and Ström, A. R. (1991). Molecular cloning and physical mapping of the *otsBA* genes encoding the osmoregulatory trehalose pathway of *Escherichia coli:* Evidence that transcription is activated by KatF (AppR). *J. Bacteriol.* **173,** 1187–1194.

Kaasen, I., McDougall, J., and Ström, A. R. (1994). Analysis of the *otsBa* operon for osmoregulatory trehalose synthesis in *Escherichia coli* and homology of the OtsA and OtsB proteins to the yeast trehalose-6-phosphate synthase/phosphatase complex. *Gene* **145,** 9–15.

Kearns, E. V., and Assmann, S. M. (1993). The guard cell–environment connection. *Plant Physiol.* **102,** 711–715.

Keleher, C. A., Redd, M. J., Schultz, J., Carlson, M., and Johnson, A. D. (1992). Ssn6-Tup1 is a general repressor of transcription in yeast. *Cell* (*Cambridge, Mass.*) **68,** 709–719.

Kennedy, B. K., Austriaco, N. R., Zhang, J., and Guarente, L. (1995). Mutation in the silencing gene *SIR4* can delay aging in *S. cerevisiae. Cell* (*Cambridge, Mass.*) **80,** 485–496.

Kidd, G., and Devorak, J. (1994). Trehalose is a sweet target for agbiotech. *Bio/Technology* **12,** 1328–1329.

Kirk, N., and Piper, P. W. (1991). The determinants of heat-shock element-directed lacZ expression in *Saccharomyces cerevisiae. Yeast* **7,** 539–546.

Kirti, P. B., Hadi, S., Kumar, P. A., and Chopra, V. L. (1991). Production of sodium-chloride tolerant *Brassica juncea* plants by in vitro selection at the somatic embryo level. *Theor. Appl. Genet.* **83,** 233–237.

Ko, C. H., and Gaber, R. F. (1991). *TRK1* and *TRK2* encode structurally related $K^+$ transporters in *Saccharomyces cerevisiae. Mol. Cell. Biol.* **11,** 4266–4273.
Ko, C. H., Liang, H., and Gaber, R. F. (1993). Roles of multiple glucose transporters in *Saccharomyces cerevisiae. Mol. Cell. Biol.* **13,** 638–648.
Kobayashi, N., and McEntee, K. (1990). Evidence for a heat shock transcription factor-independent mechanism for heat shock induction of transcription in *Saccharomyces cerevisiae. Proc. Natl. Acad. Sci. U.S.A.* **87,** 6550–6554.
Kobayashi, N., and McEntee, K. (1993). Identification of cis and trans components of a novel heat shock stress regulatory pathway in *Saccharomyces cerevisiae. Mol. Cell. Biol.* **13,** 248–256.
Kochian, L. V., Garvin, D. F., Shaff, J. E., Chilcott, T. C., and Lucas, W. J. (1993). Towards an understanding of the molecular basis of plant $K^+$ transport. Characterization of cloned $K^+$ transport cDNAs. *Plant Soil* **156,** 115–118.
Kueh, J. S. H., and Bright, S. W. J. (1982). Biochemical and genetical analysis of three proline-accumulating barley mutants. *Plant Sci. Lett.* **27,** 233–241.
Kuge, S., and Jones, N. (1994). *YAP1* dependent activation of *TRX2* is essential for the response of *Saccharomyces cerevisiae* to oxidative stress. *EMBO J.* **13,** 655–664.
Kuznetsov, V. V., Rakitin, V. Y., Borisova, N. N., and Rotschupkin, B. V. (1993). Why does heat shock increase salt resistance in cotton plants? *Plant Physiol. Biochem.* **31,** 181–188.
Lages, F., and Lucas, C. (1995). Characterization of a glycerol/$H^+$ symport in the halotolerant yeast *Pichia sorbitophila. Yeast* **11,** 111–119.
Larher, F., Leport, L., Petrivalsky, M., and Chappart, M. (1993). Effectors for the osmoinduced proline response in higher plants. *Plant Physiol. Biochem.* **31,** 911–922.
LaRosa, P. C., Handa, A. K., Hasegawa, P. M., and Bressan, R. A. (1985). Abscisic acid accelerates adaptation of cultured tobacco cells to salt. *Plant Physiol.* **79,** 138–142.
Larsson, K., Ansell, R., Eriksson, P., and Adler, L. (1993). A gene encoding *sn*-glycerol 3-phosphate dehydrogenase ($NAD^+$) complements an osmosensitive mutant of *Saccharomyces cerevisiae. Mol. Microbiol.* **10,** 1101–1111.
Latterich, M., and Watson, M. D. (1991). Isolation and characterization of osmosensitive vacuolar mutants of *Saccharomyces cerevisiae. Mol. Microbiol.* **5,** 2417–2426.
Latterich, M., and Watson, M. D. (1993). Evidence for a dual osmoregulatory mechanism in the yeast *Saccharomyces cerevisiae. Biochem. Biophys. Res. Commun.* **191,** 1111–1117.
Laüchli, A. (1984). Salt exclusion: An adaptation of legumes for crops and pastures under saline conditions. *In* "Salinity Tolerance in Plants. Strategies for Crop Improvement" (R. C. Staples and G. H. Toenniessen, eds.), pp. 171–188. Wiley, New York.
Laüchli, A., Colmer, T. D., Fan, T. W.-M., and Higashi, R. M. (1994). Solute regulation by calcium in salt-stressed plants. *In* "Biochemical and Cellular Mechanisms of Stress Tolerance in Plants" (J. H. Cherry, ed.), pp. 443–461. Springer-Verlag, Berlin.
Leigh, R. A., and Wyn Jones, R. G. (1985). Cellular compartmentation in plant nutrition: The selective cytoplasm and the promiscuous vacuole. *Adv. Plant Nutr.* **2,** 249–279.
Leigh, R. A., Ahmad, N., and Wyn Jones, R. G. (1981). Assessment of glycinebetaine and proline compartmentation by analysis of isolated vacuoles. *Planta* **153,** 34–41.
LeRedulier, D., Ström, A. R., Dandekar, A. M., Smith, L. T., and Valentine, R. C. (1984). Molecular biology of osmoregulation. *Science* **224,** 1064–1068.
Lerner, H. R., and Amzallag, G. N. (1994). The responses of plants to salinity: A working hypothesis. *In* "Biochemical and Cellular Mechanisms of Stress Tolerance in Plants" (J. H. Cherry, ed.), pp. 463–476. Springer-Verlag, Berlin.
Lesage, F., Hugnot, J.-P., Amri, E.-Z., Grimaldi, P., Barhanin, J., and Lazdunski, M. (1994). Expression cloning in $K^+$ transport defective yeast and distribution of HBP1, a new putative HMG transcriptional regulator. *Nucleic Acids Res.* **22,** 3685–3688.
Levin, D. E., and Errede, B. (1995). The proliferation of MAP kinase signaling pathways in yeast. *Curr. Opin. Cell Biol.* **7,** 197–202.

Loescher, W. H. (1987). Physiology and metabolism of sugar alcohols in higher plants. *Physiol. Plant.* **70,** 553–557.

Lone, M. I., Kue, J. S. H., Wyn Jones, R. G., and Bright, S. W. (1987). Influence of proline and glycinebetaine on salt tolerance of cultured barley embryos. *J. Exp. Bot.* **38,** 479–490.

Lopez, F., Vansuyt, G., Derancourt, J., Fourcroy, P., and Casse-Delbart, F. (1994). Identification by 2D-PAGE analysis of salt-stress induced proteins in radish (*Raphanus sativus*). *Cell. Mol. Biol.* **40,** 85–90.

Luan, S., Li, W., Rusnak, F., Assmann, S. M., and Schreiber, S. L. (1993). Immunosuppressants implicate protein phosphatase regulation of $K^+$ channels in guard cells. *Proc. Natl. Acad. Sci. U.S.A.* **90,** 2202–2206.

Lucas, C., Da Costa, M., and Van Uden, N. (1990). Osmoregulatory active sodium-glycerol co-transport in the halotolerant yeast *Debaryomyces hansenii. Yeast* **6,** 187–191.

Lurie, S., Klein, J. D., and Fallik, E. (1994). Cross protection of one stress by another: Strategies in postharvest fruit storage. *In* "Biochemical and Cellular Mechanisms of Stress Tolerance in Plants" (J. H. Cherry, ed.), pp. 201–212. Springer-Verlag, Berlin.

Lüttge, U. (1993). The role of crassulacean acid metabolism (CAM) in the adaptation of plants to salinity. *New Phytol.* **125,** 59–71.

Lynch, J., Polito, V. S., and Laüchli, A. (1989). Salinity stress increases cytoplasmic Ca activity in maize root protoplast. *Plant Physiol.* **90,** 1271–1274.

Maathuis, F. J., and Sanders, D. (1994). Mechanism of high-affinity potassium uptake in roots of *Arabidopsis thaliana. Proc. Natl. Acad. Sci. U.S.A.* **91,** 9272–9276.

Maeda, T., Wurgler-Murphy, S. M., and Saito, H. (1994). A two-component system that regulates an osmosensing cascade in yeast. *Nature (London)* **369,** 242–245.

Mager, W. H., and Varela, J. C. S. (1993). Osmostress response of the yeast *Saccharomyces. Mol. Microbiol.* **10,** 253–258.

Marchler, G., Schüller, C., Adam, G., and Ruis, H. (1993). A *Saccharomyces cerevisiae* UAS element controlled by protein kinase A activates transcription in response to a variety of stress conditions. *EMBO J.* **12,** 1997–2003.

Matsumoto, H., and Chung, G. C. (1988). Increase in proton-transport activity of tonoplast vesicles as an adaptative response of barley roots to NaCl stress. *Plant Cell Physiol.* **29,** 1133–1140.

McCue, K. F., and Hanson, A. D. (1990). Drought and salt tolerance: Towards understanding and application. *Trends Biotechnol.* **8,** 358–362.

McCulloch, S. R., Beilby, M. J., and Walker, N. A. (1990). Transport of potassium in *Chara australis.* II. Kinetics of a symport with sodium. *J. Membr. Biol.* **115,** 129–143.

McDougall, J., Kaasen, I., and Ström, A. R. (1993). A yeast gene for trehalose-6-phosphate synthase and its complementation of an *Escherichia coli otsA* mutant. *FEMS Microbiol. Lett.* **107,** 25–30.

Mendoza, I., Rubio, F., Rodriguez-Navarro, A., and Pardo, J. M. (1994). The protein phosphatase calcineurin is essential for NaCl tolerance of *Saccharomyces cerevisiae. J. Biol. Chem.* **269,** 8792–8796.

Metzler, D. E. (1977). "Biochemistry." p. 636. Academic Press, New York.

Meyerowitz, E. M. (1989). Arabidopsis, a useful weed. *Cell* **56,** 263–269.

Miralles, V. J., and Serrano, R. (1995). A genomic locus in *Saccharomyces cerevisiae* with four genes up regulated by osmotic stress. *Mol Microbiol.* **17,** 653–662.

Moran, J. F., Becana, M., Iturbe-Ormaetxe, I., Frechilla, S., Klucas, R. V., and Aparicio-Tejo, P. (1994). Drought induces oxidative stress in pea plants. *Planta* **194,** 346–352.

Munns, R. (1993). Physiological processes limiting plant growth in saline soils: Some dogmas and hypotheses. *Plant, Cell Environ.* **16,** 15–24.

Munns, R., and Termaat, A. (1986). Whole-plant responses to salinity. *Aust. J. Plant Physiol.* **13,** 143–160.

Murata, Y., Obi, I., Yoshihashi, M., Noguchi, M., and Kakutani, T. (1994). Reduced permeability to $K^+$ and $Na^+$ ions of $K^+$ channels in the plasma membrane of tobacco cells in suspension after adaptation to 50 mM NaCl. *Plant Cell Physiol.* **35,** 87–92.

Murguía, J. R., Bellés, J. M., and Serrano, R. (1995). A salt-sensitive 3′(2′),5′-bisphosphate nucleotidase involved in sulfate activation. *Science* **267,** 232–234.

Nakamura, T., Liu, Y., Hirata, D., Namba, H., Harada, S., Hirokawa, T., and Miyakawa, T. (1993). Protein phosphatase type 2B (calcineurin)-mediated, FK506-sensitive regulation of intracellular ions in yeast is an important determinant for adaptation to high salt stress conditions. *EMBO J.* **12,** 4063–4071.

Nakashima, K., Sugiura, A., Kanamaru, K., and Mizuno, T. (1993). Signal transduction between the two regulatory components involved in the regulation of the *kdpABC* operon in *Escherichia coli:* Phosphorylation-dependent functioning of the positive regulator, kdpE. *Mol. Microbiol.* **7,** 109–116.

Narasimhan, M. L., Binzel, M. L., Perez-Prat, E., Chen, Z., Nelson, D. E., Singh, N. K., Bressan, R. A., and Hasegawa, P. M. (1991). NaCl regulation of tonoplast ATPase 70-kilodalton subunit mRNA in tobacco cells. *Plant Physiol.* **97,** 562–568.

Neidhardt, F. C., Ingraham, J. L., and Schaechter, M. (1990). "Physiology of the Bacterial Cell: A Molecular Approach." Sinauer, Sunderland, MA.

Nguyen, H. T., and Joshi, C. P. (1994). Molecular genetic approaches to improving heat and drought tolerance in crop plants. *In* "Biochemical and Cellular Mechanisms of Stress Tolerance in Plants" (J. H. Cherry, ed.), pp. 279–289. Springer-Verlag, Berlin.

Nishida, E., and Gotoh, Y. (1993). The MAP kinase cascade is essential for diverse signal transduction pathways. *Trends Biochem. Sci.* **18,** 128–131.

Nissen, P. (1991). Multiphasic uptake mechanisms in plants. *Int. Rev. Cytol.* **126,** 89–134.

Niu, X., Zhu, J.-K., Narasimhan, M. L., Bressan, R. A., and Hasegawa, P. M. (1993). Plasma-membrane $H^+$-ATPase gene expression is regulated by NaCl in cells of the halophyte *Atriplex nummularia* L. *Planta* **190,** 433–438.

Nover, L. (1994). The heat stress response as part of the plant stress network: An overview with six tables. *In* "Biochemical and Cellular Mechanisms of Stress Tolerance in Plants" (J. H. Cherry, ed.), pp. 3–45. Springer-Verlag, Berlin.

Novick, P., and Botstein, D. (1985). Yeast actin mutants. *Cell* **40,** 405–416.

Ober, E. S., and Sharp, R. E. (1994). Proline accumulation in maize (*Zea mays* L.) primary roots at low water potentials. *Plant Physiol.* **105,** 981–987.

Oertli, J. J. (1968). Extracellular salt accumulation, a possible mechanism of salt injury in plants. *Agrochimica* **12,** 461–469.

Orr, W. C., and Sohal, R. S. (1994). Extension of life-span by overexpression of superoxide dismutase and catalase in *Drosophila melanogaster. Science* **263,** 1128–1130.

Panaretou, B., and Piper, P. W. (1990). Plasma-membrane ATPase action affects several stress tolerances of *Saccharomyces cerevisiae* and *Schizosaccharomyces pombe* as well as the extent and duration of the heat shock response. *J. Gen. Microbiol.* **136,** 1763–1770.

Perez-Prat, E., Narasimhan, M. L., Binzel, M. L., Botella, M. A., Chen, Z., Valpuesta, V., Bressan, R. A., and Hasegawa, P. M. (1992). Induction of a putative $Ca^{2+}$-ATPase mRNA in NaCl-adapted cells. *Plant Physiol.* **100,** 1471–1478.

Perier, F., Coulter, K. L., Liang, H., Radeke, C. M., Gaber, R. F., and Vandenberg, C. A. (1994). Identification of a novel mammalian member of the NSF/CDC48/Pas1p/TBP-1 family through heterologous expression in yeast. *FEBS Lett.* **351,** 286–290.

Perisic, O., Xiao, H., and Lis, J. T. (1989). Stable binding of *Drosophila* heat shock factor to head-to-head and tail-to-tail repeats of a conserved 5 bp recognition unit. *Cell* (*Cambridge, Mass.*) **59,** 797–806.

Pick, U. (1992). ATPases and ion transport in *Dunaliella. In* "*Dunaliella:* Physiology, Biochemistry and Biotechnology" (M. Avron and A. Ben Amotz, eds.), pp. 63–97. CRC Press, Boca Raton, FL.

Pilon-Smits, E. A. H., Ebskamp, M. J. M., Paul, M. J., Jeuken, M. J. W., Weisbeek, P. J., and Smeekens, S. C. M. (1995). Improved performance of transgenic fructan-accumulating tobacco under drought stress. *Plant Physiol.* **107,** 125–130.

Piper, P. W. (1993). Molecular events associated with acquisition of heat tolerance by the yeast *Saccharomyces cerevisiae. FEMS Microbiol. Rev.* **11,** 339–356.

Popova, L. G., and Balnokin, Y. V. (1992). $H^+$-translocating ATPase and $Na^+/H^+$ antiport activities in the plasma membrane of the marine alga *Platymonas viridis. FEBS Lett.* **309,** 333–336.

Prince, W. S., and Villarejo, M. R. (1990). Osmotic control of *proU* transcription is mediated through direct action of potassium glutamate on the transcription complex. *J. Biol. Chem.* **265,** 17673–17679.

Rains, D. W., and Epstein, E. (1967). Sodium absorption by barley roots: Role of the dual mechanisms of alkali cation transport. *Plant Physiol.* **42,** 314–318.

Ramos, J., Alijo, R., Haro, R., and Rodriguez-Navarro, A. (1994). *TRK2* is not a low-affinity potassium transporter in *Saccharomyces cerevisiae. J. Bacteriol.* **176,** 249–252.

Reuveni, M., Bennett, A. B., Bressan, R. A., and Hasegawa, P. M. (1990). Enhanced $H^+$ transport capacity and ATP hydrolysis activity of the tonoplast $H^+$-ATPase after NaCl adaptation. *Plant Physiol.* **94,** 524–530.

Robinson, L. C., Hubbard, E. J. A., Graves, P. R., DePaoli-Roach, A. A., Roach, P. J., Kung, C., Haas, D. W., Hagedorn, C. H., Goebl, M., Culbertson, M. R., and Carlson, M. (1992). Yeast casein kinase I homologues: An essential gene pair. *Proc. Natl. Acad. Sci. U.S.A.* **89,** 28–32.

Rock, C. D., and Quatrano, R. S. (1994). Insensitivity is in the genes. *Curr. Biol.* **4,** 1013–1015.

Rodriguez-Navarro, A., Blatt, M. R., and Slayman, C. L. (1986). A potassium–proton symport in *Neurospora crassa. J. Gen. Physiol.* **87,** 649–674.

Roldan, M., Donaire, J. P., Pardo, J. M., and Serrano, R. (1991). Regulation of root plasma membrane $H^+$-ATPase in sunflower seedlings. *Plant Sci.* **79,** 163–172.

Rothstein, A. (1964). Membrane function and physiological activity of microorganisms. *In* "The Cellular Functions of Membrane Transport" (J. F. Hoffman, ed.), pp. 23–39. Prentice-Hall, Englewood Cliffs, NJ.

Rudolph, H. K., Antebi, A., Fink, G. R., Buckley, C. M., Dorman, T. E., LeVitre, J., Davidow, L. S., Mao, J., and Moir, D. T. (1989). The yeast secretory pathway is perturbed by mutations in *PMR1,* a member of the $Ca^{2+}$ ATPase family. *Cell* (*Cambridge, Mass.*) **58,** 133–145.

Rush, D. W., and Epstein, E. (1981). Comparative studies on the sodium, potassium and chloride relations of a wild halophytic and a domestic salt-sensitive tomato species. *Plant Physiol.* **68,** 1308–1313.

Rushlow, K. E., Deutch, A. H., and Smith, C. J. (1984). Identification of a mutation that relieves gamma-glutamyl kinase from allosteric feed-back inhibition by proline. *Gene* **39,** 109–112.

Rutherford, S. L., and Zuker, C. S. (1994). Protein folding and the regulation of signaling pathways. *Cell* (*Cambridge, Mass.*) **79,** 1129–1132.

Saleki, R., Young, P. G., and Lefevre, D. D. (1993). Mutants of *Arabidopsis thaliana* capable of germination under saline conditions. *Plant Physiol.* **101,** 839–845.

Sanchez, Y., Taulien, J., Borkovich, K. A., and Lindquist, S. (1992). Hsp104 is required for tolerance to many forms of stress. *EMBO J.* **11,** 2357–2364.

Sanders, D. (1980). The mechanism of $Cl^-$ transport at the plasma membrane of *Chara corallina.* I. Cotransport with $H^+$. *J. Membr. Biol.* **53,** 129–136.

Sanders, D., and Hansen, U.-P. (1981). The mechanism of $Cl^-$ transport at the plasma membrane of *Chara corallina.* II. Transinhibition and the determination of $H^+/Cl^-$ binding order from a reaction kinetic model. *J. Membr. Biol.* **58,** 139–153.

Schachtman, D. P., and Schroeder, J. I. (1994). Structure and transport mechanism of a high-affinity potassium uptake transporter from higher plants. *Nature* (*London*) **370,** 655–658.

Schachtman, D. P., Schroeder, J. I., Lucas, W. J., Anderson, J. A., and Gaber, R. F. (1992). Expression of an inward-rectifying potassium channel by the *Arabidopsis KAT1* cDNA. *Science* **258,** 1654–1658.

Schroeder, J. I., and Fang, H. H. (1991). Inward-rectifying $K^+$ channels in guard cells provide a mechanism for low-affinity $K^+$ uptake. *Proc. Natl. Acad. Sci. U.S.A.* **8,** 11583–11587.

Schroeder, J. I., and Hagiwara, S. (1990). Repetitive increases in cytosolic $Ca^{2+}$ of guard cells by abscisic acid activation of nonselective $Ca^{2+}$ permeable channels. *Proc. Natl. Acad. Sci. U.S.A.* **87,** 9305–9309.

Schubert, S., and Laüchli, A. (1990). Sodium exclusion mechanisms at the root surface of two maize cultivars. *Plant Soil* **123,** 205–209.

Schuldiner, S., and Padan, E. (1993). $Na^+/H^+$ antiporters in *E. coli. In* "Alkali Cation Transport Systems in Prokaryotes" (E. Bakker, ed.), pp. 25–51. CRC Press, Boca Raton, FL.

Schüller, C., Brewster, J. L., Alexander, M. R., Gustin, M. C., and Ruis, H. (1994). The HOG pathway controls osmotic regulation of transcription via the stress response element (STRE) of the *Saccharomyces cerevisiae CTT1* gene. *EMBO J.* **13,** 4382–4389.

Sentenac, H., Bonneaud, N., Minet, M., Lacroute, F., Salmon, J. M., Gaymard, F., and Grignon, C. (1992). Cloning and expression in yeast of a plant potassium ion transport system. *Science* **256,** 663–665.

Serrano, R. (1989). Structure and function of plasma membrane ATPase. *Annu. Rev. Plant Physiol. Plant Mol. Biol.* **40,** 61–94.

Serrano, R. (1990). Recent molecular approaches to the physiology of the plasma membrane proton pump. *Bot. Acta* **103,** 230–234.

Serrano, R. (1991). Transport across yeast vacuolar and plasma membrane. *In* "The Molecular and Cellular Biology of the Yeast *Saccharomyces:* Genome Dynamics, Protein Synthesis and Energetics" (J. R. Broach, J. R. Pringle, and E. W. Jones, eds.), pp. 523–585. Cold Spring Harbor Lab. Press, Cold Spring Harbor, NY.

Serrano, R. (1994). Yeast halotolerance genes: Crucial ion transport and metabolic reactions in salt tolerance. *In* "Biochemical and Cellular Mechanisms of Stress Tolerance in Plants" (J. H. Cherry, ed.), pp. 371–380. Springer-Verlag, Berlin.

Serrano, R., and Gaxiola, R. (1994). Microbial models and salt stress tolerance in plants. *Crit. Rev. Plant Sci.* **13,** 121–138.

Sheahan, J. J., Ribeiro-Neto, L., and Sussman, M. R. (1993). Cesium-insensitive mutants of *Arabidopsis thaliana. Plant J.* **3,** 647–656.

Smith, B. J., and Yaffe, M. P. (1991). Uncoupling thermotolerance from the induction of heat shock proteins. *Proc. Natl. Acad. Sci. U.S.A.* **88,** 11091–11094.

Smith, F. A., and Walker, N. A. (1989). Transport of potassium in *Chara australis.* I. A symport with sodium. *J. Membr. Biol.* **108,** 125–137.

Smith, L. T. (1985). Characterization of a $\gamma$-glutamyl kinase from *Escherichia coli* that confers proline overproduction and osmotic tolerance. *J. Bacteriol.* **164,** 1088–1093.

Solioz, M., and Davies, K. (1994). Operon of vacuolar-type $Na^+$-ATPase of *Enterococcus hirae. J. Biol. Chem.* **269,** 9453–9459.

Sorger, P. K. (1991). Heat shock factor and the heat shock response. *Cell* (*Cambridge, Mass.*) **65,** 363–366.

Staal, M., Maathuis, F. J. M., Elzenga, J. T., Overbeek, J. H. M., and Prins, H. B. A. (1991). $Na^+/H^+$ antiport activity in tonoplast vesicles from roots of the salt-tolerant *Plantago maritima* and the salt-sensitive *Plantago media. Physiol. Plant.* **82,** 179–184.

Stein, W. D. (1990). "Channels, Carriers, and Pumps. An Introduction to Membrane Transport." Academic Press, San Diego, CA.

Stock, J. B., Stock, A. M., and Mottonen, J. M. (1990). Signal transduction in bacteria. *Nature* (*London*) **344,** 395–400.

Styrvold, O. B., and Ström, A. R. (1991). Synthesis, accumulation and excretion of trehalose in osmotically stressed *Escherichia coli* K-12 strains: Influence of amber suppressors and function of the periplasmic trehalase. *J. Bacteriol.* **173,** 1187–1192.

Sugiura, A., Nakashima, K., Tanaka, K., and Mizuno, T. (1992). Clarification of the structural and functional features of the osmoregulated *kdp* operon of *Escherichia coli. Mol. Microbiol.* **6,** 1769–1776.

Sugiura, A., Hirokawa, K., Nakashima, K., and Mizuno, T. (1994). Signal-sensing mechanisms of the putative osmosensor KdpD in *Escherichia coli. Mol. Microbiol.* **14,** 929–938.

Sumaryati, S., Negrutiu, I., and Jacobs, M. (1992). Characterization and regeneration of salt- and water-stress mutants from protoplast culture of *Nicotiana plumbaginifolia* (Viviani). *Theor. Appl. Genet.* **83,** 613–619.

Szoke, A., Miao, G. H., Hong, Z., and Verma, D. P. S. (1992). Subcellular location of $\Delta^1$-pyrroline-5-carboxylate reductase in root/nodule and leaf of soybean. *Plant Physiol.* **99,** 1642–1649.

Takase, K., Kakinuma, S., Yamato, I., Konishi, K., Igarashi, K., and Kakinuma, Y. (1994). Sequencing and characterization of the *ntp* gene cluster for vacular-type $Na^+$-translocating ATPase of *Enterococcus hirae. J. Biol. Chem.* **269,** 11037–11044.

Tarczynski, M. C., Jensen, R. G., and Bohnert, H. J. (1992). Expression of a bacterial *mtlD* gene in transgenic tobacco leads to production and accumulation of mannitol. *Proc. Natl. Acad. Sci. U.S.A.* **89,** 2600–2604.

Tarczynski, M. C., Jensen, R. G., and Bohnert, H. J. (1993). Stress protection of transgenic tobacco by production of the osmolyte mannitol. *Science* **259,** 508–510.

Thevelein, J. M. (1994). Signal transduction in yeast. *Yeast* **10,** 1753–1790.

Thompson-Jaeger, S., François, J., Gaughran, J. P., and Tatchell, K. (1991). Deletion of *SNF1* affects the nutrient response of yeast and resembles mutations which activate the adenylate cyclase pathway. *Genetics* **129,** 697–706.

Timasheff, S. N., and Arakawa, T. (1989). Stabilization of protein structure by solvents. *In* "Protein Structure: A Practical Approach" (T. E. Creighton, ed.), pp. 331–345. IRL Press, Oxford.

Trollmo, C., André, L., Blomberg, A., and Adler, L. (1988). Physiological overlap between osmotolerance and thermotolerance in *Saccharomyces cerevisiae. FEMS Microbiol. Lett.* **56,** 321–326.

Vallejo, C. G., and Serrano, R. (1989). Physiology of mutants with reduced expression of plasma membrane $H^+$-ATPase. *Yeast* **5,** 307–319.

van Zyl, P. J., Kilian, S. G., and Prior, B. A. (1990). The role of an active transport mechanism in glycerol accumulation during osmoregulation by *Zygosaccharomyces rouxii. Appl. Microbiol. Biotechnol.* **34,** 231–235.

Vernon, D. M., and Bohnert, H. J. (1992). A novel methyl transferase induced by osmotic stress in the facultative halophyte *Mesembryanthemum crystallinum. EMBO J.* **11,** 2077–2085.

Vogelien, D. L., Hickok, L. G., Augé, R. M., Stodola, A. J. W., and Hendrix, D. (1993). Solute analysis and water relations of gametophyte mutants tolerant to NaCl in the fern *Ceratopteris richardii. Plant, Cell Environ.* **16,** 959–966.

Vuorio, O. E., Kalkkinen, N., and Londesborough, J. (1993). Cloning of two related genes encoding the 56-kDA and 123-kDa subunits of trehalose synthase from the yeast *Saccharomyces cerevisiae. Eur. J. Biochem.* **216,** 849–861.

Wada, M., Satoh, S., Kasamo, K., and Fujii, T. (1989). Presence of a $Na^+$-activated ATPase in the plasma membrane of the marine raphidophycean *Heterosigma akashiwo. Plant Cell Physiol.* **30,** 923–928.

Wada, M., Urayama, O., Satoh, S., Hara, Y., Ikawa, Y., and Fujii, T. (1992). A marine algal $Na^+$-activated ATPase possesses an immunologically identical epitope to $Na^+,K^+$-ATPase. *FEBS Lett.* **309,** 272–274.

Walderhaug, M. O., Polarek, J. W., Voelkner, P., Daniel, J., Hesse, J. E., Altendorf, K., and Epstein, W. (1992). Two-component system encoded by *kdpD* and *kdpE. J. Bacteriol.* **174,** 2152–2159.

Waser, M., Hess-Bienz, D., Davies, K., and Solioz, M. (1992). Cloning and disruption of a putative NaH-antiporter gene of *Enterococcus hirae*. *J. Biol. Chem.* **267,** 5396–5400.

Watad, A.-E. A., Reuveni, M., Bressan, R. A., and Hasegawa, P. M. (1991). Enhanced net $K^+$ uptake capacity of NaCl-adapted cells. *Plant Physiol.* **95,** 1265–1269.

Watanabe, Y., Sanemitsu, Y., and Tamai, Y. (1993). Expression of plasma membrane proton-ATPase gene in salt-tolerant yeast *Zygosaccharomyces rouxii* is induced by sodium chloride. *FEMS Microbiol. Lett.* **114,** 105–108.

Weitzel, G., Pilatus, U., and Rensing, L. (1987). The cytoplasmic pH, ATP content and total protein synthesis rate during heat-shock protein inducing treatments in yeast. *Exp. Cell Res.* **170,** 64–79.

Welihuida, A. A., Beavis, A. D., and Trumbly, R. J. (1994). Mutations in *Lis1(Erg6)* confer increased Na and Li uptake in *Saccharomyces cerevisiae. Biochim. Biophys. Acta* **1193,** 107–117.

Weretilnyk, E. A., and Hanson, A. D. (1990). Molecular cloning of a plant betaine-aldehyde dehydrogenase, an enzyme implicated in adaptation to salinity and drought. *Proc. Natl. Acad. Sci. U.S.A.* **87,** 2745–2749.

Wieser, R., Adam, G., Wagner, A., Schüller, C., Marchler, G., Ruis, H., Krawiec, Z., and Bilinski, T. (1991). Heat shock factor-independent heat control of transcription of the *CTT1* gene encoding cytosolic catalase T of *Saccharomyces cerevisiae. J. Biol. Chem.* **266,** 12406–12411.

Williamson, C. L., and Slocum, R. D. (1992). Molecular cloning and evidence for osmoregulation of the $\Delta^1$-pyrroline-5-carboxylate reductase (*proC*) gene in pea (*Pisum sativum* L.). *Plant Physiol.* **100,** 1464–1470.

Wilson, T. H., and Lin, E. C. C. (1980). Evolution of membrane bioenergetics. *J. Supramol. Struct.* **13,** 421–446.

Wimmers, L. E., Ewing, N. N., and Bennet, A. B. (1992). Higher plant $Ca^{2+}$-ATPase: primary structure and regulation of mRNA abundance by salt. *Proc. Natl. Acad. Sci. U.S.A.* **89,** 9205–9209.

Winkler, K., Kienle, M., Burgert, M., Wagner, J.-C., and Holzer, H. (1991). Metabolic regulation of the trehalose content of vegetative yeast. *FEBS Lett.* **291,** 269–272.

Wyn Jones, R. G., and Pollard, A. (1983). Proteins, enzymes and inorganic ions. *In* "Encyclopedia of Plant Physiology. New Series" (A. Laüchli and A. Pirson, eds.), Vol. 15B, pp. 528–562. Springer-Verlag, Berlin.

Yamaguchi-Shinozaki, K., and Shinozaki, K. (1994). A novel cis-acting element in an *Arabidopsis* gene is involved in responsiveness to drought, low-temperature, or high-salt stress. *Plant Cell* **6,** 251–264.

Yancey, P. H., Clark, M. E., Hand, S. C., Bowlus, R. D., and Somero, G. N. (1982). Living with water stress: Evolution of osmolyte systems. *Science* **217,** 1214–1222.

Yeo, A. R., Lee, K.-S., Izard, P., Boursier, P. J., and Flowers, T. J. (1991). Short- and long-term effects of salinity on leaf growth in rice (*Oryza sativa* L.). *J. Exp. Bot.* **42,** 881–889.

Zhang, J., and Davies, W. J. (1989). Abscisic acid produced in dehydrating roots may enable the plant to measure the water status of the soil. *Plant, Cell Environ.* **12,** 73–81.

Zhou, Z., and Elledge, S. J. (1992). Isolation of *crt* mutants constitutive for transcription of the DNA damage inducible gene *RNR3* in *Saccharomyces cerevisiae. Genetics* **131,** 851–866.

Zhou, Z., and Elledge, S. J. (1993). *DUN1* encodes a protein kinase that controls the DNA damage response in yeast. *Cell* (*Cambridge, Mass.*) **75,** 1119–1127.

# Polytene Chromosomes in Mammalian Cells

Eugenia V. Zybina* and Tatiana G. Zybina†
*Laboratory of Cell Morphology and †Laboratory of Cell Pathology, Institute of Cytology of the Russian Academy of Sciences, St. Petersburg, Russia

This article deals with the structural and functional organization of polytene chromosomes in mammals. Based on cytophotometric, autoradiographic, and electron microscopic data, the authors put forward a concept of nonclassic polytene chromosomes, with special reference to polytene chromosomes in the mammalian placenta. In cells with nonclassic polytene chromosomes, two phases of the polytene nucleus cycle are described, such as the endointerphase (S phase) and endoprophase (G phase). The authors generalize that the main feature of nonclassic polytene chromosomes is that forces binding the sister chromatids are much weaker than in the Diptera classic polytene chromosomes. This concept is confirmed by comparative studies of human, mink, and fox polytene chromosomes. The final step of the trophoblast giant cell differentiation is characterized by a transition from polyteny to polyploidy, with subsequent fragmentation of the highly polyploid nucleus into fragments of low ploidy. Similarities and dissimilarities of pathways of formation and rearrangement of nonclassic polytene chromosomes in mammals, insects, plants, and protozoans are compared. The authors discuss the significance of polyteny as one of the intrinsic conditions for performance of the fixed genetic program of trophoblast giant cell development, a program that provides for the possibility of a long coexistence between maternal and fetal allogenic organisms during pregnancy.

**KEY WORDS:** Polytene chromosomes, Endoreplication, Nuclear structure, Trophoblast, Placenta.

## I. Introduction

Polytene chromosomes in Diptera, the largest chromosomes known, are a well-studied model of the interphase chromosome. In mammals, polytene chromosomes failed to be discovered for a long time, although this discovery could have made a significant contribution to advancing mammalian genetics. This article deals with the history of the discovery of polytene chromosomes in trophoblast giant cells (TGCs) of the mammalian placenta as well as with peculiarities of their structural and functional organization.

## II. Polytene Chromosomes in the Trophoblast: Morphology and Peculiarities of Cellular Reproduction

### A. Trophoblast Giant Cells

The formation of TGCs in the mammalian and human placenta has drawn the attention of many authors. Nevertheless, it was only at the end of the 1940s that the nature of their growth was questioned. Bridgman (1948) studied TGCs in the white rat placenta and noticed the lack of mitoses in these cells. No mitoses were revealed after treatment with colchicine. Based on these data, Bridgman suggested that the giant size of the trophoblast cells resulted from polyploidy.

Our goal was to ascertain whether the TGCs actually are polyploid and to determine what their nuclear ploidy level is. We used volumetry to measure polyploidy in the primary and secondary TGCs of 5- to 12-day and 6- to 18-day postcoitum (dpc) rat embryos, respectively. The primary giant cells reached levels of polyploidy of 128–256C while the secondary giant cells were 256–512C (Zybina and Tikhomirova, 1963). Cytospectrophotometric measurements of the degree of polyploidy in nuclei of the rat secondary TGCs have confirmed the results of the volumetric studies. In the course of development of the rat placenta, levels of polyploidy of the trophoblast cells increase progressively to reach the maximal value of 128–1024C (Zybina, 1963a; Zybina and Mosjan, 1967; Barlow and Sherman, 1972; E. V. Zybina *et al.,* 1975a,b; T. G. Zybina, 1985) and of 4096C (Nagl, 1972) (Table I). The progressive polyploidization of TGC nuclei is typical for early stages of development (up to 14–15 dpc in the rat and field vole); later on, the rate of polyploidization decreases markedly, to stop by 17–18 dpc (Zybina, 1963a; Zybina and Grishchenko, 1970; Zybina *et al.,* 1975a). Such high ploidy levels in the rodent TGCs first detected in the papers cited are unique for mammalian tissues.

Trophoblast giant cells in the placenta of Carnivora, American mink, and silver fox also are polyploid. Most of them, according to our cytophotometric data, have polyploidy of 8–16C but only a few nuclei reach 32–128C (Zybina *et al.,* 1989, 1992). Thus, TGCs of a number of species of rodents

TABLE I

Ploidy Levels in the Cells of the Embryonal (Trophoblast) and Maternal (Decidua) parts of the Mammalian Placenta

| Species | Ploidy level ($C$) | Reference |
|---|---|---|
| | **Primary and secondary trophoblast giant cells** | |
| Rat | 256–512–1024 | Zybina and Tikhomirova (1963), Zybina (1963a); Zybina and Mosjan (1967); T. G. Zybina *et al.* (1985) |
| | 4096 | Nagl (1972) |
| Mouse | 64–512 | Barlow and Sherman (1972); Ilgren (1981) |
| Field vole | 64–2048 | Zybina *et al.* (1975a) |
| | **Trophoblast giant cells** | |
| Rabbit | 128–512 | Zybina *et al.* (1975b) |
| Guinea pig | 4–32 | Ilgren (1980) |
| Silver fox | 8–64 | Zybina *et al.* (1989) |
| American mink | 8–64 | Zybina *et al.* (1992) |
| | **Tertiary trophoblast giant cells** | |
| Rat | 4–64 | T. G. Zybina (1987) |
| | **Trophoblast junctional zone cells** | |
| Rat | 4–32 | Zybina and Grishchenko (1970) |
| | 4–64 | Zybina *et al.* (1984a,b) |
| Field vole | 4–32 | T. G. Zybina *et al.* (1987) |
| | **Labyrinth trophoblast cells** | |
| Field vole | 4–32 | T. G. Zybina *et al.* (1987) |
| | **Decidual cells** | |
| Rat | $(4–32) \times 2^a$ | Zybina and Grishchenko (1972) |
| Mouse | 4–64 | Ansell *et al.* (1974), Ilgren *et al.* (1983) |
| | **Metrial gland granular cells** | |
| Rat | $(4–8) \times 2^a$ | Zybina *et al.* (1987) |

[a] Binucleate cells.

and, to a lesser degree, of some Carnivora, are characterized by a multifold multiplication of the chromosomal material, which is a prerequisite of polyteny. As seen from Table I, both trophoblast and decidual cells are, as a rule, polyploid. The significance of genome multiplication in placental cells is discussed in Section VI.

## B. Morphology of Polytene Chromosomes in Trophoblast Cells

In the early 1960s, when studying the cause of the giant growth of trophoblast cells in the rodent placenta, we noticed that some nuclei of TGCs contained peculiar structures that somewhat resembled polytene chromosomes in the salivary glands of Diptera (Zybina, 1961; Zybina and Tikhomirova, 1963). A comparative study of chromosomes in the TGCs of four rodent species (white rat, rabbit, mouse, and field vole) and two species of Carnivora (American mink and silver fox) has allowed us to discover their basic similarity to polytene chromosomes in Diptera.

### 1. Rat Trophoblast Cells

The initial stages of the formation of polytene chromosomes were traced in rat TGCs on nonpermanent and permanent squashed preparations stained with aceto-orcein and Feulgen reaction. In the growing nuclei of a high ploidy (16–32C), long, thin, paired strands with chromomeres start to appear; intensively stained chromocenters or chromosome regions are attached to the nucleolus. Later on, polytene-like chromosomes appear as loose bundles of a number of parallel, thin, chromosome fibrils.

Polytene chromosomes are most clearly evident in nuclei with a ploidy of 64–512C, when all chromosomes in the nucleus form rather thick and comparatively short bundles (Fig. 1). Such bundles are predominantly located under the nuclear envelope and near the nucleolus (Fig. 1c and d). The number of bundles in the nucleus is approximately equal to the haploid chromosome number. With the rise of nuclear ploidy, the number of chromosomal bundles does not increase but the bundles do enlarge. The structure of the chromosome is different at its different sites: in the pericentromeric area, chromonemes form more compact clusters while in the telomeric regions, the chromosome is split into thin fibrils that diverge radially (Fig. 1e). In a part of giant nuclei, chromonemes of the polytene chromosome decondense and sister chromatids are separated to be diffusely spread in the nucleoplasm (Fig. 1a and b). The nucleus resembles an interphase nucleus (meshwork nucleus). The chromatids remain associated only in

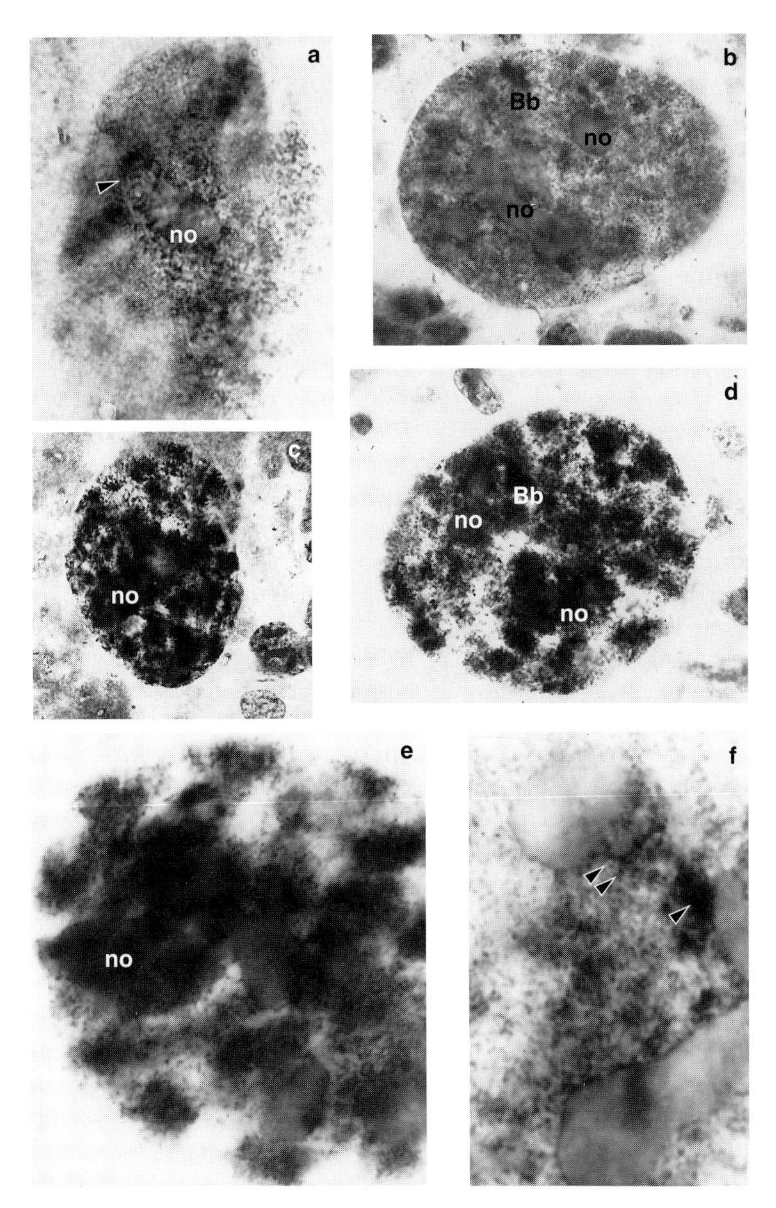


FIG. 1 Structure of nuclei of rat trophoblast giant cells. (a) In the nucleus with the decondensed chromatin, the nucleolus-organizing chromosome can be seen; near the nucleolus (no), chromomeres of adjacent chromosomes form a disklike structure (arrowhead). (b) The meshwork nucleus; chromatin is markedly decondensed; only the Barr body (Bb) is condensed. (c and d) Dense bundles of condensed chromosomes are located beneath the nuclear envelope and near the nucleolus; the size of the bundles is compatible with the Barr body but the degree of their condensation is less; the bundle size increases with polytenization. Bar: 20 $\mu$m. (e) The degree of chromoneme condensation in the polytene nucleus can be different. (f) The NOC has a heterochromatin block (arrowhead) and forms a puff (double arrowheads) near the nucleolus. Bar = 10 $\mu$m. Aceto-orecein stain.

limited regions of nucleolus-organizing chromosomes (NOCs; Fig. 1a) and probably in centromere regions.

Based on our autoradiographic data (to be discussed in Section II,C), we consider the stage of thin bundles evenly dispersed in the nucleus as the endointerphase or S phase (because of the DNA synthesis occurring at this stage), whereas the stage of bundles of polytene chromosomes is the endoprophase or G phase (Zybina, 1963b). The nucleolus in the TGC nuclei is large, round, or ovoid in shape and initially single. Subsequently it acquires lobes, becomes more and more disintegrated, and with further polytenization splits into two to five nucleoli (Fig. 1b and d–f). These nucleoli are present throughout the entire nuclear cycle, in S and G phases, and the chromatids remain closely attached in the region adjacent to the nucleolar organizer region (NOR). In the S phase bands are revealed in the NOCs. Occasionally, pufflike structures are seen near the bands. The degree of condensation of chromosome bundles may vary somewhat within the nucleus. The most condensed is one of two X-chromosomes forming the Barr body in the female cell nucleus. The compact, sharply defined Barr body is clearly detectable both in the S and G phases (Fig. 1b and d; Zybina and Mosjan, 1967; Zybina, 1970, 1977). In both nuclear cycle phases, nucleolus-organizing chromosomes are more condensed than other chromosomes of the set.

During the final period of cell differentiation, (i.e., after the end of DNA reduplication in the polytene nuclei), no bundles are formed; nuclei somewhat resemble interphase nuclei but may contain large heterochromatin clumps (Zybina, 1961). At that time sex chromatin assumes the shape of a rather loose body and therefore is difficult to identify in the nucleus.

### 2. Murine Trophoblast Cells

In a population of highly polyploid murine trophoblast cells, there are two types of nuclei: meshwork nuclei and those with bundles of condensed chromonemes (Zybina, 1977). Both types are characterized by large blocks of heterochromatin (Fig. 2). Sharply defined heterochromatin regions most commonly are oval or elongated and sometimes irregular in shape. In nuclei that contain chromoneme bundles, polytene chromosomes differ in length, degree of condensation, and attachment of chromatids to each other. Thin chromoneme threads with chromomeres are seen to run from dense heterochromatic regions (Fig. 2d). Adjacent to the nucleolus are large heterochromatic regions from which thin chromonemes radially diverge into the nucleoplasm (Fig. 2c and d). The entrance of chromosomes into the nucleolus is less evident; however, intranucleolar chromatin is clearly seen in the form of thin rings or more complex formations (Fig. 2c).

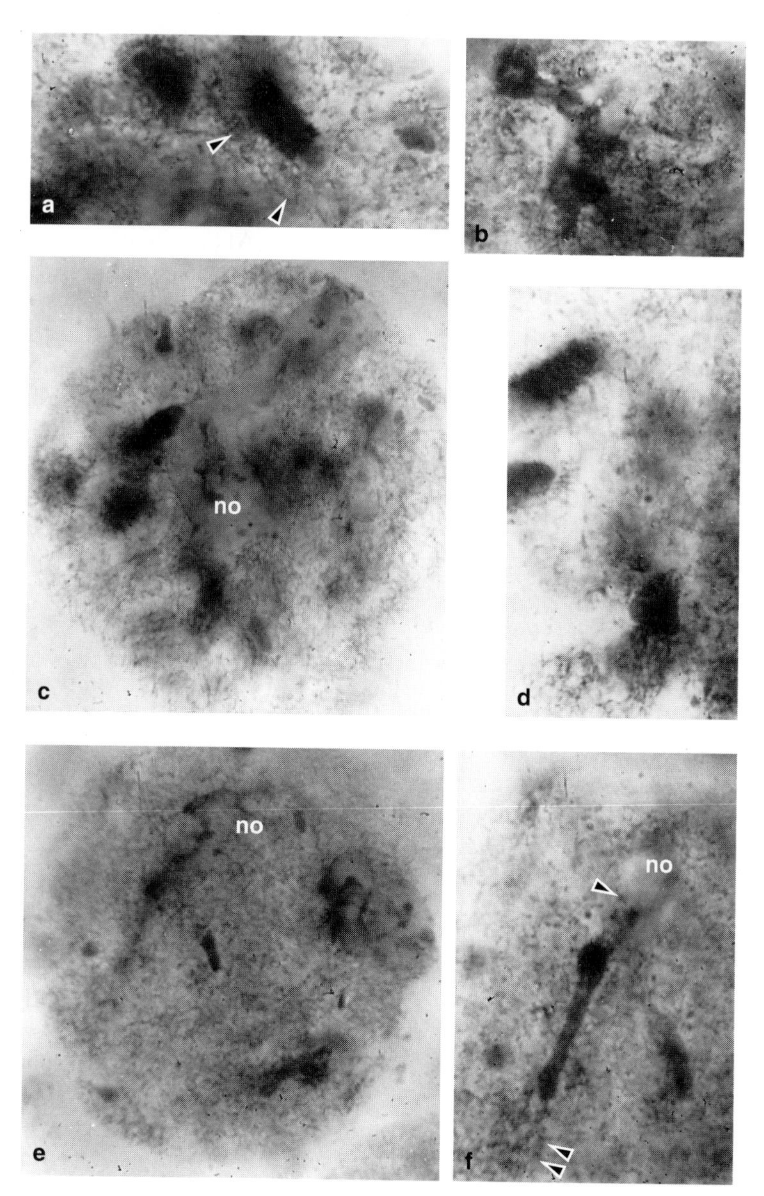


FIG. 2 Structure of polytene nuclei in mouse trophoblast secondary giant cells. (a) Thin threads resembling loops of the lampbrush chromosomes (arrowheads) emerge from some heterochromatin blocks. (b) The heterochromatin part of the chromosome has a complex outline. (c) Heterochromatin blocks of the nucleolus-organizing chromosomes are adjacent to the large nucleolus (no). (d) Polytene chromosomes consist of the large heterochromatin block and the euchromatin part in the form of diverging decondensed threads. (e) Meshwork nucleus: a significant part of the heterochromatin is located near the nucleolus. (f) Disklike structures (arrowhead) are seen in a part of the NOC adjacent to the nucleolus; chromonemes in the euchromatin part are dissociated and are revealed as separate strands (double arrowheads). Bar = 20 $\mu$m. Aceto-orcein stain.

At the stage of the meshwork nucleus in the murine trophoblast, condensed heterochromatin areas persist but become thinner, longer, and less intensively stained. Chromonemes decondense at a large distance and disperse diffusely in the nucleoplasm. This is particularly evident in NOCs: the chromatids in the euchromatic area decondense more and disintegrate at the chromosome ends to form a "fan" (Fig. 2f). At the site of the transfer from the heterochromatic block to the nucleolus, there are structures resembling bands (Fig. 2f).

At the lateral surface of heterochromatic blocks of the NOCs, a great number of thin lateral loops can be seen that protrude into the nucleoplasm at approximately right angles (Fig. 2a). These loops form a kind of a sheath around heterochromatin blocks of murine polytene chromosomes; they are composed of many thin chromosome loops that somewhat resemble the lateral loops in lampbrush chromosomes, which have been well studied in vertebrate oocytes (Baker, 1963; Callan, 1963, 1986; Kiknadze, 1966; Zybina, 1969, 1975; Zybina and Zybina, 1992). This phenomenon observed in the murine trophoblast seems to resemble puffing in polytene chromosomes of *Phaseolus coccineus* (Nagl, 1970a), which involves only peripheral parts of the chromosomes.

Polytene chromosomes in the murine TGCs are revealed for only a short time at a certain stage of cell differentiation (only at 10 and 11 dpc); subsequently, heterochromatin blocks varying in size and contour appear in the nucleus while no chromosomes can be seen (Zybina, 1977, 1986). It is probably owing to the short visualization time of the polytene chromosomes that many authors failed for a long time to discover their presence in murine trophoblasts (Sherman *et al.*, 1972; Barlow and Sherman, 1974).

Barlow and Sherman (1974) described the meshwork nuclear structure in the murine trophoblast and observed, in some cases, parallel nuclear chromatin threads forming thicker strands that somewhat resembled regions of polytene chromosomes. However, the authors did not find that this justified use of the term "polyteny" because the pattern they observed differed markedly from the classic polytene chromosomes and was uncommon in nuclei.

Snow and Ansell (1974) treated murine trophoblast cells growing in tissue culture with actinomycin D; this resulted in pronounced chromatin condensation and formation of elongated chromosome-like bodies that differed from mitotic chromosomes morphologically only in their less sharply defined contours. The authors believe these bodies represent chromosomes or their most condensed regions. The number of these bodies was approximately diploid. Since in the TGC nuclei studied, with different degrees of ploidy (from 4C to 256C), the number of the chromosome-like bodies remained about diploid, Snow and Ansell concluded that these chromosomes were polytene in nature. The polytene chromosomes these authors found in the course of several subsequent rounds of polyploidization extended their length but did not change their diameter. In the TGC nuclei of low ploidy, such bodies show a chromomeric structure similar to bands of polytenic chromosomes in Diptera.

Singh and coauthors reported several years ago that giant chromosomes could be induced from the highly polyploid nuclei of mouse trophoblast giant cells (Singh *et al.,* 1989). Two methods for condensation were used: fusion of giant cells with mitotically blocked HeLa cells, and incubation of giant cells in *Xenopus laevis* egg extracts. Such chromosomes, in the authors' opinion, may be useful for mapping and cloning mouse genes. There are also some data about the possibility of endomitosis in murine trophoblast cells at earlier stages of development. A study of the distribution of the constitutive heterochromatin in C banding-stained murine TGCs at the 7 dpc reveals stages of endomitosis (Kuhn and Therman, 1988; Kuhn *et al.,* 1991). In nuclei of different degrees of ploidy (from diploid to highly polyploid), the constitutive heterochromatin is seen either as twin dots or as larger clumps attached mainly to the nuclear envelope or to the nucleolus. There are also occasional nuclei in which a part of the heterochromatin is in the form of twin dots while the rest fuses into larger clumps. No correlation was found between the nuclear size and the pattern of the heterochromatin.

### 3. Rabbit Trophoblast Cells

In highly polyploid cell nuclei of the rabbit trophoblast, polytene chromosomes were revealed in the form of loose chromoneme bundles with small heterochromatin blocks in some chromosomes (Fig. 3a and b). In the rabbit, as in the white rat, the trophoblast has a stage in the nuclear cycle when chromosomes decondense and chromonemes dissociate at a substantial length and become separated from each other. In these cases, the nucleus looks meshy (Fig. 3a). It is even more difficult to identify the trophoblast polytene chromosomes in the rabbit than in the mouse and rat owing to difficulties in distinguishing individual bundles from each other. The easiest to identify are NOCs; these, even in the meshwork-type nuclei, can be visualized as bundles of parallel chromonemes of which chromomeres form bandlike structures (Fig. 3a). These structures are also evident in the bundle-type nuclei (Fig. 3b). In the period of terminal differentiation, prior to fragmentation of giant nuclei, the polytene chromosomes are dissociated into numerous paired chromosomes resembling the chromosomes of endomitosis. These paired chromosomes look like short twin rods attached to the nuclear envelope (Fig. 3c and d). This indicates the possibility that a significant reconstruction of the chromosomal material occurs shortly before nuclear fragmentation (Zybina, 1986).

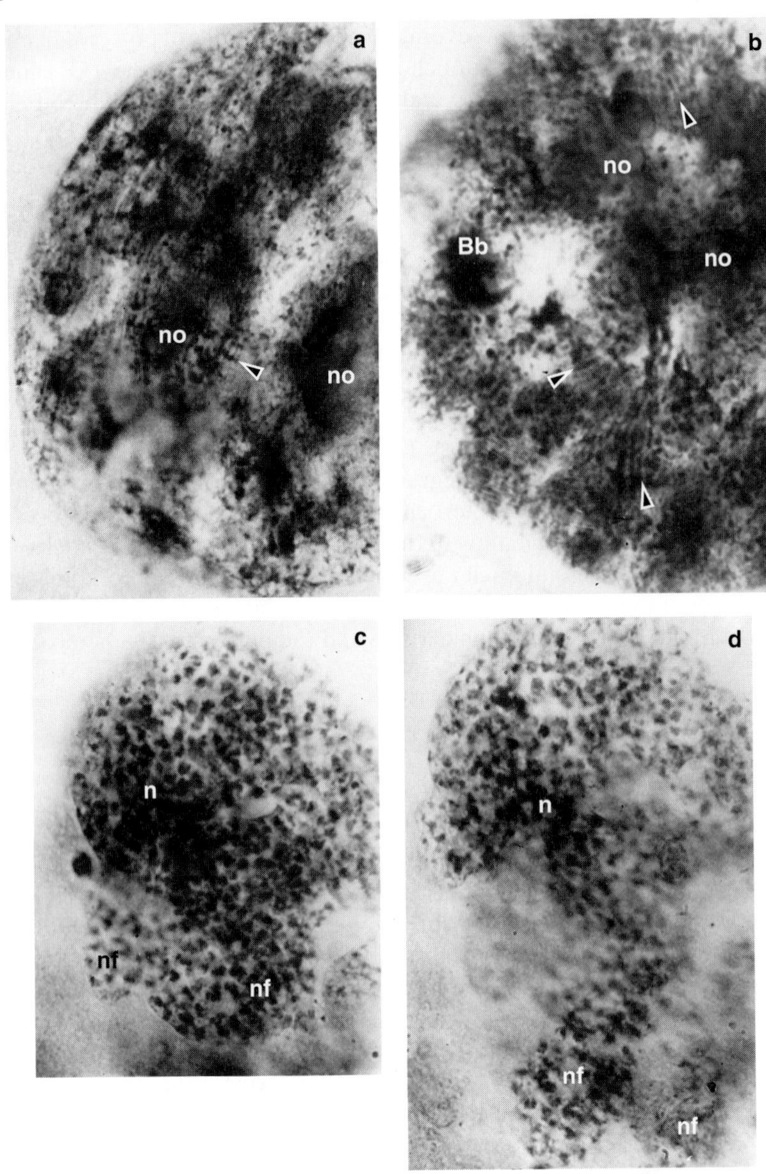


FIG. 3 Structure of the polytene (a and b) and polyploid (c and d) nuclei in the rabbit trophoblast. (a) Meshwork nucleus with the decondensed chromatin; disks (arrowhead) are seen in the nucleolus-organizing chromosome. (b) Bundles of the polytene chromosomes have different degrees of condensation, the greatest degree being observed in the Barr bodies (Bb), and disklike structures (arrowheads) are seen in some bundles. (c and d) In the initial period of giant nucleus fragmentation, paired endochromosomes are present at its periphery: n, the initial nucleus; nf, the separating nuclear fragments. Bar = 10 $\mu$m. Aceto-orcein stain.

### 4. Field Vole Trophoblast Cells

***a. Secondary Trophoblast Giant Cells*** Unlike those in the rat and rabbit trophoblast, the polytene chromosomes in the secondary TGCs in the field vole trophoblast are revealed within an even more limited period of the cell life span. We were able to discover them only in the secondary giant cells at 9 dpc. In this case, as in the trophoblasts of other rodent species studied, apart from nuclei with chromoneme threads, there are also nuclei with thin chromosome bundles that are decondensed and diffusely spread all over the nucleoplasm. Unlike TGCs in the rat, mouse, and rabbit, the TGCs in the field vole have large nucleoli of a complex shape and form numerous protrusions from which small ovoid nucleoli are separated. In polytene nuclei of female embryo TGCs, one or two large sex chromatin bodies are present (Zybina, 1986). In the nuclei of the secondary TGCs, at later stages of development (14 dpc), paired small (around 1 $\mu$m) endochromosomes or chromocenters are visualized; they form clusters under the nuclear envelope and around multiple nucleoli (Zybina, 1986).

***b. Trophoblast Supergiant Cells*** Supergiant cells of the field vole trophoblast are a particular cell population (Zybina and Zybina, 1985). These derive from the secondary TGCs, are separated from the continuous layer, and migrate a large distance inside the decidua basalis. The number of supergiant cells in each placenta is small: it usually varies from 3 to 10; these cells and their nuclei are much larger than the initial secondary TGCs.

The structure of the polytene chromosomes in supergiant cells has some peculiarities. Large bundles of the chromosomes are closely attached to the nuclear envelope, therefore the central area of the nucleoplasm looks empty (Fig. 4a). Chromonemes forming these bundles are condensed. In some chromosomes they are oriented parallel to each other, in others they are packed together at the site of their attachment to the nuclear envelope from which the chromonemes radiate into the nucleoplasm; their orientation inside the bundle often looks chaotic (Fig. 4c). A paired orientation of chromomeres in adjacent chromonemes can be seen in telomeric areas. In addition, telomeric regions of adjacent polytene chromosomes proved to be connected end to end. We failed to determine the stage of the meshwork nucleus in trophoblast supergiant cells.

The most interesting features here were discovered in the polytene NOCs. At earlier stages of development (day 9), the number of nucleoli in the supergiant nucleus initially is not high: two to five. They are round or ovoid and are attached to the nuclear envelope together with a bundle of the NOCs (Fig. 4c). The nucleoli are not homogeneous and are composed of numerous closely packed minute nucleoli (Fig. 4c), each surrounded by a Feulgen-positive "ring" (Zybina and Zybina, 1985). This is a fundamental

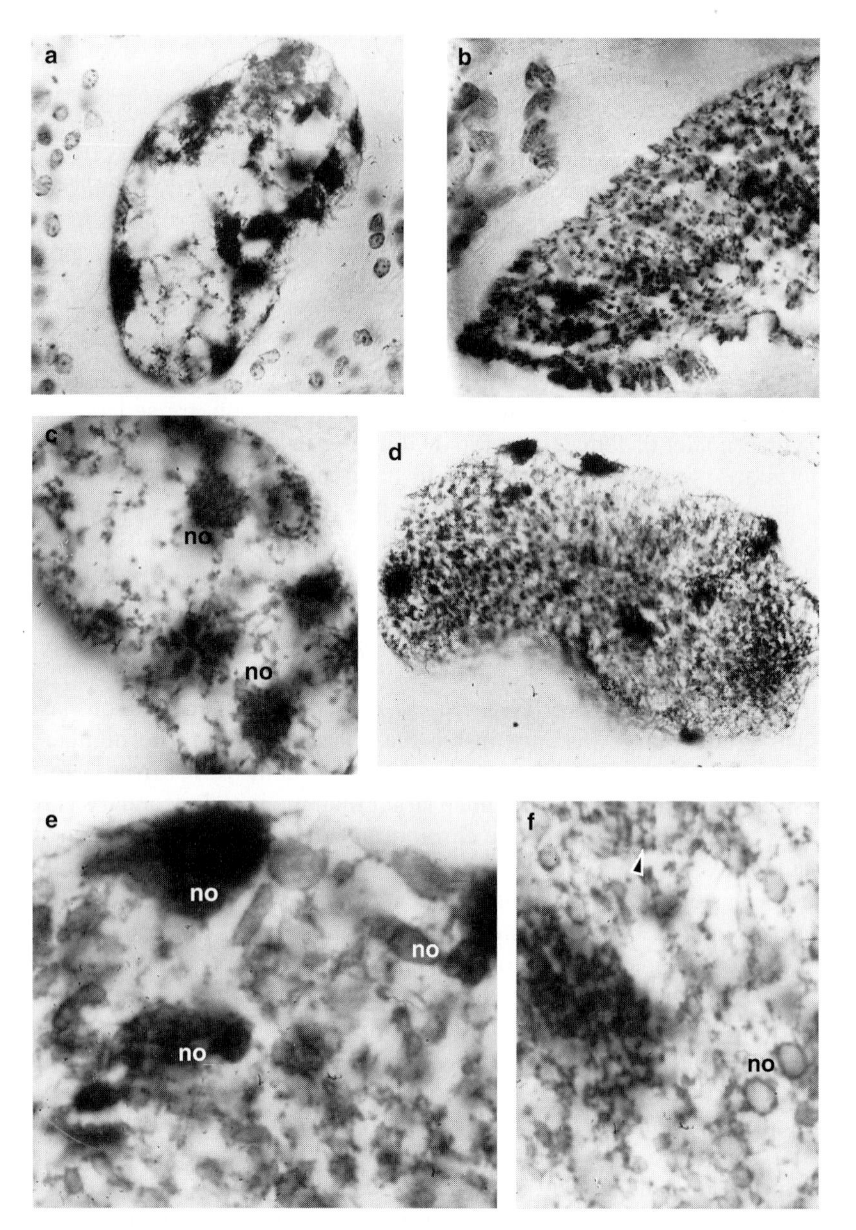


FIG. 4 Nuclei of the supergiant cells of the field vole trophoblast. (a) Chromosomes in the form of bundles are beneath the nuclear envelope. (b) Endochromosomes are seen beneath the nuclear envelope and numerous folds of the nuclear envelope are present. (c) Large nucleoli (no) are composed of numerous closely packed minute nucleoli surrounded by a Feulgen-positive ring. (d) Bundles of chromonemes in the process of dissociation; apart from large bundles in the nucleoplasm, an enormous number of small nucleoli are seen. Bar = 20 $\mu$m. (e and f) Areas of the same nucleus at a higher magnification: small nucleoli are surrounded by the Feulgen-positive ring; chromosomal threads emerge from some nucleoli; chromonemes (arrowheads) are gradually separated from the loose bundles. Bar = 10 $\mu$m. Feulgen stain.

structural difference of the nucleoli at polyteny in the trophoblast supergiant cells of the field vole compared with the giant nucleoli in the trophoblast of the rat, rabbit, and mouse. In the latter species, all NOCs of the set seem to participate in the formation of one or several nucleoli, and the NOCs show polytenic properties to the greatest degree because sister chromonemes remain coupled in the NOR at both stages of the nuclear cycle (Zybina, 1977, 1986).

However, in the field vole the coupling tendency of the sister chromonemes in the NOCs is expressed to a minimal degree; this results in a honeycomb structure of the supergiant cell nucleolus, with each minute nucleolus belonging to one sister chromatid or a small number of them. Such pronounced isolation of individual chromatids in the polytene NOCs of the field vole probably gave them the ability to split subsequently into the separately located minute nucleoli (Fig. 4d–f). Their size could correspond to, or be somewhat greater than, the nucleoli in diploid cells.

The process of separation of the giant nucleoli into individual minute nucleoli can be traced within a single nucleus. In the nucleus there are sometimes both large nucleoli composed of many meshes and loose clusters of minute nucleoli. From the latter, individual minute nucleoli are gradually separated into the nucleoplasm; they are surrounded by adjacent rings of chromatin with short diverging strands (Fig. 4e and f). The rest of the bundles of polytene chromosomes (apart from the NOCs) also are gradually split into individual chromonemes: long, thin, sometimes rather condensed chromosomal strands are consecutively separated from the polytene chromosomes (Fig. 4e). The chromosomal strands separated from the initial bundle form homogeneous aggregates in the vicinity of the nuclear envelope and are connected to it; minute nucleoli occupy the entire central part of the nucleus. This results in the complete disjunction of all chromosomal bundles into chromonemes (endochromosomes). This process is sometimes accompanied by condensation of chromonemes. As a result, the peripheral part of the nucleus is filled with elongated or condensed, rodlike, paired chromosomes (Fig. 4b) that morphologically are similar to the endomitotic chromosomes described in human tumors and trophoblast (Sarto *et al.,* 1982; Therman *et al.,* 1983; Kuhn and Therman, 1988; Therman and Kuhn, 1989) as well as in some invertebrates (Geitler, 1953; Sokolov, 1967; Kiknadze and Istomina, 1980).

### 5. Mink and Fox Trophoblast Cells

The trophoblast cell population in the American mink and silver fox placenta is composed mainly of cells with a low degree of ploidy, 2–16C, while rarely reaching 32–128C (Zybina *et al.,* 1989, 1992). It is in the cells with the maximal degree of ploidy that polytene chromosomes in the mink trophoblast were discovered as short dense bundles formed by sister chromatids and attached to the nuclear envelop (Zybina *et al.*, 1992). In smaller nuclei there are also some signs of transition to polyteny, which are particularly evident in the chromosomes bound to the nucleolus. Double-threaded or multithreaded chromosomal bundles with chromomeres are oriented parallel to each other in areas adjacent to the nucleolus. Chromomeres of the adjacent sister chromatids form structures resembling bands of polytene chromosomes. Apart from such nuclei, nuclei with a meshwork-like chromatin structure are present. The polytene chromosomes in the mink trophoblast were found not only in the developing placenta but also as early as the preimplantation blastocyst (Isakova *et al.*, 1992). In a few highly polyploid nuclei of the trophoblast cells in the fox placenta, structures similar to polytene chromosomes in the mink placenta can also be found. However, polyteny here is even less pronounced and is seen only in the chromosomes bound to the nucleolus (Zybina *et al.*, 1989).

### 6. Study of Polytene Chromosomes by Phase-Contrast Microscopy

When temporary, squashed unstained preparations of the rat trophoblast were examined in a phase-contrast microscope, the TGCs showed long chromosomes that were diffusely spread in the nucleoplasm but were predominantly located under the nuclear membrane (Fig. 5a and c–e). Since the chromosomes are interwoven, it is difficult to evaluate their length and their structure for their entire length. There are thin, dark bands in the chromosomes formed by chromomeres and large, lighter, interband regions (Zybina, 1970; Zybina and Chernogryadskaya, 1976). Apart from the nuclei with polytene chromosomes there are meshwork nuclei in which the nucleoplasm is filled with numerous convoluted fibrils with paired chromomeres (Fig. 5b). In some nuclear areas, most commonly under the nuclear membrane, adjacent fibrils can be seen to lie parallel to each other: in this case, chromomeres of adjacent chromonemes are located at the same level. It looks as if, in spite of a seemingly chaotic spreading of the decondensed chromosomal threads all over the nucleoplasm, their particular orientation in the nucleus remains. It is possible that the chromosomes are attached by their telomeric areas to the nuclear membrane and this, in some way, subsequently promotes the aggregation of the fibrils into bundles of polytene chromosomes (Fig. 5b and c). Most likely, in the course of this bundle aggregation, the chromomeres of adjacent chromonemes unite to form a bandlike structure (Zybina, 1970, 1986; Zybina and Chernogryatskaya, 1976). Thus, using phase contrast, it has become possible to reveal two states of chromosomes in the rat trophoblast cell nuclei: a compact aggregation of chromonemes to form polytene chromosomes with bands and interband

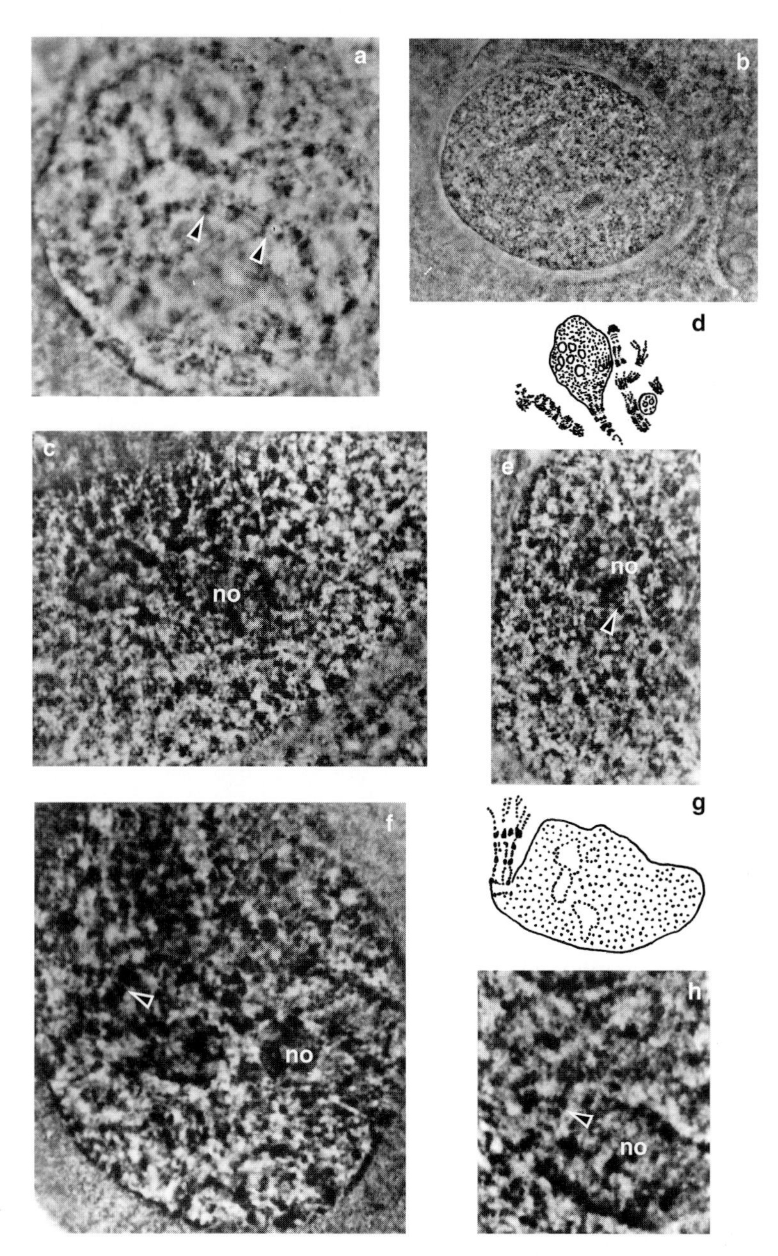


FIG. 5 Structure of polytene nuclei in the trophoblast giant cells of the rat (a–e) and rabbit (f–h) using phase-contrast microscopy. (a) Chromonemes are poorly condensed and combined in bundles; disklike structures (arrows) are seen. (b) Meshwork nucleus with decondensed chromosomes; chromonemes are combined in loose bundles attached to the nuclear envelope (arrowheads). (c) Chromonemes are condensed to a greater degree; chromoneme bundles in some areas of the nucleus are oriented to the direction of the nuclear envelope. (d and e) Disklike structures (arrowhead) are seen near the nucleolus. (d) Scheme of the nucleolus are attached chromosomes in (e). (f) Disklike structures (arrowhead) are seen in some chromosome areas. (g and h) Disklike structures are seen near the nucleolus (no). (g) Scheme of the nucleolus and a part of the attached chromosome in (h). Bar = 10 $\mu$m.

regions, and a meshwork nucleus in which the chromosomes dissociate into oligotene fibrils. It is important to note that such dissociation does not occur in some regions of the chromosomes. In the NOCs, for instance, the chromonemes remain in close contact to each other at the stage of the meshwork nucleus to maintain the polytene structure directly near nucleoli.

In the rabbit trophoblast polytene chromosomes are revealed as thin convoluted strands: a band pattern is easily seen in some short regions (Fig. 5f and h). Similar to that seen in stained preparations, some of the nuclei have a meshwork structure (Zybina and Chernogryadskaya, 1976).

It is essential to note that the morphology of polytene chromosomes in the rodent GTC nuclei may depend on the methods applied. Whereas in the squashed preparations stained with aceto-orcein or Feulgen stain the polytene chromosomes look like wide and rather short bundles of chromonemes, in phase contrast the unstained nuclei are shown to contain a network of narrow, thin, chromosome-like structures. The cause of this probably lies in morphological changes in the chromosomes in response to the Feulgen reaction, or may be caused by staining of the nucleus with aceto-orcein. Use of phase-contrast microscopy has allowed us to confirm observations in the stained preparations about the inconstancy of polyteny revealed in the nuclear cycle in the TGCs of the rat and rabbit, and to suggest the inconstancy of the attachment of sister chromatids along the length of the polytene chromosome.

## 7. Endomitosis and Polyteny in Human Trophoblast and Tumor Cells

The first communication about endomitosis and polyteny in the normal cells of the human placenta appeared in 1982 (Sarto *et al.,* 1982). Initially it was found that in the human placenta, during the first third of pregnancy, apart from cytotrophoblast cells there also are giant cells with a diameter of 5–20 μm or even somewhat larger. Only in one individual, at 65 days of pregnancy, were occasional nuclei revealed with a diameter of 50–60 μm. The morphology of these nuclei varied markedly, from the meshwork nuclei resembling interphase to the nuclei with typical dark endomitotic chromosomes. In addition, there were nuclei with bundles of parallel chromatin threads that in some areas formed structures similar to those described in polytene nuclei of the rodent trophoblast (Zybina, 1961, 1977). The authors are inclined to consider these structures a definite type of polytene chromosome similar to the rodent trophoblast polytene chromosomes (Sarto *et al.,* 1982; Therman *et al.,* 1983). According to their observations, giant cells of this type in the human placenta are more specific for trophoblast malignant tumors, but sometimes they also occur in the normally developing placenta and do not show any signs of degeneration.

Based on these data, the authors think endomitosis or polyteny occurs in some areas of the cytotrophoblast. Thus, endomitotic chromosomes were revealed in the trophoblast cells of human hydatidiform moles (Therman *et al.,* 1983; Kuhn and Therman, 1988). These chromosomes sometimes did not scatter in the course of consecutive cycles of replication and formed clusters of adjacent endochromosomes (Kuhn and Therman, 1988). A similar picture of clustered sister endochromosomes looking like rows of adjacent rods was observed in large nuclei from a cervical cancer (Kuhn and Therman, 1988; Therman and Kuhn, 1989). So far information on polyteny in the human trophoblast is fragmented. However, further accumulation of similar facts should soon resolve the problem of the existence of human polytene chromosomes.

## C. Two Stages of the Polytene Nucleus Cycle

Autoradiographic study of DNA synthesis has made it possible to elucidate the causes of variability of the structure of rat TGC giant nuclei. Four hours after [$^3$H]thymidine administration to 14-dpc rats, only the meshwork nuclei of the secondary TGC contained the label rather than nuclei with chromoneme bundles. In 24 hr, the label was present in both types of nuclei (Zybina, 1963b). Thus, these data indicate that nuclei with chromatin that is decondensed and dispersed diffusely in the nucleoplasm seem to be at the DNA synthesis stage (S phase or endointerphase) while nuclei with condensed chromosomal bundles are in the G phase (or endoprophase) when there is no DNA synthesis. Hence, the presence of polytene chromosomes in only a part of the TGC nuclei rather than in all of them is due to changes of the chromosomal structure in the polytene nucleus replication cycle: the degree of the chromoneme attachment in polytene chromosomes probably is much greater in the G phase than in the S phase. It should be noted, however, that this regularity holds true for the giant nuclei beginning only with a certain level of ploidy: not lower than 64C; at a lower degree of ploidy, typical polytene chromosome bundles do not seem to be formed yet (Zybina and Mosjan, 1967).

Because of the lack of mitosis in the TGC population, it was important to reveal different phases of the polytenic nucleus cycle to determine time characteristics of this cycle and the duration of its individual phases. Thus, using autoradiography, it has been established that at earlier developmental stages (8 dpc) in murine TGCs the entire polytene nucleus cycle is significantly shortened compared with the mitotic one; it can be as short as 6–7 hr, with the S phase lasting 4–5 hr. Later on, at 10 dpc, there was a prolongation both of the S phase (7.5 hr) and of the whole cycle (13.5 hr). In the course of mouse TGC polyploidization, the duration of the cycle increased owing to elongation of both the S and the G phase compared with diploid nuclei (Andreeva, 1964; Zavarzin, 1967). A similar regularity was revealed in the polytene nucleus cycle in the *Chironomus* larva salivary gland: with an increase of the ploidy level from $2^{10}$ to $2^{11}$, the duration of the S phase increased by 30% (Gunderina *et al.*, 1984).

These differences in the nuclear structure due to different stages of the chromosomal replication cycle have been found in polytene chromosomes of the suspensor *P. coccineus,* where the maximal polyploidy reaches 8192C (Brady, 1973a,b). It has been shown autoradiographically that in the DNA reduplication period, chromonemes of the kidney bean suspensor polytene chromosomes decondense and dissociate at a substantial length (Avanzi *et al.,* 1970; Brady, 1973b; Brady and Clutter, 1974). Hence, it is in this case, as in the case of the TGCs, that the DNA synthesis stage is characterized by a marked rearrangement of chromonemes involving their decondensation and dissociation.

Cytospectrophotometric measurements have revealed a different distribution of the DNA content in the nuclei of rat secondary TGCs, depending on the stage of the chromosomal replication cycle. Nuclei with clearly seen bundles of polytene chromosomes are characterized by sharp peaks corresponding to certain ploidy levels: 64–512C. On the other hand, no similar sharp peaks are seen in nuclei with the meshwork chromatin structure (T.G. Zybina *et al.,* 1985). The presence of many meshwork nuclei with an intermediate DNA content seems to be accounted for by the fact that they are in the DNA synthesis phase. Hence, results of the cytospectrophotometric study confirm the conclusion that nuclei with a bundle and meshwork structure belong to different stages of the polytene nucleus cycle.

## D. Markers of the Interphase Nucleus in Mammalian Cell Polytenization

### 1. Sex Chromatin

In studying mechanisms of genome multiplication, morphological characteristics of heteropyknotic sex chromosomes may help in understanding the behavior of chromosomes, particularly in cases when individual chromosomes of the set are impossible to reveal (e.g., in interphase). In this connection, the sex chromatin or the Barr body, which remains in a condensed form in interphase, can be considered a kind of a marker of the interphase nucleus; using this marker, it is possible to evaluate ways the genome can multiply. In the course of polyploidization of hepatocyte nuclei in rat and human females, the number of heteropyknotic X-chromosomes in mitosis and of Barr bodies in interphase nuclei increases parallel with the degree of ploidy (Reitalu, 1957; Ohno *et al.,* 1959).

An increase in the number of sex chromatin clumps during polyploidization has also been shown in other human tissues such as amnion, chorion, malignant tumor cells, and cardiac atrium cells, as well as in the trophoblast cells of the rat placenta junctional zone (Klinger, 1958; Martynova *et al.,* 1983; Zybina *et al.,* 1984a). Polyploid cell nuclei can have either several sex chromatin bodies the size of the diploid cell Barr body or a single sex chromatin body, but one that is larger than in diploid nuclei.

Klinger and Schwarzacher (1958, 1960) were able to determine that the DNA content in the Barr bodies corresponded to the nucleus ploidy level both in single and in multiple Barr bodies. The total DNA content in the sex chromatin bodies in human female polyploid cells of the amnion and embryonal liver has turned out to be directly proportional to the total nuclear DNA content. The authors compared three nuclear classes, with ploidy 2C, 4C, and 8C, and found the ratio of the Barr body DNA to the total nuclear DNA content to be constant. Thus, a close correlation has been shown cytophotometrically to exist between the degree of ploidy of the interphase nucleus and the DNA content in inactivated X-chromosomes.

However, morphological observations on sex chromosomes may help in some cases to decipher mechanisms of polyploidization. In the case of endomitosis, for instance, when chromosomes condense and disjoin in the centromere region, it is possible to reveal and to count heteropyknotic X-chromosomes. It is in this way that the polyploidy level was determined in different tissues of *Gerris lateralis* (Geitler, 1953) and in the testicular follicle wall cells of *Stenobothrus parallelus* (Prokofieva-Belgovskaya, 1960) and of *Schistocerca gregaria* (Kiknadze and Istomina, 1980). In the case of endoreduplication in the human fibroblast culture, the late reduplicating heteropyknotic sex chromosome is revealed as a diplochromosome. The connection between sister chromosomes is lost only in anaphase (Schwarzacher and Schnedl, 1966). It may be suggested that attraction forces holding together the sister chromonemes can act even more strongerly in the case of TGCs, in which chromosomal reproduction occurs via polyteny, than in the previous case; these forces possibly result in formation of the single complex of heteropyknotic X-chromosomes (i.e., the Barr body). Thus, whereas in endomitosis there is an increase in the number of both isopyknotic and heteropyknotic chromosomes, in the polytenization cycle, only enlargement of the single sex chromatin body occurs as well as enlargement of bundles of polytene chromosomes, without an increase in their number (Zybina, 1960, 1965; Nagl, 1972). A similar increase in the size of the Barr body together with enlargement of the nucleus has been shown in human cervical cancer cells (Therman *et al.,* 1985).

Simultaneous cytophotometric measurements of the DNA content in the Barr body and in the entire nucleus of rat TGCs have shown that an increase of the DNA in the single Barr body parallels elevation of the total nuclear DNA amount in the course of several consecutive cycles of genome multiplication from 16C to 256C (Fig. 6). The ratio of these values was constant in all ploidy degrees, and the DNA content in the Barr body amounted to about 5% of the nucleus total DNA content (Zybina and Mosjan, 1967; Zybina, 1986). Similar regularities have also been found in rabbit TGC nuclei (T. G. Zybina *et al.,* 1980; Zybina, 1986). These data are another convincing proof in favor of the polytenic nature of the inactivated X-chromosome. Regular doubling of the material in the single inactivated X-chromosome in the course of several cycles of genome multiplication also substantiates indirectly the polytenic nature of other chromosomes that either form loose bundles or remain predominantly decondensed.

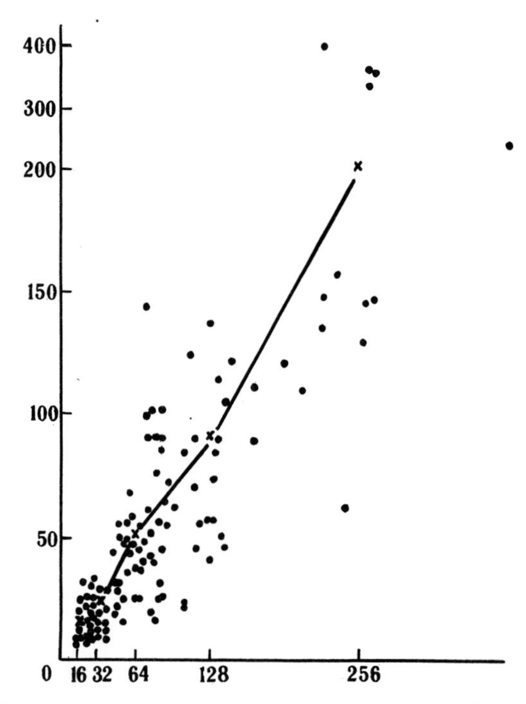


FIG. 6 Relationship between the DNA content in the Barr bodies and different ploidy degrees of the nucleus in the rat trophoblast giant cells. Lines connect the mean values of the DNA amount in the Barr bodies with different degrees of the nucleus ploidy. The abscissa shows the degree of the nucleus ploidy; the ordinate shows the Barr body DNA amount (relative units) in the same nuclei.

### 2. Nucleolus

The nucleolus, like sex chromatin, may be regarded as a kind of an interphase nucleus marker because this genome region remains visible in the interphase and its behavior reflects the state of NOCs in the course of the cell cycle: both the mitotic and endoreproduction cycle in the case of endomitosis and polyteny. In polyteny the nucleolus is constantly revealed in the nucleus (Tschermak-Woess and Hasitschka, 1953; Tschermak-Woess, 1957, 1971; Beermann, 1960, 1962, 1972; Nagl, 1970a, 1978; Kiknadze, 1972). In endomitosis, nucleoli are in most cases maintained throughout the entire nuclear cycle (Geitler, 1953; Sokolov, 1967; Nur, 1968).

In the TGC nuclei, the number of nucleoli in the course of multiple genome multiplication (up to 256–512C) in rodent trophoblasts does not change essentially but the nucleolar size increases (Zybina, 1963b, 1977, 1986). This persistence of a limited number of large nucleoli during nuclear polyploidization is another proof of the polytenic nature of TGC nuclei. In TGCs of the rat placenta junctional zone, where the initial polyploidization occurs by uncompleted polyploidizing mitoses that later are replaced by endoreplication (Zybina, 1963b, 1986; Zybina *et al.,* 1984b), the nuclei turned out to contain from one to six nucleoli, most nuclei containing two nucleoli at each ploidy level. When ploidy rises from 2C to 4C, there is a slight trend toward an increase in the number of nucleoli: the percentage of nuclei with four to six nucleoli is elevated.

On the other hand, starting with 8C, when the cells are involved in the endoreplication cycle, the number of nucleoli drops to one to three (T. G. Zybina and E. V. Zybina, 1989; T. G. Zybina *et al.,* 1994). This trend toward a decrease in the number of nucleoli that starts from the ploidy degree of 8C most likely reflects a closer association of the sister chromatids of NOCs at endoreduplication. It is interesting to note that in rat hepatocytes, which polyploidize only via uncompleted mitoses, resulting in an increase in the number of chromosomes, the trend toward an increase in the number of nucleoli was much more pronounced: from 1 to 5 nucleoli in the diploid to 3–11 in octaploid cells (Marshak *et al.,* 1994).

### 3. Heterochromatin and Other Chromosome Markers

Heterochromatin can also be an interphase nucleus marker owing to one of its properties—preservation of the condensed state in the interphase nucleus (Heitz, 1934; Brown *et al.,* 1966; Prokofieva-Belgovskaya, 1986). In mouse and rat TGCs, heterochromatin blocks are seen throughout the entire nuclear cycle, both in endointerphase and in endoprophase (Zybina, 1977, 1986). These large heterochromatin blocks are particularly visible in mouse TGCs, most of them attached directly to the nucleolus (Zybina, 1977). According to Barlow and Sherman (1974), the number of heterochromatin blocks in mouse TGCs varied from 5 to 30, the median number being 12, which is close to that in several types of mouse diploid cells: 2–24 (on average, 9). The area of chromocenters in mouse TGCs amounted to 12% of the total nucleus area; this value remained unchanged when the nucleus area increased 10 times. Hence, endopolyploidization was not accompanied by any elevation in the number of heterochromatin blocks while their enlargement did occur, which confirms the junction of sister chromatids in regions of constitutive heterochromatin in mouse TGCs.

As the heterochromatin in the mouse genome is located mainly in the pericentromere regions of all chromosomes, the data mentioned indicate a close association of chromatids in pericentromere chromatin regions. Similar results were obtained when mouse TGCs were stained with 4,6-diamino-2-phenylindole (DAPI). Brightly stained areas corresponding to the pericentromere heterochromatin were enlarged parallel to enlargement of nuclei, which indicates an association of chromatids at polyteny (Bower, 1987). In this work the number of kinetochore localization areas was also counted using indirect immunofluorescence with anti-kinetochore antibodies. As early as at the initial polyploidization from 2C to 8C, the number of flourescence areas increased only slightly to be equal on average to the haploid number of chromosomes (20) in the mouse chromosome set. These data indicate not only an association of sister chromatids in polytene chromosomes but also a somatic synapsis of homologs.

The latter suggestion is substantiated by the results of *in situ* hybridization with the satellite DNA major sequence revealed in all chromosomes of the mouse genome. At hybridization of murine primary and secondary TGCs with this satellite sequence, the number of hybridization signals showed normal distribution around the mean that was close to the haploid chromosome number (Varmuza *et al.,* 1988). These data support polytene chromatid association as well as a tendency for homologous chromosomes to synapse. To prove the polytenic nature of the TGCs, other markers were also used. Thus, on hybridization of the unique gene, $\alpha$-1-antitrypsin, with murine TGCs, there was some variability in the number of labeling zones. Of 106 nuclei studied, 22 had one labeling zone, 36 had two, and the rest of the nuclei (i.e., 45%) had more than two zones (Bower, 1987). These data confirm the polytenic association of sister chromatids in more than 50% of the nuclei and suggest the homolog synapsis in about one-quarter of the cases. Lack of the chromatid polytenic association and of the homolog synapsis in 45% of the nuclei seems to be connected with the variable character of polyteny in the nuclear cycle.

In another experiment, a plasmid containing murine $\beta$-globin DNA was used as a probe; *in situ* hybridization was performed with TGC nuclei of a murine transgenic line with the third chromosome inserted with DNA of the same plasmid. In insertion heterozygotes, only one hybridization site has been found, while in homozygotes two have been found. Similar results were obtained with some other sequences localized in different areas of mouse chromosomes (Varmuza *et al.,* 1988; Keighren and West, 1993). These data convincingly prove a polytene chromatid association in mouse TGC nuclei. Nevertheless, the presence of two labeling zones indicates a lack of conjugation in a number of regions of polytene chromosomes.

The data cited that were obtained using interphase nucleus markers confirm the polytenic nature of the TGCs. The presence, in some cases, of more than two labeling zones is connected most probably with the main peculiarity of the nonclassic form of polyteny, the nonconstant character of the attachment of chromatids along the length of the chromosome. This problem is considered in Section III,B.

The use of heterochromatin and of sequences related to pericentromere heterochromatin as markers of the interphase nucleus has made it possible not only to confirm the polytenic nature of the TGCs but also to substantiate the possibility of conjugation of homologs at polyteny in the TGCs.

### E. Conclusion

The data considered here allow several conclusions. The nuclei of the mammalian TGCs reveal the following signs of polyteny: (1) chromosomes represent bundles of the sister chromonemes that are connected to each other at some length; (2) the nuclei, despite their extremely high degree of ploidy, contain one or a few nucleoli; (3) some regions of the chromosomes have a band pattern due to a close attachment of chromomeres in adjacent chromonemes; and (4) puff-like structures sometimes occur. At the same time, the discovery of polytene chromosomes in the mammalian trophoblast is limited by the following factors: (1) the degree of nuclear polyploidy, (2) the stage of the embryo development, (3) the phase of the replication cycle of the chromosomes, and (4) the different degree of manifestation of polyteny in different chromosomes within a set and different chromosome regions. In addition, the character of the discovery of polytene chromosomes varies with the animal species.

## III. Characteristics of the Nonclassic Form of Polyteny

The study of the rodent TGC nuclear structure suggests the polytenic nature of chromosomes. It should be emphasized that, for several reasons, it is much more difficult to study trophoblast polytene chromosomes in mammals than in Diptera; nevertheless, the TGCs possess almost all the properties of classic polytene chromosomes although they are expressed to a lesser degree and sometimes are not revealed at all.

### A. Polyteny in the Trophoblast: Similarity with Diptera Polytene Chromosomes

At consecutive stages of polytenization in the growing nuclei of rat TGCs, a gradual thickening of chromosomal bundles is seen to result in their diameter becoming much larger than that of metaphase chromosomes (Zybina, 1977, 1986). In rodent trophoblasts, as in Diptera salivary gland cells, the polytene chromosomes are formed through consecutive cycles of DNA endoreduplication; as a reult, nuclei become highly polyploid: up to 512–4096C in the trophoblast of rodents (Zybina, 1963a; Barlow and Sherman, 1972; Nagl, 1972; Zybina *et al.*, 1975a,b) and 32–128C in the trophoblast of Carnivora (Zybina *et al.*, 1989, 1992), while the polyploidy level in the salivary gland cell nuclei in different Diptera species reaches 16,384–32,768C (Beermann, 1962; Rash, 1970; Valeyeva and Kiknadze, 1971). Polytene chromosomes are found in TGC nuclei at a certain, rather limited stage of cell ontogenesis, on reaching a high level of endoreproduction (higher than 32C). This situation resembles that in Chironomidae, in which formation of polytene chromosomes starts in highly polyploid (64C) nuclei (Kiknadze, 1972).

During several polytenization cycles, a chromosome is formed with $2^n$ chromatids closely attached to each other while the number of chromosomes remains unchanged (Nagl, 1978; D'Amato, 1989). Therefore, the presence of polyteny in rat and rabbit TGCs is substantiated by the observation that multiplication of the material of the single body of sex chromatin parallels the nuclear ploidy level (Zybina and Mosjan, 1967; Nagl, 1972; T.G. Zybina *et al.*, 1980) as well as by data about an increase in size of other interphase nucleus markers—heterochromatin blocks, hybridization signals of satellite DNA, areas of binding of antibodies to kinetochores, and probes to different DNA sequences in mouse chromosomes (Barlow and Sherman, 1974; Bower, 1987; Varmuza *et al.*, 1988). Of similar significance are data by Snow and Ansell (1974), who artifically induced (by adding actinomycin D to the mouse TGC culture) the formation of long chromosome-like bodies; in the course of polyploidization (from 4C to 256C), the size of these bodies increased but their number did not change and corresponded approximately to the diploid chromosome set ($2n = 40$).

A peculiarity of polytene chromosomes of Diptera consists of a decrease in their ability to complete the normal mitotic cycle of spiralization and despiralization (Ashburner, 1970). Compared with usual metaphase chromosomes, the polytene chromosomes are decondensed and persist in a state of interphase or prophase (Muller, 1935; Kiknadze, 1972). This accounts for the polytene chromosomes being much longer than the metaphase chromosomes (Beermann, 1962; Kiknadze, 1972; Pearson, 1974b). In cases when it is possible to identify polytene chromosomes, their length is seen to exceed many times the metaphase chromosomes length. In the course of endopolyploidization, the length of chromosomal bundles in the rodent TGC nuclei increases (Zybina, 1977, 1986).

Bandlike structures were discovered at the light optical level in giant chromosomes in three of four rodent species studied (Zybina, 1961, 1970, 1977, 1986; Zybina *et al.,* 1975b; Pearson, 1974b; Zybina and Chernogryadskaya, 1976); thus, this is another similarity with Diptera polytene chromosomes. However, in the rodent chromosomes, in contrast to the chromosomes in Diptera, the bandlike structures are located predominantly near the NOR (Zybina *et al.,* 1975b; Zybina and Chernogryadskaya, 1976; Zybina, 1977, 1986) and near chromocenters (Pearson, 1974b). In rat trophoblast giant chromosomes, the formation of large puffs was occasionally observed (Zybina, 1977, 1986). However, it is difficult to study the puffs because of the peculiar structure of polytene chromosomes in rodents.

Another peculiarity of Diptera polytene chromosomes is that the nuclear envelope and nucleoli persist throughout the whole cell cycle (Geitler, 1953; Ashburner, 1970; Pearson, 1974b). The nuclear envelope and nucleoli also are preserved in rodent and carnivora TGCs for the entire endoreduplication cycle (Zybina, 1977, 1986; Zybina et al., 1989, 1992).

## B. Peculiar Features of Nonclassic Polytene Chromosomes

Up to now, the similarity of polytene chromosomes in mammals and Diptera has been discussed; however, differences between them also should be reviewed. A characteristic peculiarity of mammalian polytene chromosomes is their lability, a variable character of polyteny revealed in the DNA reduplication cycle (Zybina, 1963b). When studying polytene chromosomes in rodents such as the rat, rabbit, mouse, and field vole, two stages in the polytene nucleus cycle are always found. These stages, endointerphase and endoprophase, differ in the degree to which polytene chromosomes are visualized.

The difficulty of revealing the mammalian polytene chromosomes is also due to their bundles being clearly seen only for a certain, short period of time in the course of differentiation of TGCs; after this period, they can no longer be visualized. This period in rat TGCs is known to coincide with the DNA reduplication period (Zybina, 1963b; Jollie, 1964). This is the second peculiarity of chromosomal polyteny in mammals: its temporary character in cell ontogenesis. In rat TGCs, for instance, polytene chromosome bundles can be seen at 11–14 dpc, in rabbits at 12–14 dpc, in mice at 9–11 dpc, and in field voles only at 9 dpc (Zybina, 1977, 1986). In the mink trophoblast, it is impossible to specify the exact stages of fetus development in gestation; it is only possible to state that bundles of polytene chromosomes are visualized as early as preimplantation (Isakova *et al.*, 1992) as well as middle gestation stages (Zybina *et al.*, 1992).

The third peculiarity is that sister chromatids are most commonly attached to each other in some sites rather than throughout their entire length. Therefore, the visual signs of polyteny are revealed in a different measure in different rodent species, depending on chromoneme condensation and their attachment to each other.

Another peculiarity of mammalian trophoblast polytene chromosomes is that up to now no heterochromatic region underreduplication has been found, which is often seen in Diptera polytene chromosomes (Rudkin, 1965, 1969; Mulder *et al.*, 1968). A complete replication of the whole TGC genome has been shown cytophotometrically in the mouse (Barlow and Sherman, 1972), rabbit (Zybina *et al.*, 1975b), field vole (Zybina *et al.*, 1975a), and rat (T. G. Zybina *et al.*, 1985). A study of DNA fragment reassociation kinetics has revealed the complete satellite DNA replication in mouse TGCs (Sherman *et al.*, 1972). Nevertheless, the mammalian trophoblast polytene chromosomes, as has been shown, differ significantly from the Diptera classic polytene chromosomes, which has enabled us to call the mammalian chromosomes "nonclassic polytene chromosomes." Since many data have been accumulated about the presence of polyteny in plant and animal cells, it is worth discussing to what degree the peculiarities of nonclassic polyteny in rodents are common for various cases of formation of polytene chromosomes.

Of particular interest for a comparison with mammalian polytene chromosomes are plant giant chromosomes (*Rhiesenchromosomen*) studied in much detail by the Vienna school (Geitler, 1953; Hasitschka, 1956; Tschermak-Woess, 1956, 1971, 1973; Nagl, 1962, 1970a,b, 1978, 1981, 1985). Clearly visible bands often are absent in polytene chromosomes both in plants and in rodents. It is possible to suggest that the distinct band fails to form because the corresponding chromomeres of adjacent chromonemes do not attach closely to each other and are not located exactly one above the other (Nagl, 1981). In most of the plant giant cell nuclei, as in the mammalian trophoblast, polytene chromosomes are not always visible: the possibilities of their visualization are limited by certain time periods or physiological states of the cell. Here, however, it is not always clear, in contrast to the trophoblast, what the presence of polytene chromosomes depends on (Nagl, 1978, 1981; D'Amato, 1989).

The formation of plant giant chromosomes is characterized by marked variations in the degree of the chromosome close attachment: from a complete association of chromonemes along the entire length of the chromosome in the *P. coccineus* suspensor (Nagl, 1967, 1976, 1978) to a local attachment in heterochromatic areas of the antipodes of *Papaver rhoeas* and cells of the haustorium of *Thesium alpinum* (Tschermak-Woess, 1971, 1973; Nagl, 1978, 1981). In the antipode cells of *P. rhoeas,* for instance, several nucleus types are described: (1) those with despiralized chromonemes bound in heterochromatic block areas, (2) those with many decondensed isolated chromatin fibrils, (3) those with separately located endochromosomes, and (4) those with giant chromosomes in which the chromatids are attached to each other along almost the entire chromosome length (Hasitschka, 1956; Nagl, 1981).

Nagl (1981) noticed that nuclei in the same tissue sometimes have a different structure, which can vary depending on environmental factors such as nutrition, temperature, illumination duration, and effects of chemical agents. In plants the following main regularity has been revealed: the most commonly visible polytene chromosomes are formed in those species whose chromosome sets have large heterochromatin blocks; for instance, in *Rhinanthus* and *Tropaeolum* these blocks, in the course of polytenization, promote close union of chromonemes by their stickyness.

Thus, an important similarity of the nonclassic polytene chromosomes both in mammals and in plants is that the attachment of chromonemes occurs only at some extent rather than along the whole length of the chromosome, because the chromoneme-binding forces in polytene chromosomes in angiosperms and rodents are expressed to a lesser degree than in Diptera. This extent can change in the polytenic nucleus replication cycle and in the course of cell differentiation (as shown in rodents) or under the effects of various factors (as shown in plants).

### C. Differences between Classic and Nonclassic Polyteny

A comparison of data in the literature on classic and nonclassic polyteny indicates that differences between them are not absolute. As already mentioned, the problem of homolog conjugation in polytene chromosomes in mammalian trophoblasts is yet to be solved. The number of chromosomal bundles, based on their count in rat and rabbit TGC nuclei, seems to correspond to the number of chromosomes in the haploid set (Zybina, 1977). A tendency of the homolog somatic synapsis has been revealed (Bower, 1987; Varmuza *et al.,* 1988) by using interphase nucleus markers bound to pericentromere heterochromatin in the mouse trophoblast. Use of some other markers has indicated a lack of conjugation (Barlow and

Sherman, 1974; Bower, 1987; Varmuza *et al.*, 1988). Lack of the homolog conjugation is also confirmed by the experimental data with actinomycin D mentioned earlier (Snow and Ansell, 1974): the number of compact band-shaped bodies was approximately equal to diploid. Lack of conjugation is also typical for plant polytene chromosomes (Avanzi *et al.*, 1970; Nagl, 1978; D'Amato, 1989).

However, the homolog conjugation that is so characteristic of Diptera classic polytene chromosomes (Beermann, 1962; Prokofieva-Belgovskaya, 1986) is not always found in cases of classic polyteny. Thus, a lack of conjugation of homolog chromosomes is found in polytene chromosomes of the salivary glands of Collembola (Cassagnau *et al.*, 1979; Dallai, 1979; Deharveng, 1982) as well as in *Harmandia laevi* (Zhimulev and Lychev, 1972). In these chromosomes, a close attachment of sister chromonemes does not always occur; their structure can change markedly in the course of cell differentiation, with the chromosomes dissociating into separate chromonemes (White, 1948, 1977; Matuszewski, 1965; Henderson, 1967a; Zhimulev and Lychev, 1972; Zhimulev and Belyaeva, 1976).

Therefore, other features typical of mammalian and plant polytene chromosomes, such as noncomplete union of sister chromonemes along the length of the chromosome, lack of a clear band pattern, and temporary character of the chromoneme union in the polytene chromosome, are also peculiar to polytene chromosomes in some Diptera species. Classic polytene chromosomes can lose some signs of polyteny at certain ontogenetic stages. Thus, in the larva abdominal epidermis and enocyte nuclei of *Calliphora erythrocephala,* at a certain ontogenetic stage, typical polytene chromosomes dissociate into oligotene fibrils that form a meshwork in the nucleoplasm and attach only in short regions so that it is not possible to identify individual chromosomes along their entire length (Pearson, 1974a,b). Some authors (Mainx, 1949; Zhimulev, 1992) present many examples of marked variations of the classic polytene chromosome morphology in various tissues of Diptera and some other insects because of different degrees of the sister chromatid conjugation. These authors classify several types of the polytene chromosome morphology and structure, depending on the degree of the chromatid synapsis: (1) classic polytene chromosomes when chromatids exhibit a close synapsis, with an exact union in homolog regions and, as a result, a clearly seen band pattern; the number of the chromosomes is haploid. Chromosomes of this type are present in salivary glands and other organs of larvae of many families; (2) the chromatid synapsis is weaker, the chromosomes are loose, and the clear uninterrupted band pattern remains only in the area of large bands. Examples include some species of Cecidomyidae (White, 1948; Henderson, 1967a,b) and Collembola (Cassagnau, 1971); (3) "pompon-like" chromosomes with an almost entirely lost band pattern that is maintained only in some regions of the chromosome;

the polytene chromosomes themselves are visible only as large chromocenters or dense bundles from which chromatids diverge radially. Examples include Collembola (Cassagnau, 1971; Cassagnau, 1974; Dalens, 1976); (4) hidden polyteny. Only individual chromocenters are seen in highly polyploid nuclei. Such nuclei occur in *Drosophila* ovarian nurse cells (Hammond and Laird, 1985).

It is important to note that different types of polytene chromosome structure may occur in the same tissue and transform from one into another at different ontogenetic stages or under the effects of various factors (Bier, 1957, 1958; Ribbert and Bier, 1969; Ribbert, 1979; Zhimulev, 1992). These data indicate that the close conjugation of sister chromatids along the entire length of the chromosome is not a mandatory event for classic polytene chromosomes. Therefore, the presence of a sharp band pattern along the entire length of the chromosome may hardly be considered a necessary condition for polyteny. The structure of classic polytene chromosomes also can change during the DNA replication cycle. In the *Chironomus plumosus* larva, sharply defined classes of DNA are present only in nuclei with closely packed polytene chromosomes, whereas most of the intermediate DNA values are seen in nuclei with loose packing of chromosomes. Il'yinskaya and Selivanova (1982) explain this phenomenon by noting that the intermediate DNA values correspond to the S phase while nuclei with more condensed chromosomes correspond to the G phase.

Thus, owing to the recent accumulation of a large number of new facts, the concept of polyteny has become more extensive and its content changed. Whereas previously this phenomenon was mainly regarded as one form of chromosome organization, at present polyteny may be considered, apart from this, a peculiar shortened variant of the cell cycle that manifests itself morphologically as different ways of packing of united sister chromonemes in the interphase nucleus (Nagl, 1978; Brodsky and Uryvaeva, 1981, 1985; D'Amato, 1989). At present it is already known that the degree of association of chromonemes can change markedly under the effects of various factors as well as during various periods of the cell and organism life cycle. Further studies will probably elucidate the cause and the biological significance of these changes in the nuclear structure.

## IV. Ultrastructural Organization of the Polytenic Nucleus in Trophoblast Giant Cells

### A. Polytene Chromosomes and Nucleolus-Like Bodies

Electron microscopy studies confirm light microscopy data on the presence of polytene chromosomes in TGCs of some rodents. In electron micrographs of rat trophoblast cells, two types of nuclei are revealed that seem to correspond to stages in the polytene nucleus cycle: (1) nuclei with decondensed chromatin and with signs of polyteny that as a rule are not revealed, and (2) nuclei with condensed chromatin areas and structures resembling polytene chromosome bands (Zybina, 1980c,d, 1986). In nuclei of the first type, at earlier stages of the development of placenta (12–14 dpc), thin chromatin threads are spread diffusely throughout the nucleoplasm (Fig. 7). Perinuclear and perinucleolar chromatin layers are poorly distinguished. A distinct orientation of the chromatin threads in a certain direction is sometimes seen; it is particularly evident near NOCs. Small fibrillogranular nucleolus-like bodies (NLBs) with a diameter of about 0.5 $\mu$m are sometimes revealed in the nucleoplasm.

In nuclei of the second type, there are large, condensed chromatin blocks. Perinuclear and perinucleolar chromatin is clearly seen. Whereas the chromatin blocks in the nucleoplasm look to be spread chaotically, they are arranged in a definite order near the nucleolus: large, elongated chromatin lumps are oriented parallel to each other and to the perinucleolar chromatin layer (Fig. 8a). These rows of condensed chromatin clumps cannot be ruled out as representing structures analogous to bands. These interrupted bands in trophoblast nuclei are 0.2–0.3 $\mu$m thick. Large chromatin clumps forming the band are sometimes separated into smaller clumps, probably due to dissociation of a single chromoneme bundle into several individual bundles. Thin chromatin fibrils in the space between the bandlike structures are evidently oriented from one band to another or from the perinucleolar chromatin to the closest of the parallel bands (Fig. 8a). In their ultrastructure, size, and spatial arrangement, the bandlike TGC structures resemble bands of polytene chromosomes in Diptera (Beermann, 1972; Perov *et al.*, 1976); however, the parallel arrangement of large chromatin clumps in the trophoblast is seen in more limited regions along the length of the chromosomes than in Diptera.

In the 17-dpc placenta, when the chromosomal bundles are not visualized anymore and DNA synthesis stops almost completely, the nuclear ultrastructure changes significantly compared with that at earlier stages. Nevertheless, even in this period, two types of nuclei are present that differ in their chromatin state and the morphology of presumed products of chromosomal activity: some nuclei are composed of a decondensed chromatin while in others the chromatin is condensed only partially. At this developmental stage, nuclei with a partially condensed chromatin (Fig. 8b) have a greater NLB variability than at the preceding stage or compared to nuclei with decondensed chromatin. There are many interchromatin and perichromatin granules, rough fibrillar NLBs (with 20-nm-diameter fibrils), as well as numerous fibrillogranular round NLBs (with 20-nm-diameter granules and 8-nm-diameter fibrils) or large elongated aggregates of small

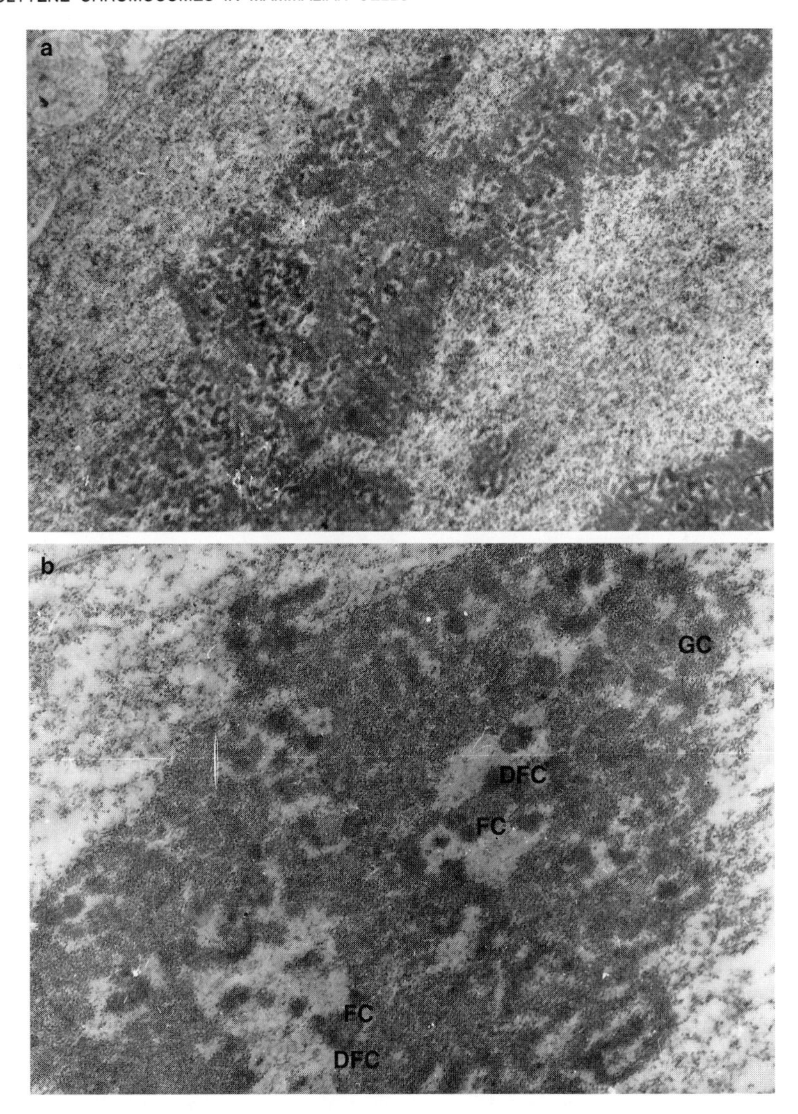


FIG. 7 Ultrastructure of nuclei with decondensed chromatin in rat giant trophoblast on day 12 postcoitum. (a) The decondensed chromatin is distributed throughout the nucleus rather homogeneously; the perinuclear chromatin layer is poorly defined; the nucleolus has a nucleoloneme structure. (b) The nucleolus contains fibrillar centers (FC) surrounded by the dense fibrillar component (DFC) and a pronounced agglomerate of the granular component (GC). Bar = 1 $\mu$m.

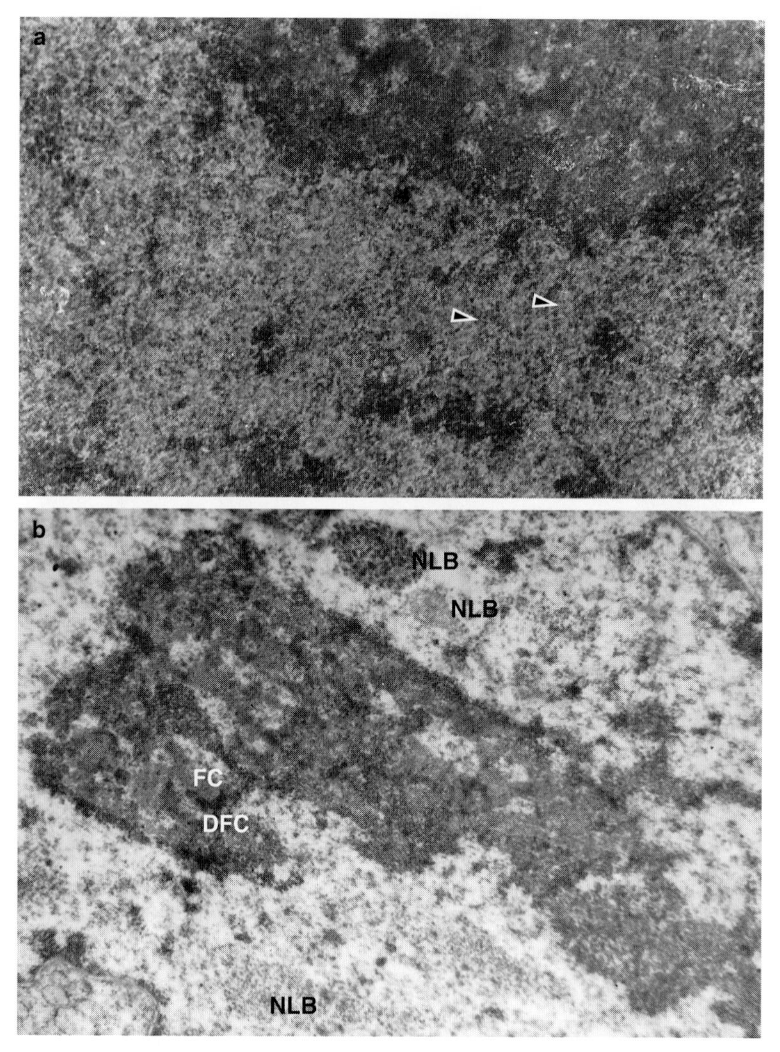


FIG. 8 Ultrastructure of nuclei with decondensed chromatin in rat giant trophoblast on day 13 (a) and day 17 (b) postcoitum. (a) Clearly seen are the condensed perinucleolar chromatin layer and zone with the decondensed chromatin threads oriented from the condensed chromatin nucleoplasmic aggregates to the nucleolus (arrowheads); the condensed chromatin clusters resemble a disk; nucleolar fibrillar centers are poorly defined. (b) Clearly seen in the nucleus is the condensed perinuclear chromatin, the latter forming strands that penetrate the nucleolus through the lacunae; fibrillar centers (FC) and the dense fibrillar component (DFC) are clearly seen but the amount of granular component falls markedly; the nucleoplasm contains finely granular and roughly granular nucleolus-like bodies (NLB). Bar = 0.5 $\mu$m.

granules with the same diameter. This stage is also characterized by the appearance of new structures: thin, fibrillar, round NLBs with a 2-$\mu$m diameter, composed of densely packed thin threads. There are also fibrillo-granular bodies with large granules, 40–60 nm in diameter, resembling coiled bodies. Formation of all the above-mentioned types of the bodies was also noticed in lampbrush chromosomes in oocytes of the mouse (Chouinard, 1973, 1975a) and of the golden hamster (Zybina and Grishchenko, 1977). Secondary nucleoli in oocytes of *Acheta* (Jaworska and Lima-de-Faria, 1973) and "granular coils" in oocytes of *Chrysopa* (Gruzova and Parfenov, 1993) have a structure and size similar to those in the coiled bodies.

As to the nature of NLBs, they are known to contain basic and acidic proteins and traces of RNA (Zybina, 1968; Arronet, 1973; Gruzova and Parfenov, 1993). They are supposed to contain products of chromosomal transcriptional activity, including activity of additional ribosomal cistrons (Sandler and Pavan, 1972; Stahl *et al.,* 1975) as well as small nuclear RNAs (snRNAs) (Raska *et al.,* 1994). They also contain, according to some authors (Wolgemuth-Jarashow *et al.,* 1977), amplified DNA. Formation of NLBs in the regions that are not identical to NOR has been described in both heterochromatic and euchromatic polytene chromosome segments in polytenic nuclei of the kidney bean suspensor (Nagl, 1970b, 1973a). A great number of micronucleoli (i.e., bodies that are not identical to the main nucleolus) appear in areas of intercalar heterochromatin in *Drosophila* salivary gland nuclei with classic polytene chromosomes (Ananiev and Barsky, 1985). This seems to compensate for underreduplication of heterochromatic regions containing the main NOR (Prokofieva-Belgovskaya, 1986).

At any rate, NLBs are either products of the activity of individual chromosome loci or a form of the newly synthesized RNA accumulation. Therefore, the diversity of the NLB forms probably reflects the expression of different chromosome loci at different stages of the TGC cycle and differentiation. At 17 dpc, the so-called nuclear bodies appear in nuclei with partially condensed chromatin (Zybina, 1980d, 1986). They are composed of an external layer of concentrically arranged fibrils, which surrounds aggregates of granules of different diameters. These bodies are known to be seen in various actively functioning cells (e.g., in viral infections or carcinogenesis) and to contain DNA and ribonucleoprotein (RNP), glycogen, and lipids (Monneron and Bernhard, 1969; Recher *et al.,* 1972; Smetana, 1974).

In summary, it may be said that in the course of TGC differentiation, a cyclicity has been observed in the changes in chromatin state from diffuse to more condensed; this cyclicity is accompanied by a periodicity in the appearance of supposed products of chromosomal activity (Table II). During the entire period of differentiation, polytene chromosome transcrip-

TABLE II

Cycle of Structural and Functional Changes in Trophoblast Giant Cell Nuclei in the Course of Differentiation

| | Polytenization | | Completion of polytenization | | | |
|---|---|---|---|---|---|---|
| State of nucleus | Endointerphase | Endoprophase | Endointerphase | Endoprophase | Polygenomic nucleus | Fragmentation of the nucleus |
| Morphology of the nucleus | 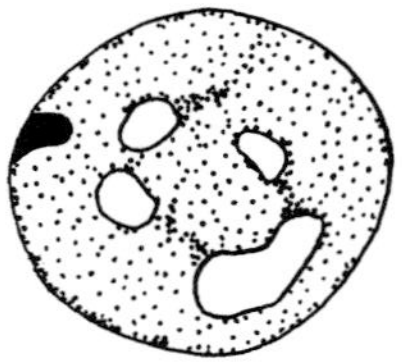 |  | 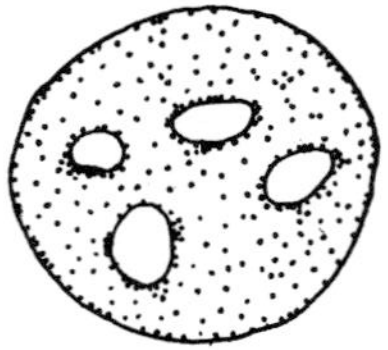 |  |  |  |
| Chromosome structure | Meshwork nucleus, chromatid association near the nucleolus | Bundles | Meshwork nucleus, chromatid association near the nucleolus | Chromatin is partially condensed | Paired endochromosomes | Condensed chromatin attached to the nuclear membrane |
| Barr bodies | Distinct | Distinct, more condensed | Less distinct | Less distinct | Not observed | Sometimes present |
| DNA synthesis | Present | No synthesis | Present in some nuclei | No synthesis | Not studied | No synthesis |
| State of nucleolus (ultrastructurally) | Maximal development of GC. Pch is decondensed. FCs are distinct | Decrease of GC. Pch is condensed. FCs are less distinct | Progressive decrease of GC. Pch is decondensed. FCs are less distinct | GC scarse or absent. Pch is condensed. FCs are distinct | Not studied | GC scarse or absent. Pch is condensed |
| Predominant products of chromosome activity (ultrastructurally) | Interchromatin granules. Fibrillogranular NLB | Interchromatin and perichromatin granules. Fibrillar and fibrillogranular NLB | Interchromatin granules. Fibrillogranular NLB | Interchromatin and perichromatin granules. Fibrillar and fibrillogranular NLB. Coiled bodies. Nuclear bodies | Not studied | Interchromatin granules. Fibrillar NLB |

Abbreviations: FC, fibrillar center; GC, granular component; NLB, nucleolus-like bodies; Pch, perinucleolar chromatin.

tional activity is maintained (Zybina, 1963b), as well as production of different NLBs characteristic for each stage of the cell cycle and ontogenesis (Zybina, 1986).

## B. The Nucleolus and Its Components

To characterize the TGC polytene nucleus cycle, it is of interest to follow changes in the interrelation of nucleolar components since the nucleolus reflects, to a significant degree, the total level of the cell functional activity (Busch and Smetana, 1970; Schwarzacher and Wachtler, 1983; Hernandez-Verdun, 1986). In interphase cell nucleoli, the following main structural components can be distinguished: fibrillar centers (FCs), dense fibrillar component (DFC), granular component (GC), perinucleolar and intranucleolar chromatin, and nucleolar lacunar space. The FCs represent round clusters of fibrils, 2–5 nm in diameter, which contain a number of proteins, including RNA polymerase I and topoisomerase I, as well as DNA in a despiralized state ready for transcription (Hernandez-Verdun and Derenzini, 1983; Hernandez-Verdun, 1991). The DFC most commonly surrounds the FCs with a dense layer or is distributed in the nucleolus as free strands composed of fibrils 4–8 nm in diameter. It contains rDNA cistrons (Wachtler *et al.,* 1992), newly synthesized preribosomal RNA, and various proteins participating in rRNA synthesis and processing.

The FCs and surrounding DFC seem to correspond, most likely, to despiralized parts of NORs in which transcription of rDNA occurs (Mirre and Knibiehler, 1981; Knibiehler *et al.,* 1982; Hernandez-Verdun and Derenzini, 1983; Hernandez-Verdun *et al.,* 1984; Vagner-Capodano *et al.,* 1984). The GC represents areas rich in preribosomal particles at different stages of maturation (Miller and Beatty, 1969) and also contains various proteins participating both in synthesis and in processing of rRNA (Hernandez-Verdun *et al.,* 1993; Raska and Dundr, 1993).

The nucleolar lacunar space represents light intranucleolar zones inside the nucleolus, the material of these zones being indistinguishable from nucleoplasm. At the nucleolar surface and inside the lacunar space, perinucleolar and intranucleolar chromatin is formed by DNP fibrils 20–30 nm in diameter. It is structurally bound to FCs and, based on *in situ* hybridization data, contains rDNA (Thiry and Thiry-Blaise, 1989). At 12 dpc, nucleoli in TGC nuclei with decondensed chromatin are nucleolonemal and composed mainly of a fibrillogranular material; numerous lacunae are often connected directly with the nucleoplasm (Fig. 7) and are, as a rule, situated near the DFC areas. Fibrillar centers surrounded by the DFC invaginate into the lacunar space.

At this stage of the polytenic nucleus cycle, the layer of the condensed perinucleolar chromatin around the nucleolus is absent; nevertheless, it is possible to trace thin chromatin fibrils penetrating from the nucleoplasm into the open nucleolar lacunar space and often being attached to DFC areas (Zybina, 1981a). In TGC nuclei with decondensed chromatin, often there are bunches or chains of FCs bound to each other via the DFC. It is probable that such FC aggregates correspond to the fibrillar complexes resulting from genome doubling as it has been described at the transition of the pig kidney cell culture from the $G_0$ to $G_2$ phase (Zatsepina *et al.*, 1988a). Although the FCs are spread throughout the entire space of the TGC nucleolus, the distal FCs are usually attached to the very surface of the nucleolus and protrude into the nucleoplasm. Thus, each FC complex turns out to be connected with peripheral nucleolar areas. In one slice three or four FC complexes can be revealed, each containing from 3 to 12–15 FCs, 260–360 nm in diameter (E.V. Zybina and T.G. Zybina, 1989). The presence of several separate FC groups in the nucleolus may be assumed to reflect an arrangement of sister chromatid homologous areas in the polytene NOCs. However, the nature of FCs in TGC nucleoli still needs detailed investigation.

In polytene nuclei with condensed chromatin, the granular component is expressed to a lesser degree. The nucleolar structure is somewhat more compact, with a condensed perinucleolar chromatin layer around the nucleolus (Fig. 8a). At this stage of the polytene nucleus cycle, penetration of the perinucleolar chromatin strands into nucleolar lacunar areas is revealed better than at the stage with the decondensed chromatin. The FCs at this stage are seen less distinctly (Zybina, 1981a, 1986).

Thus, within two phases of the polytene nucleus cycle, the NOR state changes markedly. At the endointerphase, these chromosomal regions are markedly decondensed and are revealed inside nucleolar lacunae as thin chromatin threads and probably a developed system of single FC and FC complexes. At the endoprophase, NOR areas penetrating inside the nucleolus are probably more condensed. Similar changes of the NOR state occur in onion root meristem cells during the transition from interphase to prophase and, on the contrary, during the return of telophasic nuclei to interphase (Lafontain and Lord, 1974; Chouinard, 1975b, 1981). In prophase and telophase, NOR areas in the spiralized state can be observed in contact with the nucleolus whereas in interphase, chromatin is composed of thin, barely visible fibrils owing to despiralization of these areas. The same tendency in TGCs is expressed to a lesser degree. The NOR state can be supposed to change also in the course of the shortened cell cycle (polytenization) when the mechanism of mitosis is completely absent but the degree of chromatin condensation changes cyclically.

The same tendency is indicated by a change in the interrelation between granular and fibrillar components in the cycle of the TGC polytenic nucleus. At the supposed S stage, the proportion of the GC in the nucleolar mass is prominent while it decreases at the G stage. It may be assumed that the marked decrease in the number of preribosomal granules reflects a decrease in NOR transcriptional activity in the polytene nucleus cycle. This is confirmed by autoradiographic study of RNA synthesis in rat TGC nucleoli. The level of nucleolar RNA synthesis evaluted by [$^{14}$C]adenine incorporation is about 25% higher in the S phase than in the G phase (Zybina, 1963b, 1977).

An analogy with the mitotic cycle is also seen with respect to the granular component. Thus, nucleoli of the Chinese hamster cell culture in the S phase contain many granules; however, the nucleolar volume falls markedly in the $G_2$ phase, not long before the start of the mitotic prophase (Noel *et al.*, 1971). During rat TGC differentiation (17 dpc), when DNA synthesis stops and RNA synthesis in the nucleus decreases somewhat, two types of nucleolar structure may be observed, which, as at previous stages, correspond to different degrees of chromatin condensation. At the same time, the nucleolar morphology as a whole changes markedly, which seems to be connected with a progressive chromatin condensation in the nucleus. The GC gradually decreases in both meshwork nuclei and in nuclei with a partially condensed chromatin. The latter always contain a perinuclear chromatin ring as well as, in slices of better quality, chromatin threads penetrating inside the nucleolus via the lacunar space (Fig. 8b). It is interesting to note that whereas earlier stages of TGC differentiation are characterized by the FCs clearly seen at the meshwork nucleus stage, FC visualization at later developmental stages is approximately equal in the nuclei both with the more structured and with the less structured chromatin (Zybina, 1981b). This is due, in our opinion, to a progressive FC fusion into larger formations surrounded by the DFC. In the active nucleolus, the number of FCs is known to be 5–10 times higher than the number of NORs (Zatsepina *et al.*, 1988b), probably because of "division" of the NOR into several zones of the active rDNA transcription. However, as has been shown in erythroblasts, inactivation of the nucleolus is accompanied by a progressive fall in the number of FCs, probably as a result of the FC fusion into larger structures (Hernandez-Verdun, 1986; Zatsepina *et al.*, 1988b). It cannot be ruled out that FC fusion in the TGC nuclei also can be due to a gradual inactivation and condensation of NORs. Thus, the changes of nucleolar components in the course of the TGC differentiation described here indicate inhibition of the transcriptional NOR activity. This has been confirmed by autoradiographic studies of RNA synthesis in TGC nucleoli (Zybina, 1963b).

# V. Depolytenization and Fragmentation of Trophoblast Giant Cells as the Terminal Stage of Their Differentiation

## A. General

Rodent placenta secondary TGCs remain viable almost until the end of gestation and mainly degenerate just before parturition. This sometimes is preceded by dissociation of the giant nuclei into smaller nuclear fragments. In the mouse and field vole placentas, this process starts in occasional giant nuclei as early as 10 dpc (Zybina *et al.,* 1979; E. V. Zybina, 1986; T. G. Zybina, 1990) and in the rat placenta, at 13 dpc (Zybina and Grishchenko, 1970; Zybina, 1986). This process is particularly intensive during the last several days of gestation. Sometimes the entire mass of the giant nucleus breaks down almost simultaneously into many (several tens) ovoid fragments that are closely attached to each other and that preserve the shape of the initial nucleus. The nuclear envelope forms numerous short projections and folds invaginating into the initial nucleus. In other cases, the nuclear fragments are detached from the periphery of the nucleus. Both the initial nucleus and the nuclear fragments are stained intensively with Feulgen reaction, the clumps of the condensed chromatin being located predominantly at the periphery of the nuclear fragments (Fig. 3c and d). Such clumps resemble morphologically rodlike endomitotic chromosomes.

## B. Genome Distribution of Chromosomal Material in Nuclear Fragments

The question arises as to how chromosomes are distributed in nuclear fragments. If this is a random process, the nuclear fragments should contain a nonbalanced number of chromosomes in nonviable cells. Therefore it was important to trace separate chromosomes and entire genomes during fragmentation to confirm their whole-genome distribution in individual fragments.

Interesting results on the distribution of NOCs have been obtained when studying fragmentation of mouse TGC nuclei. The initial nucleus of the giant cell contains as a rule several large nucleoli while the nuclear fragments have one or several small nucleoli (Zybina and Tikhomirova, 1963; Zybina, 1986). The mouse polytene NOCs are characterized by the presence of large heterochromatin blocks in the vicinity of the nucleolus (Zybina, 1977, 1986). The fragmentation of the giant nucleus is also accompanied by dissociation of heterochromatin blocks attached to the nucleolus. Ovoid, small, but clearly visible heterochromatin clumps appear in the newly formed nuclear fragments near small nucleoli, their number and arrangement in the nuclear fragments resembling that in the initial giant nucleus (Zybina *et al.,* 1979; Zybina, 1986). In the course of fragmentation, the large Barr body becomes loose and dissociates into many small bodies. Small nuclear fragments located in the vicinity of the initial giant nucleus contain one or several small chromatin clumps, of which the size and localization correspond to that of the Barr body (Fig. 9; Zybina and Mosjan, 1967; Zybina, 1986). All these data support the dissociation of polytene

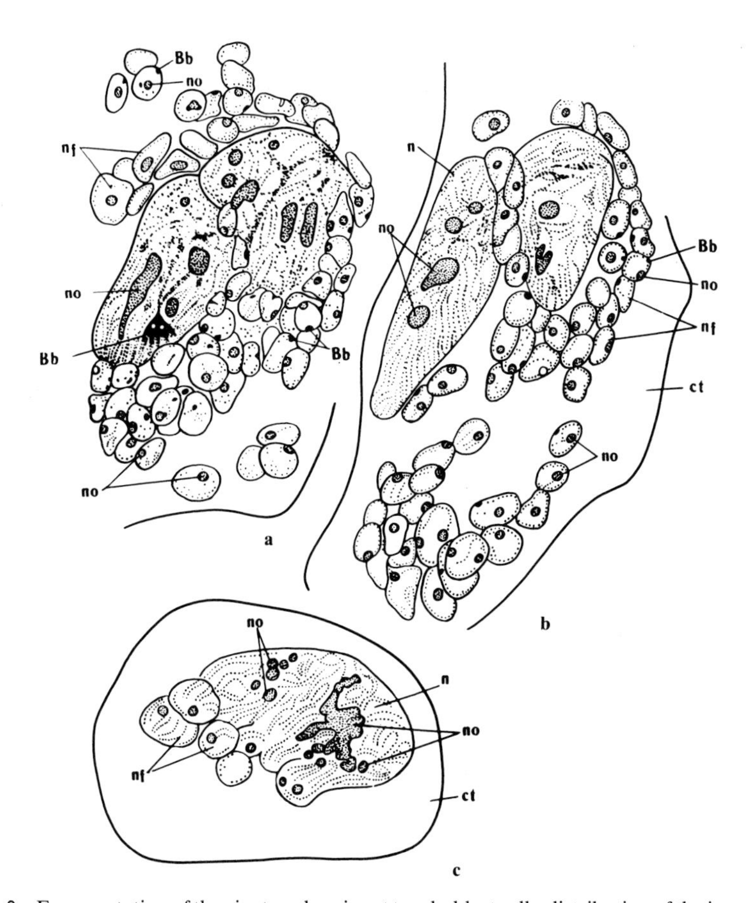


FIG. 9 Fragmentation of the giant nucleus in rat trophoblast cells; distribution of the interphase nuclear markers (nucleolus and Barr bodies) at the time of nuclear fragmentation. (a) The large Barr body (Bb) in the initial nucleus and the small Barr bodies in the nuclear fragments (nf). (b and c) Fragmentation of large nucleoli (no) into smaller ones and distribution of the latter into the newly formed nuclear fragments. n, Nucleus; ct, cytoplasm.

chromosomes into individual chromatids and a regular chromosome distribution in nuclear fragments in the course of fragmentation of the giant nucleus. In this way the nucleoli, heterochromatin blocks, and Barr body play a role as markers, indicating the distribution of individual chromosome sets of a lower ploidy in each fragment.

Cytophotometric measurements of DNA in the nuclear fragments have revealed a clear pattern in their whole-genome DNA distribution. Histograms of DNA distribution in the nuclear fragments of mouse, rabbit, and rat TGCs have shown rather distinct classes corresponding to 1C, 2C, 4C, 8C, 16C, and 32C (Zybina *et al.*, 1975b, 1979). An even more distinct whole-genome character of DNA in the nuclear fragments has been revealed with dissociation of secondary TGCs in the field vole *Microtus subarvalis* (Fig. 10). Histograms show distinct peaks corresponding to the theoretically expected DNA content values of 2C, 4C, 8C, and 16C. Only a few fragments had an amount of DNA intermediate between these classes. This indicates convincingly that at fragmentation of the giant nucleus, distribution of the chromosomal material in most nuclear fragments does not occur randomly and each fragment gets several (2*n*, 4*n*, 8*n*) entire genomes (Zybina, 1990). The whole-genome character of the DNA distribution in the field vole trophoblast is more evident than in other rodent species studied (rabbit, rat, and mouse). In our opinion, it is reasonable to believe that this is due to an easier dissociation of the field vole polytene chromosomes into endochromosomes, as discussed in Sections II,B and III,C. It should be noted that at fragmentation the nuclei formed are predominantly of the same ploidy classes as in the initial period of polyploidization (i.e., 2C, 4C, 8C, and 16C). This suggests the existence of a mechanism of division of the highly endopolyploid nucleus that is different from mitosis; this mechanism allows each daughter nucleus to get $2^n$ entire genomes. Such a distribution should be facilitated by separation of chromosomal sets of different ploidy divisible by $2^n$ inside the initial endopolyploid nucleus (i.e., without the disappearance of the nuclear envelope).

### C. Transformation of Polytene Chromosomes during Transfer into the Polygenome Nucleus

The transformation of genetic material in the initial giant nucleus to produce depolytenization prior to fragmentation is directly bound, in all probability, to the lability of nonclassic polytene chromosomes. Apart from this, the manifestation of the polyteny lability may be quite diverse: from a slight conjugation of homolog chromosomes and a noncomplete attachment of chromonemes, to their complete dissociation. To better understand the transformation of polytene chromosomes in a period preceding fragmenta-

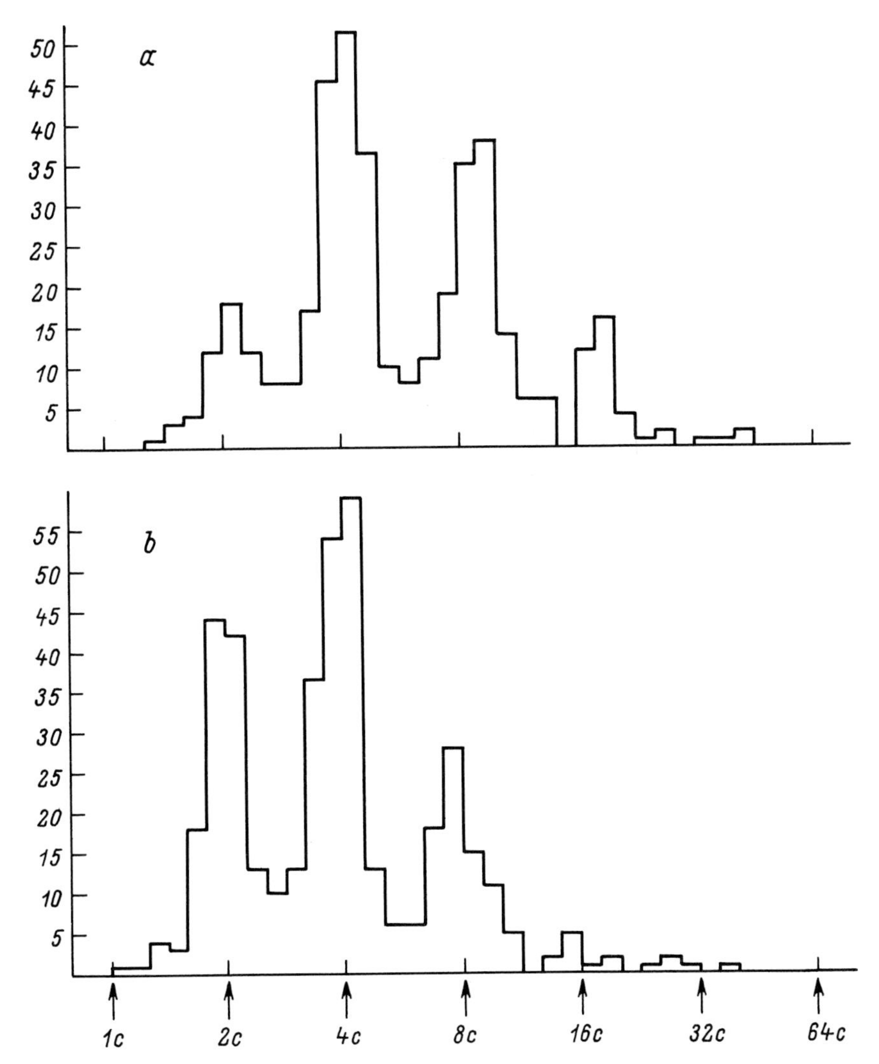


FIG. 10 Distribution of field vole trophoblast nuclear fragments with respect to DNA amount on day 15 (a) and day 17 (b) dpc. postcoitum. The abscissa shows ploidy degree, *C* (logarithmic scale); the ordinate shows the number of nuclei.

tion, it is worth reviewing a number of examples indicating the possibility of a partial or complete dissociation of polytene chromosomes into elementary chromosome fibrils in various plant and animal species, including Diptera.

The possibility of the dissociation of extensive regions of polytene chromosomes into oligotene fibrils may be considered a reflection of their structural variability. This process, at the same time, seems to be a step toward depolytenization. In salivary gland polytene chromosomes in Cecidomyidae, there is sometimes no close attachment of sister chromonemes to each other. These chromosomes undergo deep structural changes during cell differentiation, with the chromosomes dissociating into oligotene fibrils and individual chromonemes (White, 1948; Matuszewski, 1965; Henderson, 1967a; Zhimulev and Lychev, 1972). In different areas of the salivary gland in *Dasyneura urtica* (Cecidomyidae), the degree of dissociation of polytene chromosomes is different: in the cells of the basal reservoir, typical polytene chromosomes are preserved; in the transitory region, the polytene chromosome splits into separate fibrils at both ends; and in the distal region cells, the condensed chromonemes are attached only at the heterochromatic sites (White, 1948). With further progression of these processes, polytene chromosomes most likely can split completely to form chromosomes of a lesser degree of polyteny. For instance, polyploid polytene nuclei (i.e., nuclei with the number of polytene chromosomes equal to approximately 60–120*n*) appear in the coelomic cavity of *D. urtica* larvae under the effect of parasitizing wasp embryos (Matuszewski, 1964). Henderson (1967a) considers this a possible case of a complete dissociation of polytene chromosomes into oligotene fibrils. A similar picture has been observed in the salivary gland cell nuclei of *Lestodiplosis* (Cecidomyidae), in which the number of polytene chromosome sets is equal to 8–16*n* (Henderson, 1967b).

There are rare papers dealing with the morphology of the process of dissociation of polytene chromosomes into endochromosomes. When studying polytene chromosomes in the salivary glands of *Dasyneura affinis,* White (1948) found cells with polytene and "polyploid" nuclei (i.e., nuclei with numerous small chromosomes). Among the zones of cells with polytene and "polyploid" nuclei, there is an area of cells with a transitional type of nuclei. White's opinion, dissociation of polytene chromosomes can proceed gradually: when two polytene chromosomes of the markedly changed loosened type are still present in the nucleus, two others in the same nucleus dissociate completely into diffuse chromatin strands that most likely represent endochromosomes.

As to the possibility of dissociation of polytene chromosomes into individual chromosomes (or endochromosomes), of interest are data about the induction of cell divisions in cotyledon cells of the pea, *Pisum sativum,* with a high amount of DNA (Markes and Davies, 1979). In the course of cell culture formation, either polyploid cells, 4–32*n,* or cells with polytene chromosomes have been revealed; in the latter case there was a chromoneme condensation resulting in transformation of the usual bundles characteristic for plant giant chromosomes to peculiar aggregates of numerous spiralized chromonemes. With further transformations, individual chromosomes are gradually separated from these aggregates, which subsequently produces scattering of chromosomes. This results in the formation of a figure resembling the metaphase plate. This figure differs from the polyploid mitosis metaphase plate in that homolog chromosomes are long located near each other. Whereas in the previous examples the process of polytene chromosome dissociation was accompanied by some degree of condensation of chromosomes (the formation of endomitotic chromosomes), in the cell culture of pea cotyledons this process was preceded by an almost complete chromoneme condensation. These examples demonstrate that some degree of condensation (or a tendency for progressive spiralization) is a possible prerequisite for the dissociation of the polytene chromosome into chromonemes (endochromosomes). In some rare cases the depolytenization can result in the separation of some genomes inside the giant nucleus and even in a division of the nucleus into fragments of a much lower ploidy degree.

An example of whole-genome chromosome distribution is somatic reduction in the larval hindgut cells of the mosquito of the genus *Culex* (Berger, 1938). Chromosomes in these cells show some signs of polyteny. During metamorphosis, polyploid mitoses with 64*n* chromosomes appear in these cells. During several consecutive, so-called reductional mitoses (without interphase), the chromosomes divide to form octa- and tetraploid nuclei. Reductional division has also been described in the larval epidermis cells of the mosquito *Aedes aegypti* (Risler, 1959). Other known examples of whole-genome chromosome distribution are seen in invertebrates and Protozoa. In trophic polyploid cell nuclei of the seminal follicle wall in different species of locusts, chromosomes of endopolyploid nuclei have been observed to separate into two or several groups (Kiknadze and Tuturova, 1970; Istomina, 1976; Istomina and Vysotskaya, 1977). Predominant were chromosomal groups corresponding to tetraploid, triploid, diploid, and haploid sets.

All the examples cited here indicate that in animal and plant cells there is in some cases a complete dissociation of polytene chromosomes into endochromosomes as well as a whole-genome chromosome separation inside the endopolyploid nucleus. This separation sometimes is completed by somatic reduction. It may be suggested that nuclear fragmentation in the rodent trophoblast is preceded by similar transformations that make possible the nonmitotic division of the giant nucleus into numerous nuclei of a lower ploidy.

Since these transformations are connected with dissociation of polytene chromosomes into oligotene fibrils and subsequently into endochromosomes, we consider it worthwhile to review facts about some species of Diptera showing a longitudinal splitting of polytene chromosomes into strands of a lower degree of ploidy. Thus, in Collembola each homolog can split in some limited chromosome regions into two, four, and eight strands of approximately equal thickness and with equal distribution of bands and puffs (Cassagnau, 1976; Dalens, 1976, 1978; Dallai, 1979). This may indicate some "hierarchy" in the polytene chromosome structure, i.e., a possibility that the chromosomes dissociate not in a random way but in the same order in which they were formed in the course of nucleus endopolyploidization.

Dissociation of polytene chromosomes into oligotene fibrils also occurs when puffs and Balbiani rings are formed. Thus, one of the polytene chromosomes in *Acritopus lucidus* in a nucleolar region splits into four large bundles; a band pattern is still preserved in the bundles, and the bands split subsequently into thinner fibrils (Mechelke, 1953). In the process of puff formation, polytene chromosomes can split dichotomically several times (Beermann and Clever, 1964).

Electron microscopy has shown the Balbiani rings in *Chironomus tentans* to contain thin strands that probably result from a local splitting of the polytene chromosome; these strands divide into thinner bundles of despiralized fibrils of which the length is approximately equal to one gene (Ericsson *et al.,* 1989). The number of these thin fibrils is estimated to be five to eight, but not greater than eight. It looks like the chromosome dissociates not randomly but into fibrils composed of 8 sister chromatids. Had the division been random, it would have been highly probable to see, for instance, 9, 10, or more fibrils originating from one bundle. It cannot be ruled out that these bundles are combined into higher order bundles (16*n,* 32*n,* 64*n,* etc), which provides the possibility of dissociation (local or complete) of the polytene chromosome into chromosomes of a lower level of ploidy. Some data about the character of conjugation of sister chromatids in polytene chromosomes show the same tendency. In the case of a partial conjugation in polytene chromosomes of ovarian nurse cells of *C. erythrocephala,* there is an alternation of synaptic and asynaptic zones. The latter can represent a bundle of 16 oligotene fibrils preserving a band pattern (Bier, 1960). Electron microscopy observation of artificially dispersed polytene chromosomes of *Drosophila melanogaster* allowed visualization of bundles of fibrils oriented parallel to the chromosome long axis. There were four such bundles and each contained denser structures that according to their localization seemed to correspond to band remnants (Ananiev and Barsky, 1985). In our opinion, these data indicate that inside each polytene chromosome there is a structural orientation that provides for the possibility of the local or complete dissociation of the chromosome into $2^n$ polytenic bundles.

The most convincing example of the whole-genome distribution of the chromosomal material is fragmentation of the polyploid nucleus in Protozoa (Grell, 1953; Raikov, 1978, 1982). Many Radiolariae are characterized by the presence of highly polyploid nuclei, the division of which produces segregation of genomes, and ony the so-called primary nucleus becomes polyploid. Division of the primary nucleus in the radiolarian *Aulacantha*

(Phaeodaria) results in an immense number of large chromosomes which, as Grell (1953) suggests, are chains of composite chromosomes, each such chain probably being a haploid genome (subnuclear organization). During endomitosis preceding division, a reproduction of the composite chromosomes occurs, without their dissociation into single chromosomes. This seems to be what provides for a mechanism of subsequent depolyploidization: in the course of sporogenesis the primary polygenomic nucleus dissociates into numerous secondary nuclei within the cytoplasm. Later, the cytoplasm is fragmented into several multinuclear spheres. After several series of divisions of the secondary nuclei, zoospores are formed, each receiving one composite chromosome. All this complicated process is regarded as dissociation of the individual primary nucleus into separate genomes via their segregation, and depolyploidization of the nucleus proceeds via several stages (Raikov, 1978). Macronuclei of some Infusoria also seem to have a so-called subnuclear organization. During division of the macronucleus of the infusorium *Colpoda,* segregation of genomes also takes place (Raikov and Ammerman, 1976; Frenkel, 1978). Hence, the phenomena connected with depolyploidization and segregation of genomes occur, albeit rarely, in different taxonomic groups. It cannot be excluded that this mechanism appeared and was fixed in the evolution of animals and plants as one mechanism of cell division. However, it is most likely of significantly lesser importance than mitosis.

The mechanisms of depolytenization and isolation of genomes in the period preceding fragmentation of TGC nuclei remained up to now hypothetical since it has been possible to trace several stages in only a few rodent species. Comparative analysis of this phenomenon in various animal and plant species has enabled us to delineate the process of reconstruction of the chromosomal material prior to, and in the course of, the breakdown of the giant nucleus in the trophoblast.

The initial stage, dissociation of polytene chromosomes, can be convincingly traced in field vole trophoblast supergiant cells. In this case, condensation of chromonemes also precedes depolytenization: the chromosomal bundles just prior to dissociation look like aggregates of somewhat spiralized chromosomal strands. Dissociation of polytene chromosomes starts with the movement of external chromonemes, one after another, out of the bundle surface. The separate endochromosomes formed subsequently condense to form short, rodlike, paired chromosomes (Zybina and Zybina, 1985). The whole complex of morphological and cytospectrophotometric data on the differentiation of rodent TGCs allows us to suggest the following sequence of the reconstruction of chromosomal material, which probably results in the whole-genome distribution of the chromosomes into nuclear fragments.

Initially, the polytene nuclei contain chromosomes in the form of chromoneme bundles that are more closely attached to each other in heterochromatic regions. Later stages of differentiation are accompanied by looser connections of chromonemes in the bundles. In the period preceding fragmentation when, according to Zybina (1963b), DNA synthesis stops completely, full dissociation of the bundles into chromonemes seems to take place; this occurs also in heterochromatic regions, for instance, in mouse trophoblast perinucleolar chromocenters (Zybina, 1986). Subsequently, the polytene chromosomes dissociate into endochromosomes because this is seen before fragmentation of rabbit and field vole TGCs. Hence, the fragmentation seems to be preceded by a complex reconstruction of the genetic material in the initial nucleus; as a result, the polytene nucleus becomes polygenomic and individual genomes probably are separated and segregated, with division into fragments (Table II; Zybina, 1986). In favor of this interpretation are the data cited earlier on the tendency toward whole-genome DNA distribution into nuclear fragments as well as the distribution of several markers of interphase nucleus: nucleoli (field vole, rat), heterochromatic NOC blocks (mouse), and Barr bodies (rat).

It is not yet clear how individual genomes are separated in the nucleus after dissociation of polytene chromosomes and what the mechanism of subsequent aggregation of all chromosomes forming one genome is. It is interesting that in the TGCs as well as in a number of other cases, both genome multiplication and depolyploidization occur without disappearance of the nuclear envelope. However, this envelope undoubtedly plays some role, at least in the processes of genome segregation and nucleus fragmentation.

### D. Role of Membraneous Structures in Nuclear Fragmentation and Cytoplasmic Compartmentalization

Studies of the ultrastructure of the giant nucleus during fragmentation have shown that the nuclear envelope and its derivatives play an active role in this process. Deep folds of the nuclear envelope are formed to divide the nucleus into several lobes. The envelope of giant nucleus undergoing division contains multiple pores, both at the external surface of the nucleus and in the deep nuclear envelope invaginations (Ollerlich and Carlson, 1970; Zybina, 1979, 1980a, 1986). Fragmentation of the initial giant nucleus results in the formation of a multinuclear cell (polykaryocyte). All small nuclei are closely attached to each other, divided only by narrow cytoplasmic strips. Nuclear fragments are then gradually separated from each other. Each of them is surrounded by a nuclear envelope composed of two membranes and containing numerous pores. Chromatin in the nuclear fragments is condensed and is located mainly beneath the nuclear envelope and near nucleolus. Numerous nuclear fragments of one polykaryocyte have the same degree of chromatin condensation (Fig. 11a). Ovoid small nucleoli are homogeneous and are composed of moderately electron-dense fibrils. In the course of fragmentation, intranuclear membraneous structures, such as annulate lamellae, aggregates of pore complexes, take part in the nuclear division (Zybina, 1980b, 1986); they initially play a role in increasing the surface of the growing endopolyploid nucleus (Jollie, 1969; Ollerlich and Karlson, 1970).

Ultrastructural study of polykaryocytes has produced unexpected results. It has turned out that in the polykaryocyte, a separation of cytoplasmic areas around individual nuclear fragments takes place; this separation occurs with the aid of membranes, derivatives of the nuclear envelope (Fig. 11b). Compartmentation of the cytoplasm areas in this case is an example of the close connection of the nuclear envelope with cytoplasmic organoids (Zybina and Rumyantsev, 1980; Zybina, 1986). Narrow cisterns emerging from the nuclear envelope external membrane represent a smooth endoplasmic reticulum which, in turn, participates in formation of double membranes separating the cytoplasmic zone around individual nuclei or their aggregates inside the trophoblast polykaryocytes (Figs. 11b and 12). It should be emphasized that the double membranes surrounding the fragment of one nucleus belong topologically to the double nuclear envelope of another nuclear fragment. Thus, the separated area, including the nuclear fragment and cytoplasmic zone, differs from the cell in that it is surrounded by the double rather than by the elementary membrane (i.e., the space between the membranes surrounding this cell fragment seems to belong topologically to the perinuclear space of another nucleus if no communication with the extracellular space appears). Desmosome-like structures appear in flat cisterns of the smooth endoplasmic reticulum. The separation of such areas around nuclear fragments through the endoplasmic reticulum cisterns seems to occur gradually: first, the cytoplasmic areas are separated around aggregates of nuclear fragments, then the fragments are separated from each other and spread throughout the cytoplasm of the polykaryocyte, with subsequent isolation of the cytoplasmic areas also around individual nuclei (Zybina and Rumyantsev, 1980; Zybina, 1986). The morphology and character of orientation of the cytoplasmic organoids inside the cytoplasmic compartments hardly differ from that in usual interphase nuclei. No signs of degeneration are present.

The isolation of cellular areas around individual nuclei has an analogy in polykaryocytes of other organisms. In frog adrenal cells, the formation of plasmalemma between two daughter cells during nuclear fragmentation results from the aggregation of the smooth endoplasmic reticulum channels (Pehlman, 1968). A complex system of nuclei (macro- and micronuclei)

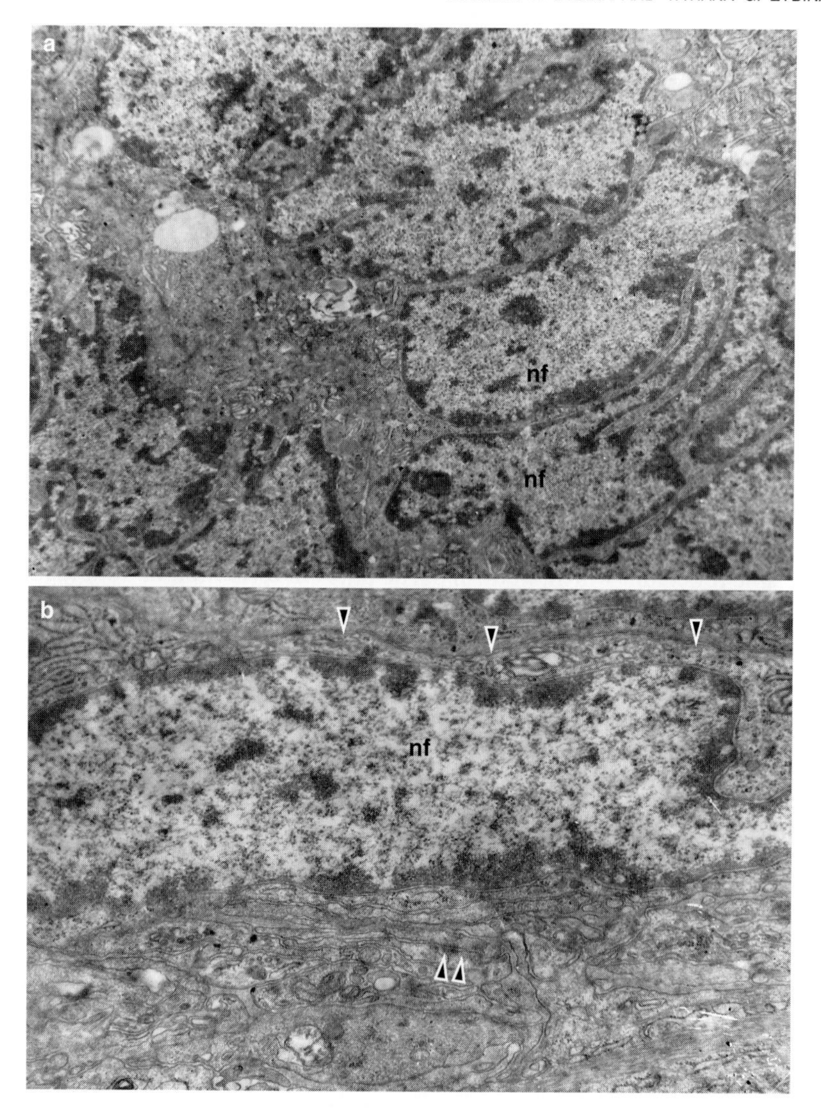


FIG. 11 Nuclei of trophoblast giant cells during fragmentation on day 17 postcoitum. (a) Area of the polykaryocyte; nuclear fragments (nf) adjoin each other; the condensed chromatin is located mainly beneath the nuclear envelope. (b) The nuclear fragment with a cytoplasmic zone is separated from the rest of the nuclei by a flat cistern of the smooth cytoplasmic reticulum (arrowhead); the external membrane of the nuclear envelope in the lower part of the nuclear fragment forms outpouches in the form of flat, branching cisterns that form in some areas desmosome-like structures (double arrowheads). Bar = 1 $\mu$m.

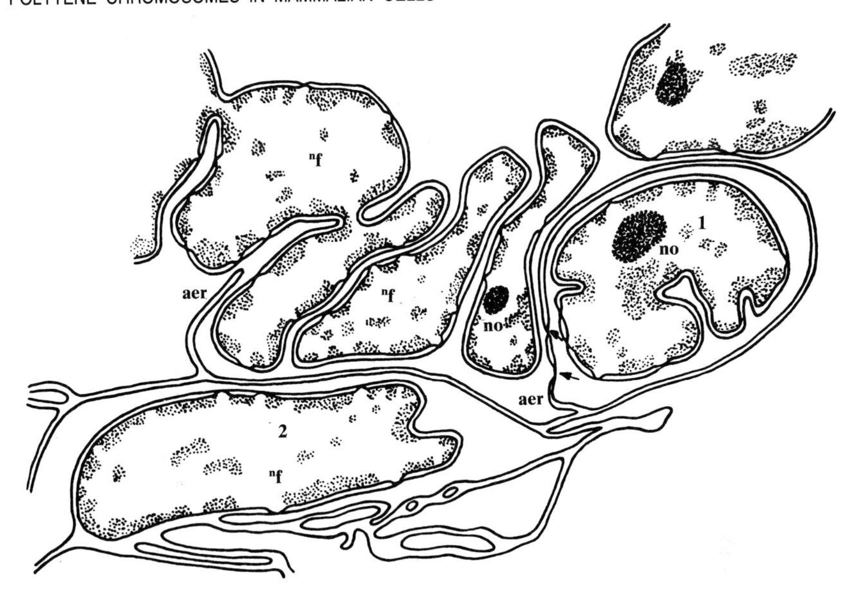


FIG. 12 Scheme of formation of the plasma membrane at the separation of cellular territories around the nuclear fragments (nf) inside the trophoblast multinuclear giant cell. Narrow cisterns emerging from the nuclear envelope external membrane at fragment 2 represent agranular cytoplasmic reticulum (aer), which, in turn, participates in formation of the double membranes separating the cytoplasmic zone around fragment 1; arrows show desmosome-like structures; no, nucleoli.

united by their double membrane has been revealed in several lower Infusoria (Raikov and Kovaleva, 1977; Kovaleva *et al.*, 1979). The entire nuclear complex in *Trachelorhaphis totevi* is surrounded by a lacuna of the smooth endoplasmic reticulum whose proximal membrane is adjacent to the outer surfaces of all the macronuclei and also gives rise to deep invaginations between individual macronuclei (Kovaleva *et al.*, 1979). In both cases, in the trophoblast and in the macronucleus, the separation occurs only when the nuclei are oligoploid or diploid and unable to divide mitotically. The compartments are formed with the aid of the endoplasmic reticulum double membranes, which never have ribosomes. The functional significance of this phenomenon is not yet clear.

In the cytoplasm of the separating intracellular compartments, 10-nm-thick microfilament bundles have been found that are well developed in the terminal period of the trophoblast nucleus fragmentation. One end of the microfilaments is bound to the external nuclear envelope while the other end is bound to the cytoplasm membranous organoids (Zybina and Rumyantsev, 1980). It cannot be ruled out that the microfilaments participate, in a way, in fragmentation of the giant nucleus. This seems possible, particularly if one considers that the trophoblast giant cells have no centriolar apparatus but do have centrosome-like bodies composed of intermediate filaments (Glasser *et al.,* 1984). The intermediate microfilaments forming the centrosome-like bodies and, possibly, bound to them cannot be excluded as playing a role in the regular nonmitotic fragmentation of the nucleus. There are data on the participation in this fragmentation of other mitotic apparatus elements as well. This has been revealed during nuclear fragmentation in the callus of the tobacco plant, *Nicotiana glauca* (Nuti Ronchi *et al.,* 1973). The exact role of microfilaments and microtubules in fragmentation of the giant nucleus is yet unknown.

## VI. Significance of Polyteny in Trophoblast Giant Cells

To appreciate the significance of the fact that it is in placenta trophoblast cells that polyteny in mammals has been found, specific functions and unique properties of these cells are reviewed here. The trophoblast provides for implantation of the blastocyst into the endometrium and plays an important role in the close interconnection between the embryo and maternal organism in the formation of the placenta (Mossman, 1937; Boyd and Hamilton, 1970; Tachi and Tachi, 1979; Kaufman, 1983; Welsh and Enders, 1987). Owing to an active invasive and phagocytic role for primary and secondary TGCs in rodents, maternal tissues are destroyed, which promotes implantation of the embryo into the endometrium (Alden, 1948; Orsini, 1954; Zybina and Tikhomirova, 1963; Enders and Schlafke, 1969, 1971; Pijnenborg *et al.,* 1974; Zybina, 1976a,b, 1986; Denker, 1993).

In the human placenta, initial invasion by the syncytium gives way to subsequent invasive behavior of the cytotrophoblast (Pijnenborg *et al.,* 1980; Aplin, 1991). Through the phagocytosis and digestion of maternal cells, the trophoblast cells perform a trophic function that is particularly active in the first half of pregnancy, until the complete formation of the placenta (Zybina and Tikhomirova, 1963; Dickson, 1969; Zybina, 1976a,b, 1986; Jollie, 1981; Welsh and Enders, 1987; Bevilaqua and Abrahamson, 1988; Aplin, 1991; Hoffman and Wooding, 1993).

The trophoblast also has an important defensive function in that it protects the allogenic embryo from the maternal rejection reaction (Kolb *et al.,* 1984; Goldsobel *et al.,* 1986; Lala *et al.,* 1988; Stewart and Mukhtar, 1988; Lin *et al.,* 1993; Mikhailov *et al.,* 1994). The trophoblast cells secrete a great number of hormones that presumably regulate the course of pregnancy: progesterone (Sherman, 1983: Hoffman and Wooding, 1993), placental lactogen (Gosseye and Fox, 1984; Hall and Talamantes, 1984; Renegar *et al.,* 1990; Soares *et al.,* 1991), chorionic gonadotropin (Gosseye and Fox, 1984; Billingsley and Wooding, 1990; Aplin, 1991; Kurman *et al.,* 1984), and so on.

It should be emphasized that the trophoblast cells compensate for many functions that cannot be performed by the developing embryo because it lacks definitive organ systems (Govorka, 1970; Carlson, 1981; T. G. Zybina, 1988). Most likely, the early formation of different trophoblast cell functions requires rapid differentiation and, at the same time, the intensive growth of these cells. Endoreduplication resulting in polyteny is the mechanism promoting the combination of these processes. Polyteny is a shortened cell cycle in which multiplication of chromosomal material and cell growth occurs without cell mitosis, the cells growing by reaching a high level of ploidy rather than by dividing (Nagl, 1978, 1981; Brodsky and Uryvaeva, 1981, 1985; D'Amato, 1989). Therefore endoreduplication allows the accumulation of a large cellular mass and a quick start of tissue-specific functions (Brodsky and Uryvaeva, 1985).

Probably, in this form of genome multiplication, differentiation often is irreversible. Therefore, endoreduplication resulting in a high degree of genome multiplication is most commonly characteristic of tissues with a precise program of development. Rodent TGCs are an example of irreversible differentiation. The placenta is a provisional organ that has a definite life span corresponding to the total duration of pregnancy in each animal species. Therefore, the entire complex sequence of events occurring at implantation and formation of the placenta seems to be connected with strongly genetically programmed time parameters. Thus, TGCs in the rodent placenta are characterized by a programmed life span and apoptosis. In experiments with transplantation of the rat placenta into different organs such as the liver, renal capsule, and genital tract mesenterium, as well as in tissue culture, the trophoblast life span turned out to be constant—22 days (Jollie, 1960; Dorgan and Schultz, 1971). The capability of TGCs for invasion and phagocytosis is strictly determined; this takes place in the case both of normal implantation (Enders and Schlafke, 1967; Zybina, 1976a,b, 1986; Welsh and Enders, 1987; Bevilaqua and Abrahamson, 1988) and of blastocyst transplantation in other organs (Fawcett *et al.,* 1947; Kirby, 1960, 1963; McLaren and Tarkowski, 1963). The trophoblast in the human placenta also has a specific differentiation program according to which the invasive behavior ends 4–5 months after implantation (Aplin, 1991). The number of chromosome endoreduplication cycles is also programmed: on average there are 6–11 cycles in rodents (Zybina, 1986). Both at normal implantation and at blastocyst transplantation under the renal capsule to the nonpregnant female mouse, TGC nuclei reach the same level of ploidy—32–512C (Barlow and Sherman, 1972; Ilgren, 1981). It is interesting to note that probably there is an interrelation between the duration of manifestation of TGC invasive properties and time of the multifold genome multiplication in these cells (Zybina, 1976a, 1986; E. V. Zybina and T. G. Zybina, 1993) when signs of polyteny are revealed (Table II).

It should be noted that some conditions necessary for the giant cells to perform their program of development have already been elucidated. Among these conditions are the nature of the contact of trophoblast cells with each other and the effects of the inner cell mass (Gardner, 1972; Rossant and Offer, 1977; Ilgren, 1980, 1981, 1983). Studies on murine trophoblast cells cultivated *in vitro* have shown that the degree of cell gigantism depends on the conditions of tissue culturing, which provide for different degrees of intercellular contact. Thus, a trophoblast growing in a suspension culture as spheroid cellular aggregates maintains many cellular contacts, and the trophoblast cells actively divide mitotically to remain mainly diploid. Meanwhile, a trophoblast explanted on a tissue culture gradient forms a monolayer and contains less numerous cellular contacts, a great number of giant cells appearing with highly polyploid nuclei (up to 256C).

An even greater percentage of highly polyploid cells is observed when the trophoblast is cultivated as single cells so that intercellular contacts either are minimal or are completely absent (Ilgren, 1981, 1983). These data are in agreement with *in vivo* observations on the trophoblast cells invading the endometrial stroma. These cells are separated from the main cell layer and completely lose contact with each other. They migrate individually deeply into the stroma, sometimes to reach the myometrium, and are characterized by a particularly marked gigantism; for instance, cells of the obplacental part of the rabbit placenta (Zybina *et al.,* 1975; T. G. Zybina *et al.,* 1980; Hoffman and Wooding, 1993) and supergiant cells of the field vole (E. V. Zybina and T. G. Zybina, 1985). Most likely, changes in the character of intercellular contacts and the behavior of the nuclear envelope and cytoskeleton play an essential role in realization of invasive and phagocytic properties of trophoblast cells (Enders and Schlafke, 1969; Steer, 1971; Tachi and Tachi, 1979; Denker, 1993).

Another example of early cell differentiation involving endoreduplication are embryonal sac cells in angiosperm plants (the haustorium and suspensor). They, like trophoblast cells, enter endoreduplication from the moment of isolation from the embryo and begin to function much quicker and much earlier than the embryonal tissues themselves (Nagl, 1973b, 1978); it is in these cells that polytene chromosomes have been found. In cells of the suspensor of *P. coccineus* the transcriptional activity per DNA unit turned out to be double that of embryo diploid cells (Clutter *et al.,* 1974). This seems to promote rapid cell growth and their quick differentiation, the latter being especially important for highly active cells in tissues and organs with a short life span, such as suspensor cells in plants (Nagl, 1978, 1981), trophocytes in insect ovaries (King and Burnett, 1959; Basile, 1969), and rodent trophoblast. This is true for many provisional organs, including Diptera larval salivary glands.

The high level of polyploidy and transition to the endoreduplication cycle that rules out mitoses can be of defensive significance in establishing close relations between allogenic maternal and embryonic organisms. Immunological mechanisms, for instance, the natural killer activity of trophoblast and decidual cells (Kolb *et al.,* 1984; Van Vlasselaer and Vanderputte, 1984; Boehm *et al.,* 1989), as well as some other mechanisms, play important roles in the defensive function of the trophoblast. In the course of implantation and placentation, the trophoblast and decidual cells come into close contact and often are damaged thereby; this damage probably can also involve the chromosomal apparatus. It should be noted that all cell populations, both in the maternal and in the embryonal part of the placenta, are polyploid (Table I). Multifold genome doubling can protect the interacting cells from the undesirable effects of changes in the hereditary apparatus (Uryvaeva, 1979). It is known that mutation frequency can rise with the introduction of foreign DNA into the organism (Guershenzon *et al.,* 1975). From that point of view, the existence of an interdependence between the type of TGC reproduction and the capability for phagocytosis seems reasonable. As a rule, TGCs able to phagocytose not only anuclear erythrocytes but also nucleus-containing cells (such as decidual, epithelial, and endothelial cells and leukocytes) pass to the endoreduplication cycle without disappearance of the nuclear envelope. Hence, there is no contact of two genomes. This kind of cell reproduction seems to protect, to a greater degree, the chromosomal apparatus from possible action of the allogenic DNA penetrating into the trophoblast cell (Zybina, 1986).

In the past few years, new data have appeared that suggest that processes occurring at the level of genes and possibly individual chromosomes also can participate in formation of an immunological barrier between fetal and maternal organisms. Thus, in the cells of extraembryonic tissues that are directly adjacent to the maternal tissues (the placental trophoblast and derivatives of the primary entoderm in the yolk sac), predominant inactivation of the paternal X-chromosome has been observed (Takagi and Sasaki, 1975; Wake *et al.,* 1976; West and Chapman, 1978; Frels *et al.,* 1979; Frels and Chapman, 1980; Papaioannou *et al.,* 1981; Takagi *et al.,* 1982). Meanwhile in most tissues of the adult organism, activity of both the maternal and the paternal X-chromosome can take place (West *et al.,* 1977; West and Chapman, 1978). Preferential inactivation of the paternal X-chromosome cannot be ruled out as somewhat decreasing the immunological incompatibility of the invasive trophoblast cells and yolk sac cells with the endometrium, both in the initial implantation period and in the course of the formation and functioning of the placenta. Therefore, it may be suggested that the predominant expression of the gene localized in the maternal X-chromosome can facilitate the trophoblast invasion into the maternal tissues.

This is confirmed by finding the expression of one of the homeobox genes, *Pem,* both in a great number of extraembryonic tissues, including the trophoblast, and in highly invasive malignant cells, for instance, in T lymphomas, as well as in several transformed and immortalized tissues (Lin *et al.,* 1994). Expression of this gene is particularly prominent in the actively invading ectoplacental cone cells and in the secondary TGCs. Unlike all other homeobox genes studied, the *Pem* gene is localized in the X-chromosome, the temporal and spatial pattern of this gene expression coinciding with the corresponding pattern of the preferential inactivation of the paternal X-chromosome (Sobis *et al.,* 1991; Lin *et al.,* 1994). These authors assume that the X-linked maternal genes have a stronger effect than the paternal genome on implantation and placenta formation. We believe that it is the cells with the inactivated paternal X-chromosome that may be immunologically privileged and have a greater possibility for invasion. Therefore, further investigation of the expression of the X-linked genes, like the *Pem* gene, will contribute to an elucidation of the mechanisms of genetic control of the immunological compatibility of the trophoblast cells and endometrium in the course of placenta formation.

As to the problem of the importance of polyteny in the mammalian trophoblast, it should be noted that the most frequent occurrence of the inactivated X-chromosome clearly correlates in time with manifestation of the invasive properties and appearance of polytene chromosomes (Table II). It is in the implantation and placentation period that up to 90% of trophoblast cells contain the well-formed Barr body. By the time of the gradual cessation of the invasive activity, the Barr bodies, like polytene chromosomes, become less clearly defined and at last disappear (Zybina, 1960, 1961, 1986; Zybina and Tikhomirova, 1963; Zybina and Mosjan, 1967). Thus, there are sufficient grounds for believing that the program of trophoblast cell development involves a great number of closely correlated mechanisms of different types, which are able to prevent the dramatic consequences of close interaction of the cells of allogenic organisms in the course of functioning of the placenta.

## VII. Conclusion

A comparative study of TGC chromosomes in a number of rodent species has made it possible to establish that in the period of active TGC functioning in the developing placenta, nuclei reach extremely high (for mammals) levels of polyploidy (512–4096C), and their chromosomes acquire some similarity to polytene chromosomes in Diptera. However, mammalian polytene chromosomes are even more similar to the polytene chromosomes in the haustorium and suspensor of angiosperm plants. A comparison of the mammalian polytene chromosomes with those in Diptera and in higher plants has allowed us to put forward a concept of a nonclassic form of polyteny, of which the peculiar feature is its nonpermanent character in the nuclear cycle and in the course of cell differentiation. The degree of sister chromatid attachment in nonclassic polyteny can vary from complete association along the entire length of the chromosome to a connection only in the region of the pericentromere heterochromatin. Based on the available data, a conclusion may be drawn that the phenomenon of polyteny is to be regarded, not only as a form of organization of chromosomes, as it has been considered, but also as a peculiar type of cell cycle that lacks the mitotic division stage. The transition to the shortened cycle of endoreduplication that results in polyteny is of great importance for the rapid growth and differentiation of provisional organs, including the placenta, which provide for the development of the embryo until its own organ systems can start functioning.

## Acknowledgments

The authors sincerely thank Prof. I. I. Kiknadze, Prof. I. B. Raikov and Prof. V. Ya. Brodsky for participation in discussions of many aspects of our generalizations. Translation and editing of the manuscript by Dr. L. Z. Pevzner is gratefully acknowledged. Preparation of the manuscript was supported by the governmental program "Priority Trends in Genetics" and by the Russian Fund for Fundamental Investigations. We appreciate the kind permission of the Publisher of Tsitologia (St. Petersburg, Russia) to reproduce (in this chapter) Figures 1–12 from our papers previously published in Tsitologia.

## References

Alden, R. (1948). Implantation of the rat egg. III. Origin and development of primary trophoblast giant cells. *Am. J. Anat.* **83,** 143–169.

Ananiev, E. V., and Barsky, V. E. (1985). Elementary structures in polytene chromosomes of *Drosophila melanogaster. Chromosoma* **93,** 104–112.

Andreeva, L. F. (1964). DNA synthesis and cell population kinetics study in the giant cells and cellular trophoblast of placenta. *In* "Cell Cycle and Nucleic Acid Metabolism in the Course of Cell Differentiation" (L. N. Zhinkin and A. A. Zavarzin, eds.), pp. 136–147. Nauka, Moscow.

Ansell, J. D., Barlow, P. W., and McLaren, A. (1974). Binucleate and polyploid cells in the decidua of the mouse. *J. Embryol. Exp. Morphol.* **31,** 223–227.

Aplin, J. D. (1991). Implantation, trophoblast differentiation and haemochorial placentation: Mechanistic evidence *in vivo* and *in vitro. J. Cell Sci.* **99,** 681–692.

Arronet, V. N. (1973). Morphological changes of nuclear structures in the oogenesis of reptiles (Lacertidae, Agamidae). *J. Herpetol.* **7,** 163–193.

Ashburner, M. (1970). Function and structure of polytene chromosomes during insect development. *Adv. Insect Physiol.* **7,** 1–95.

Avanzi, S., Cionini, P., and D'Amato, F. (1970). Cytochemical and autoradiographic analyses on the embryo suspensor cells of *Phaseolus coccineus. Caryologia* **23,** 605–638.

Baker, T. G. (1963). A quantitative and cytological study of germ cells in human ovaries. *Proc. R. Soc. London* B158, 417–433.

Barlow, P. W., and Sherman, M. J. (1972). The biochemistry of differentiation of mouse trophoblast: Studies on polyploidy. *J. Embryol. Exp. Morphol.* **27,** 447–465.

Barlow, P. W., and Sherman, M. J. (1974). Cytological studies on the organization of DNA in giant trophoblast nuclei of the mouse and the rat. *Chromosoma* **47,** 119–131.

Basile, R. (1969). Nucleic acid synthesis in nurse cells of *Rhynchosciara angelae* (Nonato and Pavan, 1951). *Genetics* **61,** 261–273.

Beermann, W. (1960). Der Nucleolus als lebenwichtiger Bestandteil des Zellkernes. *Chromosoma* **11,** 263–296.

Beerman, W. (1962). Riesenchromosomen. *Protoplasmatologia* **6,** 1–161.

Beermann, W. (1972). Chromosomes and genes. *In* "Developmental Studies on Giant Chromosomes. Results and Problems in Cell Differentiation" (W. Beerman, ed.), pp. 1–33, Springer, Heidelberg, Germany.

Beermann, W., and Clever, U. (1964). Chromosome puffs. *Sci. Am.* **10,** 50–58.

Berger, C. A. (1938). Multiplication and reduction of somatic chromosome groups as a regular developmental process in the mosquito *Culex pipiens. Carnegie Contrib. Embryol.* **27,** 209–232.

Bevilaqua, E. M. A. F., and Abrahamson, P. A. (1988). Ultrastructure of trophoblast giant cell transformation during invasive stage of implantation of the mouse embryo. *J. Morphol.* **198,** 341–351.

Bier, K. (1957). Endomitose und Polytanie in den Nahrzellkernen von *Calliphora erythrocephala* Meigen. *Chromosoma* **8,** 493–522.

Bier, K. (1958). Bezeihungen zwischen Wachstumsgeschwindigkeit, endometaphasischer Kontaktion und der Bildung von Riesenchromosomen in den Nahrzellen von *Calliphora. Z. Naturforsch.* **13**b, 85–93.

Bier, K. (1960). Der Karyotyp von *Calliphora erythrocephala* Meigen unter besonderer Berucksichtigung der Nahrzellkernchromosomen in gebundelten und gepaarten Zustand. *Chromosoma* **11,** 335–364.

Billingsley, S. A., and Wooding, F. B. R. (1990). An immunogold, cryoultrastructural study of sites of synthesis and storage of chorionic gonadotropin and placental lactogen in human syncytiotrophoblast. *Cell Tissue Res.* **261,** 375–382.

Boehm, K. D., Kelley, M. F., Ilan, J., and Ilan, J. (1989). The interleukin 2 gene is expressed in the syncytiotrophoblast of the human placenta. *Proc. Nat. Acad. Sci. USA* **86,** 656–660.

Bower, D. J. (1987). Chromosome organization in polyploid mouse trophoblast nuclei. *Chromosoma* **95,** 76–80.

Boyd, J. D., and Hamilton, W. J. (1970). "The Human Placenta." W. Heffer and Sons, Cambridge.

Brady, T. (1973a). Feulgen cytophotometric determination of the DNA content of the embryo proper and suspensor cells of *Phaseolus coccineus. Cell Differ.* **2,** 65–75.

Brady, T. (1973b). Cytological studies on the suspensor polytene chromosomes of *Phaseolus*: DNA content and synthesis, and the ribosomal cistrons. *Caryologia* **25** (Suppl.), 233–272.

Brady, T., and Clutter, M. (1974). Structure and replication of *Phaseolus* polytene chromosomes. *Chromosoma* **45,** 63–79.

Bridgman, J. (1948). A morphological study of the development of the placenta of the rat. II. Histological and cytological study of the development of the chorioallantoic placenta of the white rat. *J. Morphol.* **83,** 195–224.

Brodsky, V. Ya., and Uryvaeva, I. V. (1981). Cell Polyploidy. "Proliferation and Differentiation." Nauka, Moscow.

Brodsky, V. Ya., and Uryvaeva, I. V. (1985). "Genome Multiplication in Growth and Development. Cambridge University Press, Cambridge.

Brown, C. W., Jacobs, P., Buckton, K., and Tough, I. M. (1966). "Chromosome Studies on Adults." Cambridge University Press, London.

Busch, H., and Smetana, K. (1970). "The Nucleolus." Academic Press, New York.

Callan, H. G. (1963). The nature of lampbrush chromosomes. *Int. Rev. Cytol.* **15,** 1–33.

Callan, H. G. (1986). "Lampbrush Chromosomes." Springer-Verlag, Berlin.

Carlson, B. (1981). "Patten's Foundation of Embryology." McGraw-Hill, New York.

Cassagnau, P. (1971). Les chromosomes salivares polytene chez *Bilobella grassei* (Denis), Collembola: Neanuridae. *Chromosoma* **35,** 57–83.

Cassagnau, P. (1974). Les chromosomes polytene de *Neanura monticola* Cassagnau (Collembola). I. Polymorphisme ecologique du chromosome X. *Chromosoma* **46,** 343–363.

Cassagnau, P. (1976). La variabilite des chromosomes polytenes chez *Bilobella aurantiaca* Caroli (Collembola: Neanuridae) et ses rapports avec la biogeographie et l'ecologie de l'espece. *Arch. Zool. Exp. Gen.* **117,** 511–572.

Cassagnau, P., Dallai, R., and Deharveng, L. (1979). Le polymorphisme des chromosomes polytenes de *Lathriopyga longiseta* Caroli (Collembole, Neanuridae). *Caryologia* **32,** 461–483.

Chouinard, L. A. (1973). An electron-microscope study of the extranucleolar bodies during growth of the oocyte in the prepubertal mouse. *J. Cell Sci.* **12,** 183–211.

Chouinard, L. A. (1975a). A light- and electron-microscope study of the oocyte nucleus during development of the antral follicle in the prepubertal mouse. *J. Cell Sci.* **17,** 589–615.

Chouinard, L. A. (1975b). An electron-microscope study of the intranucleolar chromatin during nucleologenesis in root meristematic cells of *Allium cepa. J. Cell Sci.* **19,** 85–101.

Clutter, M., Brady, T., Walbot, V., and Sussex, I. (1974). Macromolecular synthesis during plant embryogeny. *J. Cell Biol.* **63,** 1097–1102.

Dalens, H. (1976). Variations de structure des chromosomes polytenes cours au sein deux populations pyreneennes de *Bilobella aurantiaca* Caroli (Collemboles). *C.R. Acad. Sci. Paris Ser. D* **282,** 1817–1820.

Dalens, H. (1978). Structure des chromosomes polytenes dans des populations de *Bilobella aurantiaca* Caroli (Collemboles) de la bordure sud du Massif Central. *C.R. Acad. Sci. Paris Ser. D* **287,** 273–275.

Dallai, R. (1979). Polytene chromosomes in some *Bilobella aurantiaca* (Collembola) Italian populations. *Boll. Zool.* **46,** 231–249.

D'Amato, F. (1989). Polyploidy in cell differentiation. *Caryologia* **42,** 183–211.

Deharnveng, L. (1982). Polymorphism of polytene chromosomes in *Bilobella aurantiaca* (Insecta: Collembola). Study of a population from Sierra de Credos (Central Spain). *Chromosoma* **85,** 201–214.

Denker, H.-W. (1993). Implantation: A cell biological paradox. *J. Exp. Zool.* **266,** 541–558.

Dickson, A. D. (1969). Cytoplasmic changes during the trophoblastic giant cell transformation of blastocysts from normal and ovariectomized mice. *J. Anat.* **105,** 371–380.

Dorgan, W. I., and Schultz, R. L. (1971). An *in vitro* study of programmed death in rat placental giant cells. *J. Exp. Zool.* **178,** 497–512.

Enders, A. C., and Shlafke, S. (1967). A morphological analysis of the early implantation stages in the rat. *Am. J. Anat.* **120,** 185–226.

Enders, A. C., and Schlafke, S. (1969). Cytological aspects of trophoblast–uterine interaction in early implantation. *Am. J. Anat.* **125,** 1–30.

Enders, A. C., and Schlafke, S. (1971). Penetration of the uterine epithelium during implantation in the rabbit. *Am. J. Anat.* **132,** 219–240.

Ericsson, C., Mehlin, H., Bjorkroth, B., Lamb, M. M., and Daneholt, B. (1989). The ultrastructure of upstream and downstream regions of an active Balbiani ring gene. *Cell* **56,** 631–639.

Fawcett, D. W., Wislocki, G., and Waldo, C. M. (1947). The development of mouse ova in the anterior chamber of the eye in the abdominal cavity. *Am. J. Anat.* **81,** 413–444.

Frels, W. I., and Chapman, V. M. (1980). Expression of the maternally derived X-chromosome in the mural trophoblast of the mouse. *J. Embryol. Exp. Morphol.* **56,** 179–190.

Frels, W. I., Rossant, J., and Chapman, V. M. (1979). Maternal X-chromosome expression in mouse chorionic ectoderm. *Dev. Genet.* **1,** 123–132.

Frenkel, M. A. (1978). The DNA content in micronuclei and chromatin aggregates of dividing macronucleus within reproduction cysts of ciliate *Colpoda steini. Tsitologia* **20,** 465–469.

Gardner, R. L. (1972). An investigation of inner cell mass and trophoblastic tissue following their isolation from the mouse. *J. Embryol. Exp. Morphol.* **28,** 279–312.

Geitler, L. (1953). Endomitose und endomitotische Polyploidisierung. *Protoplasmatologia* **60,** 1–89.

Glasser, S. R., Soares, M. J., and Julian, J. (1984). Morphological and functional differentiation of rat trophectoderm *in vitro*: Organization of microtubules, expression of cytokeratins and rat placental lactogen in midgestation trophoblast giant cells. Int. Symp. Mammal. Reprod. Early Development, August 23–24, 1984, Tokyo.

Goldsobel, A., Ank, B., Spina, C., Giorge, J., and Stiehm, E. R. (1986). Phenotypic and cytotoxic characteristics of the immune cells of the human placenta. *Cell Immunol.* **97,** 335–343.

Gosseye, S., and Fox, H. (1984). An immunohistological comparison of the secretory capacity of villous and extravillous trophoblast in the human placenta. *Placenta* **5,** 329–348.

Govorka, E. (1970). "Human Placenta." Polish State Medical Press, Warsaw.

Grell, K. G. (1953). Die Chromosomen von *Aulacantha scolymantha* Haeckel. *Arhc. Protistenk.* **99,** 1–54.

Gruzova, M. N., and Parfenov, V. N. (1993). Karyosphere in oogenesis and intranuclear morphogenesis. *Int. Rev. Cytol.* **144,** 1–52.

Guershenson, C. M., Alexandrov, Yu.N., and Maluta, C. C. (1975). "Mutageneous effect of DNA and Viruses in *Drosophila.*" Naukova Dumka, Kiev.

Gunderina, L. I., Sherudilo, A. I., and Mitina, R. D. (1984). The replication cycle of DNA during chromosome polytenization in the salivary gland cells of *Chironomus thummi.* III. The duration of DNA synthesis period. *Tsitologia* **26,** 925–935.

Hall, J., and Talamantes, F. (1984). Immunocytochemical localization of mouse placental lactogen in mouse placenta. *J. Histochem. Cytochem.* **32,** 379–382.

Hammond, M. P., and Laird, C. D. (1985). Chromosome structure and DNA replication in nurse and follicle cells of *Drosophila melanogaster. Chromosoma* **91,** 267–278.

Hasitschka, G. (1956). Bildung von Chromosomenbundeln nach Art der Speicheldrusenchromosomen, spiralisierte Ruhekernchromosomen und andere Struktureigentumlichkeiten in den endopolyploiden Riesenkernen der Antipoden von *Papaver rhoeas. Chromosoma* 8, 87–113.

Heitz, E. (1934). Uber $\alpha$- und $\beta$-Heterochromatine sowie Konstanz und Bau der Chromomeren bei *Drosophila. Biol. Zbl.* **54,** 588–699.

Henderson, S. A. (1967a). The salivary gland chromosomes of *Dasyneura crataegi* (Diptera: Cecidomyiidae). *Chromosoma* **23,** 38–58.

Henderson, S. A. (1967b). A second example of normal coincident endopolyploidy and polyteny in salivary gland nuclei. *Caryologia* **20,** 181–186.

Hernandez-Verdun, D. (1986). Structural organization of the nucleolus in mammalian cells. *Methods Achiev. Exp. Pathol.* **12,** 26–62.

Hernandez-Verdun, D., and Derenzini, M. (1983). Non-nucleosomal configuration of chromatin in nucleolar organizer regions of metaphase chromosome *in situ. Eur. J. Cell Biol.* **31,** 360–365.

Hernandez-Verdun, D., Derenzini, M., and Bouteille, M. (1984). Relationship between the Ag-NOR proteins and ribosomal chromatin *in situ* during drug-induced RNA synthesis inhibition. *J. Ultrastruct. Res.* **88,** 55–65.

Hernandez-Verdun, D., Roussel, P., and Gautier, T. (1993). Nucleolar proteins during mitosis. *In* "Chromosomes Today" (A. T. Sumner and A. C. Chandley, eds.), v. 11, pp. 79–90. Chapman and Hall, London.

Hoffman, L., and Wooding, F. B. P. (1993). Giant and binucleate trophoblast cells in mammals. *J. Exp. Zool.* **266,** 559–577.

Ilgren, E. B. (1980). On the control of trophoblastic growth in the guinea pig. *J. Embryol. Exp. Morphol.* **60,** 405–418.

Ilgren, E. B. (1981). On the control of the trophoblastic giant-cell transformation in the mouse: Homotypic cellular interaction and polyploidy. *J. Embryol. Exp. Morphol.* **62,** 183–202.

Ilgren, E. B. (1983). Control of trophoblastic growth. *Placenta* **4,** 307–328.

Ilgren, E. B., Evans, E. P., and Burtenshaw, M. D. (1983). Origin of the multinucleate decidual cell of the mouse. *Cytologia* **48,** 313–322.

Ilynskaya, N. B., and Selivanova, G. V. (1982). The structure of polytene chromosomes at various stages of endoreduplication. I. The DNA content and and nuclear structure of salivary glands in the fourth instar larvae of *Chironomus. Tsitologia* **24,** 144–155.

Isakova, G. K., Zybina, T. G., and Zybina, E. V. (1992). Polyteny in the mink preattachment embryo cells. *Dokl. Akad. Nauk* **326,** 900–902.

Istomina, A. G. (1976). Endomitosis peculiarities in several grasshopper species. Candidate (Ph.D.) thesis. 24 p. Institute of Cytology and Genetics. Novosibirsk, USSR.

Istomina A. G., and Vysotskaya, L. V. (1977). Investigation of amitosis in polyploid cells of the inner parietal layer of the testicular follicle in some grasshopper species. *In* "Proceedings of All Union Symposium on Somatic Polyploid," pp. 47–48. Yerevan, USSR.

Jaworska, H., and Lima-de-Faria, A. (1973). Amplification of ribosomal DNA in Acheta. IV. Ultrastructure of two types of nucleolar components associated with ribosomal DNA. *Hereditas* **74,** 169–186.

Jollie, W. P. (1960). The persistence of trophoblast on extrauterine tissues in rat. *Am. J. Anat.* **106,** 109–115.

Jollie, W. P. (1964). Radioautographic observations on variations in desoxyribonucleic acid synthesis in rat placenta with increasing gestational age. *Am. J. Anat.* **114,** 161–167.

Jollie, W. P. (1969). Nuclear and cytoplasmic annulate lamellae in trophoblast giant cells of rat placenta. *Anat. Rec.* **165,** 1–14.

Jollie, W. P. (1981). Age changes in the fine structure of trophoblast giant cells. *Anat. Embryol.* **162,** 105–119.

Kaufman, M. H. (1983). The origin, properties and fate of trophoblast in the mouse. *In* "Biology of Trophoblast" (Y. W. Loke and A. Whyte, eds.), pp. 23–68. Elsevier, Amsterdam.

Keighren, M., and West, J. D. (1993). Analysis of cell ploidy in histological sections of mouse tissues by DNA–DNA *in situ* hybridization with digoxigenin-labelled probes. *Histochem. J.* **25,** 30–44.

Kiknadze, I. I. (1966). The changes of nuclear structures in the course of ovogenesis in the mink. *Tsitologia* **8,** 384–387.

Kiknadze, I. I. (1972). "Functional Organization of Chromosomes." Nauka, Leningrad.

Kiknadze, I. I., and Istomina, A. G. (1980). Endomitosis in grasshopper. I. Nuclear morphology and synthesis of DNA and RNA in the endopolyploid cells of the inner parietal layer of the testicular follicle. *Eur. J. Cell Biol.* **21,** 122–133.

Kiknadze, I. I., and Tuturova, K. F. (1970). On the transcription capacity of chromosomes of endomitotic cells in the testicle of *Chrysochraon dispar dispar. Tsitologia* **12,** 844–855.

King, T., and Burnett, A. (1959). Audioradiographic study of uptake of tritiated glycine, thymidine and uridine by fruit fly ovaries. *Science* **129,** 1674–1675.

Kirby, D. R. S. (1960). Development of mouse eggs under the kidney capsule. *Nature* (*London*) **187,** 707–708.

Kirby, D. R. S. (1963). The development of mouse blastocytes transplanted to the scrotal and cryptorchid testis. *J. Anat.* **97,** 119–130.

Klinger, H. P. (1958). The finer structure of the sex chromatin body. *Exp. Cell Res.* **14,** 207—210.

Klinger, H. P., and Schwarzacher, H. J. (1958). Amount of sex chromatin in female tissues is correlated with degree of tissue ploidy. *Nature* (*London*) **181,** 1150–1152.

Klinger, H. P., and Schwarzacher, H. J. (1960). The sex chromatin and heterochromatin bodies in human diploid and polyploid nuclei. *J. Biophys. Biochem. Cytol.* **8,** 345–364.

Knibiechler, M. C., Mirre, C., and Rossett, R. (1982). Nucleolar organizer structure and activity in nucleolus without fibrillar centres: The nucleolus in an established *Drosophila* cell line. *J. Cell Sci.* **57,** 351–364.

Kolb, I. P., Chaouat, G., and Chassoux, D. (1984). Immunoactive products of the placenta. III. Suppression of natural killing activity. *J. Immunol.* **132,** 2305–2310.

Kovaleva, V. G., Vinnikova, N. V., and Raikov, I. B. (1979). Electron microscopic study of the nuclear apparatus of the lower ciliate *Tracheloraphis totevi. Tsitologia* **21,** 280–286.

Kuhn, E. K., and Therman, E. (1988). The behavior of heterochromatin in mouse and human nuclei. *Cancer Genet. Cytogenet.* **34,** 143–151.

Kuhn, E. K., and Therman, E., and Susman, B. (1991). Amitosis and endocycles in early cultured mouse trophoblast. *Placenta* **12,** 251–261.

Kurman, R. J., Main, C. S., and Chen, H.-C. (1984). Intermediate trophoblast: A distinctive form of trophoblast with specific morphological, biochemical and functional features. *Placenta* **5,** 349–370.

Lafontaine, J. G., and Lord, A. (1974). A correlated light- and electron-microscope investigation of the structural evolution of the nucleolus during the cell cycle in plant meristematic cells (*Allium porrum*). *J. Cell Sci.* **16,** 63–93.

Lala, P. K., Kennedy, T. G., and Parhar, R. S. (1988). Suppression of lymphocyte alloreactivity by early gestational human decidua. II. Characterization of the suppressor mechanisms. *Cell Immunol.* **116,** 411–422.

Lin, H., Mossmann, T. R., Guilbert, L., Tuntipopipat, S., and Wegnann, T. G. (1993). Synthesis of T helper-2-type cytokines at the maternal–fetal interface. *J. Immunol.* **151,** 4562–4573.

Lin, T.-P., Labosky, P. A., Grabel, L. B., Kozak, C. A., Pitman, J. L., Kleeman, J., and MacLeod, C. L. (1994). The *Pem* homeobox gene is X-linked and exclusively expressed in extraembryonic tissues during early murine development. *Dev. Biol.* **166,** 170–179.

Mainx, F. (1949). The structure of the giant chromosomes in some Diptera. Proc. VIII Int. Congr. Genetics. *Hereditas* 19, Suppl. pp. 622–623.

Markes, G. E., and Davies, D. R. (1979). The cytology of cotyledon cells and the induction of giant polytene chromosomes in *Pisum sativum. Protoplasma* **101,** 73–80.

Marshak, T. L., Dungenova, R. E., Sedkova, N. A., and Brodsky, V. Ya. (1994). Relation between ribosomal RNA synthesis and number of different type nucleoli in the rat hepatocytes. *Tsitologia* **36,** 252–258.

Martynova, M. G., Antipanova, E. M., and Rumyantsev, P. P. (1983). Studies of DNA content, sex chromatin bodies and nucleoli in the myocyte nuclei of normal and hypertrophied human heart atricles. *Tsitologia* **25,** 614–619.

Matuszewski, B. (1964). Polyploidy and polyteny induced by a Hymenopteran parasite in *Dasyneura urtica* (Diptera, Cecidomyiidae). *Chromosoma* **15,** 31–35.

Matuszewski, B. (1965). Transition from polyteny to polyploidy in salivary glands of Cecidomyiidae. *Chromosoma* **16,** 22–34.

McLaren, A., and Tarkowski, A. K. (1963). Implantation of mouse eggs in the peritoneal cavity. *J. Reprod. Fertil.* **6,** 385–392.

Mechelke, F. (1953). Reversible Strukturmodificationen der Speicheldrusenchromosomen von *Acritopus lucidus. Chromosoma* **5,** 511–543.

Mikhailov, V. M., Podporina, A., and Malygin, A. (1994). Immunological functions of granulated cells of endometrium. *In* "The 8th International Conference of the International Society of Differentiation" (E. Tahara, ed.), p. 193. Hiroshima.

Miller, O. L., and Beatty, B. R. (1969). Visualization of nucleolar genes. *Science* **164,** 955–957.

Mirre, C., and Knibiehler, B. (1981). Ultrastructural autoradiographic localization of the rRNA transcription sites in the quail nucleolar component using two RNA antimetabolites. *Biol. Cell* **42,** 73–78.

Monneron, A., and Bernhard, W. (1969). Fine structure organization of the interphase nucleus in some mammalian cells. *J. Ultrastruct. Res.* **27,** 266–268.

Mossman, H. W. (1937). Comparative morphogenesis of the fetal membranes and accessory uterine structures. *Contrib. Embryol. Carneg. Inst.* **26,** 129–246.

Mulder, M. P., Duijn, P., and Gloor, H. J. (1968). The replicative organization of DNA in polytene chromosomes of *Drosophila hydei. Genetica* **39,** 385–428.

Muller, H. J. (1935). On the dimensions of chromosomes and genes in Dipteran salivary glands. *Am. Nature* **69,** 405–411.

Nagl, W. (1962). Uber Endopolyploidie, Restitutionkernbildung und Kernstructuren im Suspensor von Angiospermen und einer Gymnosperme. *Osterr. Bot. Zeitschr.* **109,** 432–495.

Nagl, W. (1967). Die Riesenchromosomen von *Phaseolus coccineus* L.: Baueigentumlichkeiten. Strukturmodifikationene, zusatzliche Nucleolen und Vergleich mit den mitotischen Chromosomen. *Oster. Bot. Zeitschr.* **114,** 171–182.

Nagl, W. (1970a). Temperature-dependent functional structures in the polytene chromosomes of *Phaseolus,* with special reference to the nucleolus organizers. *Cell Sci.* **6,** 87–107.

Nagl, W. (1970b). Differentielle RNS-Synthese on pflanzlichen Riesenchromosomen. *Ber. Dtsch. Bot. Ges.* **83,** 301–309.

Nagl, W. (1972). Giant sex chromatin in endopolyploid trophoblast nuclei of the rat. *Experientia* (*Basel*) **28,** 217–218.

Nagl, W. (1973a). Photoperiodic control of activity of the suspensor polytene chromosomes in *Phaseolus vulgaris. Z. Pflanzephys.* **70,** 350–357.

Nagl, W. (1973b). The angiosperm suspensor and the mammalian trophoblast: Organ with similar cell structure and function. *Bull. Soc. Bot. Fr. Mem.* **119,** 289–302.

Nagl, W. (1976). Nuclear organization. *Annu. Rev. Plant. Physiol.* **27,** 39–69.

Nagl, W. (1978). "Endopolyploidy and Polyteny in Differentiation and Evolution." North-Holland, Amsterdam.

Nagl, W. (1981). Polytene chromosomes of plants. *Int. Rev. Cytol.* **73,** 21–53.

Nagl, W. (1985). Chromosomes in differentiation. *In* "Advances in Chromosome and Cell Genetics" (A. K. Sharma, and A. Sharma, eds.), pp. 135–172. Oxford and IBH, New Delhi.

Noel, J. S., Dewey, W. C., Abel, J. H., Jr., and Thompson, R. P. (1971). Ultrastructure of the nucleolus during the Chinese hamster cell cycle. *J. Cell Biol.* **49,** 830–847.

Nur, U. (1968). Endomitosis in the mealy bug *Planococcus citri* (Homoptera: Coccoidea). *Chromosoma* **24,** 202–209.

Nuti Ronchi, V., Bennici, A., and Martini, G. (1973). Nuclear fragmentation in dedifferentiating cells of *Nicotiana glauca* pith grown in vitro. *Cell Differ.* **2,** 77–85.

Ohno, S., Kaplan, W. D., and Kinosita, R. (1959). Formation of sex chromatin by a single X-chromosome in liver cells of *Rattus norvegicus. Exp. Cell Res.* **182,** 415–418.

Ollerlich, D., and Carlson, E. (1970). Ultrastructure of intranuclear annulate lamella in giant cells of rat placenta. *J. Ultrastruct. Res.* **30,** 411–422.

Orsini, M. (1954). The trophoblastic giant cells and endovascular cells associated with pregnancy in the hamster, *Cricetus auratus. Am. J. Anat.* **94,** 273–321.

Papaioannou, V. G., West, J. D., Bucher, T., and Linke, I. M. (1981). Nonrandom X-chromosome expression early in mouse development. *Dev. Genet.* **2,** 305–315.

Pearson, M. J. (1974a). Polyteny and the functional significance of the polytene cell cycle. *J. Cell Sci.* **15,** 457–479.

Pearson, M. J. (1974b). The abdominal epidermis of *Callifora erythrocephala* (Diptera). I. Polyteny and growth in the larval cells. *J. Cell Sci.* **16,** 113–131.

Pehlmann, F. W. (1968). Die amitotische Zellteilung. Eine electronenmicroskopische Untersuching an Interrenalzellen von *Rana temporaria. Zellforsch. Mikrosk. Anat.* **84,** 516–548.
Perov, N. A., Kiknadze, I. I., and Chentsov, Yu. S. (1976). The puff ultrastructure in polytene chromosomes of *Chironomus thummi. Tsitologia* **18,** 840–846.
Pijnenborg, R., Robertson, W. B., and Brosens, I. (1974). The arterial migration of trophoblast in the uterus of golden hamster, *Mesocricetus auratus. Reprod. Fertil.* **40,** 269–280.
Pijnenborg, R., Dixon, G., Robertson, W. B., and Brosens, I. (1980). Trophoblastic invasion of human decidua from 8–18 weeks of pregnancy. *Placenta* **1,** 3–19.
Prokofieva-Belgovskaya, A. A. (1960). Nuclear cycle and differentiation of somatic cells. *In* "Problems of Cytology and General Physiology" (Yu.I. Polyansky, ed.), pp. 215–253. Nauka, Moscow.
Prokofieva-Belgovskaya, A. A. (1986). "Heterochromatic Chromosome Regions." Nauka, Moscow.
Raikov, I. B. (1978). "Protozoa Nucleus: Morphology and Evolution." Nauka, Leningrad.
Raikov, I. B. (1982). "The Protozoan Nucleus. Morphology and Evolution." Cell Biology Monographs, Vol. 9. Springer-Verlag, Vienna.
Raikov, I. B., and Ammermann, D. (1976). The macronucleus of ciliates: Recent advances. *In* "Caryology and Genetics of Protozoa" (I. B. Raikov, ed.), pp. 64–90. Nauka, Leningrad.
Raikov, I. B., and Kovaleva, V. G. (1977). The fine structure of nuclear complex of the lower marine *Tracheloraphis phoenicopterius. Tsitologia* **19,** 350–355.
Rasch, E. H. (1970). Two-wavelength cytophotometry of *Sciara* salivary gland chromosomes. *In* "Introduction of Quantitative Cytochemistry" (G. L. Wied and G. F. Bahr, eds.), Vol. 2, pp. 335–355, Academic Press, New York.
Raska, I., and Dundr, M. (1993). Compartmentalization of cell nucleus: Case of nucleolus. *In* "Chromosomes Today" (A. T. Sumner and A. C. Chandley, eds.), Vol. 11, pp. 101–119. Chapman and Hall, London.
Raska, I., Dundr, M., and Koberna, K. (1994). The nature of coiled bodies. *Cell Biol. Int.* **18,** 480.
Recher, L., Parry, N., Whitescarver, I., and Briggs, L. (1972). Lead-positive nuclear structures and their behavior under the effect of various drugs. *J. Ultrastruct. Res.* **38,** 398–410.
Reitalu, J. (1957). Observation of the so-called sex-chromatin in man. *Acta Genet. Med. Gem.* **6,** 393–402.
Renegar, R. H. Southard, I. N., and Talamantes, F. (1990). Immunohistochemical colocalization of placental lactogen II and relaxin in the golden hamster (*Mesocricetus auratus*). *J. Histochem. Cytochem.* **38,** 935–940.
Ribbert, D. (1979). Chromomeres and puffing in experimentally induced polytene chromosomes of *Calliphora erythrocephala. Chromosoma* **74,** 269–298.
Ribbert, D., and Bier, K. (1969). Multiple nucleoli and enhanced nucleolar activity in the nurse cells of the insect ovary. *Chromosoma* **27,** 178–197.
Risler, H. (1959). Polyploide und somatische Reduction in der Larvenepidermis von *Aedes aegypti* L. (Culicidae). *Chromosoma* **10,** 184–209.
Rossant, J., and Offer, L. (1977). Properties of extraembryonic ectoderm isolated from postimplantation mouse embryos. *J. Embryol. Exp. Morphol.* **39,** 183–194.
Rudkin, G. T. (1965). Nonreplication DNA in giant chromosomes. *Genetics* **52,** 470.
Rudkin, G. T. (1969). Non-replicating DNA in *Drosophila. Genetics* (Suppl.) **61,** 227–238.
Sandler, L., and Pavan, C. (1972). Heterochromatin in development of normal and infected cells. *In* "Cell Differentiation" (R. Harris, P. Allin, and D. Viza, eds.), pp. 162–175. Munksgaard, Copenhagen.
Sarto, G. E., Stubblefield, P. A., and Therman, E. (1982). Endomitosis in human trophoblast. *Hum. Genet.* **62,** 228–232.
Schwarzacher, H. G., and Schnell, W. (1966). Endoreduplication in human fibroblast cultures. *Cytogenetics* **4,** 1–18.

Schwarzacher, H. G., and Wachtler, F. (1983). Nucleolus organizer regions and nucleoli. *Hum. Genet.* **63,** 89–99.

Sherman, M. I. (1983). Endocrinology of rodent trophoblast cells. *In* "Biology of Trophoblast" (Y. W. Loke and A. Whyte, eds.), pp. 401–467. Elsevier, Amsterdam.

Sherman, M. I., McLaren, A., and Walker, P. M. B. (1972). Mechanism of accumulation of DNA in giant cells of mouse trophoblast. *Nature New Biol.* **238,** 175–176.

Singh, P. B., Mills, A. D., Kothary, R., Surani, M. A. H., Amos, B., White, H., Sheenan, M., Laskey, R., and Johnson, R. (1989). Experimentally induced giant chromosomes in the mouse. *Genet. Res.* **53,** 228.

Smetana, K. (1974). The ultrastructural morphology and formation of nucleolar ribonucleoprotein structures. *Acta Fac. Med. Univ. Brun.* **49,** 155–179.

Snow, M. H., and Ansell, I. D. (1974). The chromosomes of giant trophoblast cells of the mouse. *Proc. R. Soc. London* **B187,** 93–98.

Soares, M. I., Faria, T. N., Roby, K. F., and Deb, S. (1991). Pregnancy and prolactin family of hormones: Coordination of anterior pituitary, uterine, and placental expression. *Endocrine Rev.* **12,** 402–423.

Sobis, H., Verstuyf, A., and Vandeputte, M. (1991). Histochemical differences in expression of X-linked glucose-6-phosphate dehydrogenase between ectoderm- and entoderm-derived embryonic and extraembryonic tissues. *J. Histochem. Cytochem.* **39,** 569–574.

Sokolov, I. I. (1967). Endomitotic polyploidy in testicular epithelial cells of spiders (Araneina). Communication I. *Tsitologia* **9,** 152–161.

Stahl, A. Luciani, I. M., Devictor, M., Capodano, A. M., and Gagne, R. (1975). Constitutive heterochromatin and micronucleoli in human oocyte at the diplotene stage. *Humangenetik* **26,** 315–327.

Steer, H. W. (1971). Implantation of the rabbit blastocyst: The adhesive phase of implantation. *J. Anat.* **109,** 215–227.

Stewart, I. J., and Mukhtar, D. D. J. (1988). The killing of mouse trophoblast cells by granulated metrial grand cells *in vitro. Placenta* **9,** 417–423.

Tachi, S., and Tachi, C. (1979). Ultrastructural studies on maternal–embryonic cell interaction during experimentally induced implantation of rat blastocysts to the endometrium of the mouse. *Dev. Biol.* **68,** 203–223.

Takagi, N., and Sasaki, M. (1975). Preferential inactivation of the paternally derived X-chromosome in th extraembryonic membranes of the mouse. *Nature* (*London*) **250,** 640–642.

Takagi, N., Sugawara, O., and Sasaki, M. (1982). Regional and temporal changes in pattern of X-chromosome replication during the early post implantation development of the female mouse. *Chromosoma* **85,** 275–286.

Therman, E., and Kuhn, M. (1989). Mitotic modification and aberration in cancer. *CRC Clin. Rev. Oncogenesis* **1,** 393–405.

Therman, E., Sarto, G. E., and Buchler, D. A. (1983). The structure and origin of giant nuclei in human cancer cells. *Cancer Genet. Cytogenet.* **9,** 9–18.

Therman, E., Denniston, C., Nieminen, U., Buchler, D. A., and Thimonen, S. (1985). X chromatin, endomitosis, and mitotic abnormalities in human cervical cancer. *Cancer Genet. Cytogenet.* **16,** 1–11.

Thiry, M., and Thiry-Blaise, L. (1989). *In situ* hybridization at the electron microscope level: An improved method for precise localization of ribosomal DNA and RNA. *Eur. J. Cell Biol.* **50,** 235–243.

Tschermak-Woess, E. (1956). Karyologische Pflanzenanatomie. *Protoplasma* **46,** 798–834.

Tschermak-Woess, E. (1957). Uber der regelmaßige Auftreten von "Riesenchromosomen" in Cholazahaustorium von *Rhinanthus. Chromosoma* **8,** 523–544.

Tschermak-Woess, E. (1971). "Der Zellkern." Springer, Heidelberg.

Tschermak-Woess, E. (1973). Somatische Polyploidie bei Pflanzen. *In* "Grundlagen der Cytologie" (G. C. Hirsch, H. Ruska, and P. Sitte, eds.), pp. 189–204. Fischer, Stuttgart.

Tschermak-Woess, E., and Hasitschka, G. (1953). Veranderungen der Kernstruktur der Endomitose, rhythmisches Kernwachstum und verschiedenes Heterochromatin bei Angiosperm. *Chromosoma* **5,** 574–614.

Tschermak-Woess, E.. and Hasitschka-Jenschke, G. (1963). Das Verhalten von B-Chromosomen besonderer Ausbildung in den endopolyploiden Riesenkernen des chalasalen Endospermhaustoriums von *Rhinanthus. Osterr. Bot. Ztschr.* **110,** 468–480.

Uryvaeva, I. V. (1979). Polyploidizing mitosis and biological significance of liver cell polyploidy. *Tsitologia* **21,** 1427–1437.

Vagner-Capodano, A. M., Henderson, S. A., Lissitzky, S, and Stahl, A. (1984). The relationships between ribosomal genes and fibrillar centers in thyroid cells cultivated *in vitro. Biol. Cell* **51,** 11–22.

Valeeva, F. S., and Kiknadze, I. I. (1971). Changes in the amount of DNA following the polytenization of nuclei in salivary glands of Chironomids. *Ontogenez* (*Moscow*) **2,** 406–410.

Van Vlasselaer, P., and Vanderputte, M. (1984). Immunosuppressive properties of murine trophoblast. *Cell Immunol.* **83,** 422–432.

Varmuza, S., Prideaux, V., Kothary, R., and Rossant, J. (1988). Polytene chromosomes in mouse trophoblast giant cells. *Development* **102,** 127–134.

Wachtler, F., Schoefer, C., Mosgoeller, W., Weipoltshammer, K., Schwarzacher, H., Guichaoua, M., Hartung, M., Stahl, A., Berge-Lefranc, J. L., Gonzales, I., and Silvester, J. (1992). Human ribosomal RNA gene repeats are localized in the dense fibrillar component of nucleoli: Light and electron microscopic *in situ* hybridization in human Sertoli cells. *Exp. Cell Res.* **198,** 135–143.

Wake, N., Takagi, N., and Sasaki, M. (1976). Nonrandom inactivation of X-chromosome in the rat yolk sac. *Nature* (*London*) **262,** 580–581.

Welsh, A. O., and Enders, A. C. (1987). Trophoblast–decidual cell interaction and establishment of maternal blood circulation in the parietal yolk sac placenta of the rat. *Anat. Rec.* **217,** 203–219.

West, J. D., and Chapman, V. M. (1978). Variation for X-chromosome expression in mice detected by electrophoresis of phosphoglycerate kinase. *Genet. Res.* **32,** 91–102.

West, J. D., Frels, W. L., Chapman, V. M., and Papaioannou, V. E. (1977). Preferential expression of the maternally derived X-chromosome in the mouse yolk sac. *Cell* **12,** 873–882.

White, M. J. D. (1948). The salivary gland chromosomes of several species. *J. Morphol.* **82,** 53–80.

White, M. J. D. (1977). "Animal Cytology and Evolution." Cambridge University Press, Cambridge.

Wolgemuth-Farashow, D. J., Jagiello, G. M., and Henderson, A. S. (1977). The localization of rDNA in small, nucleolus-like structures in human diplotene oocyte nuclei. *Hum. Genet.* **36,** 63–68.

Zatsepina, O. V., Hozak, P., Babadjanyan, D., and Chentsov, Yu. (1988a). Quantitative ultrastructural study of nucleolus-organizing regions at some stages of cell cycle ($G_0$-period, $G_2$-period, mitosis). *Biol. Cell.* **62,** 211–218.

Zatsepina, O. V., Chelidze, P. V., and Chentsov, Yu. S. (1988b). Changes in the number and volume of fibrillar centres with the inactivation of nucleoli at erythropoesis. *J. Cell Sci.* **91,** 439–448.

Zavarzin, A. A. (1967). "DNA Synthesis and Kinetics of Cell Populations in the Mammalian Ontogenesis." Nauka, Leningrad.

Zhimulev, I. F. (1992). "Polytene Chromosomes: Morphology and Structure." Nauka, Novosibirsk.

Zhimulev, I. F., and Belyaeva, E. S. (1976). Changes of the structure of *Drosophila melanogaster* polytene chromosomes after a long-term cultivation of larval salivary glands in the imago abdomen. *Tsitologia* **18,** 5–9.

Zhimulev, I. F., and Lychev, V. A. (1972). Functioning of salivary gland chromosomes in *Harmandia laevi* larvae (Diptera, Cecidomyiidae). I. Morphological changes of giant cell chromosomes in ontogenesis. *Ontogenez* (*Moscow*) **3,** 194–201.

Zybina, E. V. (1960). Sex chromatin in the trophoblast of early embryos of white rat. *Dokl. Akad. Nauk. USSR* **130,** 633–635.

Zybina, E. V. (1961). Endomitosis and polyteny of giant cells of the trophoblast. *Dokl. Akad. Nauk. USSR* **140,** 1177–1180.

Zybina, E. V. (1963a). Cytophotometric determination of DNA content in nuclei of giant cells of trophoblasts. *Dokl. Akad. Nauk. USSR* **153,** 1428–1431.

Zybina, E. V. (1963b). Autoradiographic and cytophotometric investigation of nucleic acids (DNA and RNA) in the endomitotic cycle of giant trophoblast cells. *In* "Morphology and Cytochemistry of Cells" (I. Sokolov, ed.), pp. 34–44. Nauka, Moscow.

Zybina, E. V. (1965). Sex chromatin in giant cells of trophoblast and in cells of early rabbit embryos. *Fed. Proc. Transl.* Suppl. **24,** 868–876.

Zybina, E. V. (1968). The structure of nucleus and nucleolus during ovogenesis of mice. *Tsitologia* **10,** 36–42.

Zybina, E. V. (1969). Behavior of the chromosome–nucleolar apparatus during the growth period of the rabbit oocytes. *Tsitologia* **11,** 25–30.

Zybina, E. V. (1970). Features of polyploidization patterns of trophoblast cells. *Tsitologia* **12,** 1084–1091.

Zybina, E. V. (1975). An electron microscope study of the lampbrush chromosomes and their activity products during the rabbit's oogenesis. *Tsitologia* **17,** 875–880.

Zybina, E. V. (1976a). Phagocytic activity and reproduction of the trophoblast cells in the white mouse foetus. *Tsitologia* **18,** 683–688.

Zybina, E. V. (1976b). Peculiarities of phagocytosis in mitotic and endomitotic chromosome replication in the trophoblast of the grey vole. *Tsitologia* **18,** 834–838.

Zybina, E. V. (1977). The structure of polytene chromosomes in the mammalian trophoblast cells. *Tsitologia* **19,** 327–337.

Zybina, E. V. (1979). Folding pattern of the nuclear envelope and investigation of its derivates in the nuclei of trophoblast polyploid cells of the rat placenta. *Tsitologia* **21,** 1253–1258.

Zybina, E. V. (1980a). Division of polyploid nuclei by fragmentation in the rat giant trophoblast cells. II. Formation of deep invaginations in the nuclear envelope as the beginning fragmentation. *Tsitologia* **22,** 10–13.

Zybina, E. V. (1980b). Division of polyploid nuclei by fragmentation in the rat trophoblast giant cells. III. The involvement of intracellular membrane structures in the division. *Tsitologia* **22,** 15–19.

Zybina, E. V. (1980c). Electron microscopic investigation of the endopolyploid nuclei in the giant cells of the rat trophoblast. I. Fine structure of the polytene chromosomes and the products of their activity at various stages of the endoreplicative cycle. *Tsitologia* **22,** 1284–1289.

Zybina, E. V. (1980d). Electron microscopic investigation of the endopolyploid nuclei in the giant cells of the rat trophoblast. II. Changes of the fine structure of the polytene nucleus during differentiation. *Tsitologia* **22,** 1379–1386.

Zybina, E. V. (1981a). Electron microscope investigation of the endopolyploid nuclei in the giant cells of the rat trophoblast. III. Fine structure of the nucleolus and its fibrillar centre at various stages of the endoreplication cycle. *Tsitologia* **23,** 5–11.

Zybina, E. V. (1981b). Electron microscope investigation of the endopolyploid nuclei in the giant cells of the rat trophoblast. IV. Changes of the fine structure of the nucleolus during cell differentiation. *Tsitologia* **23,** 129–132.

Zybina, E. V. (1986). "The trophoblast Cytology." Nauka, Leningrad.

Zybina, E. V., and Chernogryadskaya, N. A. (1976). A study of polyploid nuclei of the giant trophoblast cells of some rodent species by phase contrast microscopy. *Tsitologia* **18,** 161–165.

Zybina, E. V., and Grishchenko, T. A. (1970). Polyploid cells of the trophoblast in different regions of white rat placenta. *Tsitologia* **12,** 585–595.

Zybina, E. V., and Grishchenko, T. A. (1972). Spectrophotometrical estimation of ploidy level in decidual cells of the endometrium of the white rat. *Tsitologia* **14,** 284–290.

Zybina, E. V., and Grishchenko, T. A. (1977). The ultrastructure of nucleolus-like bodies in the golden hamster's oocytes at the diplonema stage. *Tsitologia* **19,** 1231–1237.

Zybina, E. V., and Mosjan, I. A. (1967). Sex chromatin bodies during endomitotic polyploidization of trophoblast cells. *Tsitologia* **9,** 265–272.

Zybina, E. V., and Rumyantsev, P. P. (1980). Formation of a complex plasmatic membrane and microfilament bundles during the completion of nuclear fragmentation in the trophoblast giant polycariocytes. *Tsitologia* **22,** 890–898.

Zybina, E. V., and Tikhomirova, M. M. (1963). The question of endomitotic polyploidization in the giant cells of the trophoblast. *In* "Morphology and Cytochemistry of Cells" (I. Sokolov, ed.), pp. 53–63. Nauka, Moscow.

Zybina, E. V., and Zybina, T. G. (1985). Polyteny and endomitosis in supergiant trophoblast cells of *Microtus subarvalis. Tsitologia* **27,** 402–410.

Zybina, E. V., and Zybina, T. G. (1989). A comparative study of the interrelation of different nucleolar components in several rat trophoblast cell populations during differentiation. II. Nucleoli of glycogen cells and secondary giant cells. *Tsitologia* **31,** 1428–1434.

Zybina, E. V., and Zybina, T. G. (1992). Changes in the arrangements of chromosomes and nucleoli related to functional peculiarities of developing mammalian oocytes during meiotic prophase. *Tsitologia* **34,** 3–32.

Zybina, E. V., and Zybina, T. (1993). Polyteny and protective role of genome multiplication in giant trophoblast cells in the rodent placenta. *Placenta* **14,** A84.

Zybina, E. V., Kudryavtseva, M. V., and Kudryavtsev, B. N. (1975a). Morphological and cytofluorometric study of the common vole giant cells trophoblast. *Tsitologia* **17,** 254–260.

Zybina, E. V., Kudryavtseva, M. V., and Kudryavtsev, B. N. (1975b). Polyploidization and endomitosis in giant cells of rabbit trophoblast. *Cell Tissue Res.* **160,** 525–537.

Zybina, E. V., Kudryavtseva, M. V., and Kudryavtsev, B. N. (1979). The distribution of chromosome material during giant nuclei division by fragmentation in the trophoblast of rodents. Morphological and cytophotometrical study. *Tsitologia* **21,** 12–20.

Zybina, E. V., Zybina, T. G., and Stein, G. I. (1984a). Peculiarities of trophoblast cell reproduction in the rat's placenta connective zone. I. Determination of the degree of ploidy and the number of Barr's bodies in the interphase nuclei. *Tsitologia* **26,** 525–530.

Zybina, E. V., Zybina, T. G., and Stein, G. I. (1984b). Peculiarities of trophoblast cell reproduction in the rat's placenta connective zone. II. Determination of ploidy degree in the mitotic figures. *Tsitologia* **26,** 531–535.

Zybina, E. V., Zybina, T. G., and Dalmane, A. R. (1987). The metrial gland cells: Relationship with trophoblast cells and reproduction peculiarities. *Tsitologia* **29,** 771–781.

Zybina, E. V., Zybina, T. G., Zhelezova, A. I., Kiknadze, I. I. , and Stein, G. I. (1989). Trophoblast cells in the placenta of silver fox: their peculiarities and polyploidization. *Tsitologia* **31,** 492–495.

Zybina, E. V., Zybina, T. G., Isakova, G. K., Kiknadze, I. I. , and Stein, G. I. (1992). Polyploidy and polyteny in the mink trophoblast cells. *Tsitologia* **34,** 53–57.

Zybina, T. G. (1987). Polyploidization dynamics of tertiary giant trophoblast cells in the rat's placenta. *Tsitologia* **29,** 1012–1019.

Zybina, T. G. (1988). Glycogen-containing cells in the maternal and fetal parts of the rat and field vole placentas. *Tsitologia* **30,** 1180–1187.

Zybina, T. G. (1990). DNA in the nuclear fragments that appear in the course of fragmentation of the secondary giant trophoblast cells in the field vole. *Tsitologia* **32,** 806–810.

Zybina, T. G., and Zybina, E. V. (1989). A quantitative study of nuclear Ag-positive areas revealed by silver staining in interphase nuclei of trophoblast cells from the rat placenta connective zone. *Tsitologia* **31,** 1292–1305.

Zybina, T. G., Zybina, E. V., Kudryavtseva, M. V., and Kudryavtsev, B. N. (1980). A cytophotometrical study of sex chromatin in the giant cell nuclei of rabbit trophoblast. *Tsitologia* **22,** 1037–1045.

Zybina, T. G., Zybina, E. V., and Stein, G. I. (1985). Nuclear DNA content in the secondary giant cells of rat trophoblast at different phases of cell cycle of polytene nucleus. *Tsitologia* **27,** 957–960.

Zybina, T. G., Zybina, E. V., and Stein, G. J. (1987). Characteristic features of differentiation and polyploidization of trophoblast cells in the connective zone and labyrinth of placenta of the field vole, *Microtus subarvalis. Tsitologia,* **29,** 549–559.

Zybina, T. G., Zybina, E. V., Stein, G. I., Severova, E. L., and Dyban, A P. (1994). A quantitative study of Ag-stained nucleolar zones and other nucleolar parameters in the cambial rat trophoblast cells in the course of differentiation. *Tsitologia* **36,** 641–654.

# Ependymins: Meningeal-Derived Extracellular Matrix Proteins at the Blood–Brain Barrier

Werner Hoffmann* and Heinz Schwarz†
*Institut für Molekularbiologie und Medizinische Chemie, Otto-von-Guericke-Universität, D-39120 Magdeburg, Germany
†Max-Planck-Institut für Entwicklungsbiologie, D-72076 Tübingen, Germany

Ependymins represent regeneration-responsive piscine glycoproteins and in many teleost fish they appear as the predominant cerebrospinal fluid constituents. Thus far, no homologous sequences have been characterized unambiguously in mammals. Sialic acid residues of the N-linked carbohydrate moiety of ependymins are responsible for their calcium-binding capacity. Ependymins from some species bear the L2/HNK-1 epitope typical of many cell adhesion molecules.

After their synthesis in fibroblast-like cells of the inner endomeningeal layer, soluble ependymins are widely distributed via the cerebrospinal fluid system. Furthermore, ependymins presumably cross the intermediate endomeningeal barrier layer by way of a transcellular transport phenomenon (transcytosis). A bound form of ependymins is associated with collagen fibrils of the extracellular matrix typically found around cerebral blood vessels. Here, they might modulate the endothelial barrier function. Generally, ependymins are thought to represent a new class of possibly antiadhesive extracellular matrix proteins playing a role in specific cell contact phenomena (e.g., during regeneration).

**KEY WORDS:** Ependymins, Cell adhesion, Meninges, Transcytosis, Cerebrospinal fluid, Blood–brain barrier, Extracellular matrix, Arachnoid, Cell migration.

## I. Introduction

Ependymins represent a prominent group of secretory proteins in the brain of many teleost fish. Historically, they have been discovered twice in the goldfish when investigating two functionally different perspectives (memory and regeneration assays). Interestingly, in both cases, increased synthesis of these glycoproteins monitored by radioactive *in vivo* labeling led to their detection.

In 1976, Shashoua reported on two proteins in goldfish brain showing enhanced expression in fish that learned a new pattern of swimming (Shashoua, 1976a,b). Based on their initial immunohistochemical localization at (sub)ependymal zones (Benowitz and Shashoua, 1977; Shashoua, 1977a), these proteins were termed ependymins $\beta$ and $\gamma$ [apparent molecular mass on sodium dodecyl sulfate-polyacrylamide gel electrophoresis (SDS-PAGE) about 35 and 30 kDa].

Challenged by this unusual discovery, ependymins were often discussed to be implicated in processes of learning and memory (Shashoua, 1986, 1991; Nelson and Alkon, 1989). Consequently, many attempts have been undertaken to support these speculations. For example, infusion of ependymin antibodies was reported to block memory consolidation (Shashoua, 1977a, 1986; Shashoua and Moore, 1978; Schmidt, 1986, 1989; Piront and Schmidt, 1988), and conditioning experiments seemed to influence the concentration as well as the expression of ependymins (Schmidt, 1987; Shashoua and Hesse, 1989; Schmidt *et al.,* 1992). Furthermore, in an initial hypothesis, the molecular function of ependymins was tentatively explained by a polymerization into insoluble fibrils after local $Ca^{2+}$ depletion (Shashoua, 1986, 1988, 1991; Shashoua *et al.,* 1990). However, there is no molecular evidence supporting this hypothesis thus far.

In 1988, ependymins were rediscovered when studying regeneration of the goldfish optic nerve (Thormodsson *et al.,* 1988). It was observed that the synthesis of these glycoproteins increases during regeneration (Schmidt and Shashoua, 1988; Thormodsson *et al.,* 1992a).

In the present article we focus on the site of synthesis (i.e., meningeal fibroblasts) and on more recent localization studies, in particular at the electron microscopy level. Based on these results, we discuss new concepts and future perspectives for the molecular function of ependymins.

## II. From Gene to Protein: Molecular Analysis

### A. Gene Structure

Molecular characterization of ependymin genes from various cypriniform (zebrafish, goldfish, and carp; Rinder *et al.,* 1992; Adams and Shashoua, 1994) and a salmoniform fish (Atlantic salmon; Müller-Schmid *et al.,* 1992) revealed that the ependymin genes of these species have a similar structure comprising six exons (Figs. 1 and 2).

Analysis of the promoter regions of the ependymin genes from the zebrafish and the Atlantic salmon did not result in canonical regulatory sequences other than a TATA box. Furthermore, there are no significant similarities in the 5′-flanking sequences of the ependymin genes from these two species. However, by transient expression studies in zebrafish embryos the 2.0-kb upstream promoter region of the zebrafish ependymin gene has been shown to contain all necessary elements directing a correct temporal and spatial expression pattern (Rinder *et al.,* 1992).

In agreement with results from cDNA cloning (see Fig. 2), Southern blot analysis indicated the probable existence of two ependymin genes per haploid genome in the goldfish and the rainbow trout (Königstorfer *et al.,* 1989b; Müller-Schmid *et al.,* 1992). This reflects the quasitetraploid genome of these species (Ohno *et al.,* 1967; Bailey *et al.,* 1969), which might originate from a genome duplication about 70 million years ago (Uyeno and Smith, 1972). In contrast, the zebrafish contains only a single ependymin gene (Sterrer *et al.,* 1990).

## B. Structure of Precursor

Molecular cloning from three different orders of teleost fish (Cypriniformes, Salmoniformes, and Clupeiformes) revealed unambiguously that ependymins are synthesized via precursors containing an amino-terminal, cleavable, hydrophobic signal sequence typical of secretory proteins (Königstorfer *et al.,* 1989a,b; Sterrer *et al.,* 1990; Müller-Schmid *et al.,* 1992, 1993; Adams and Shashoua,1994). These precursors have a deduced molecular mass of about 23.7–24.5 kDa, which is reduced after cleavage by signal peptidase to about 21.6–22.3 kDa. The amino acid sequences obtained are highly

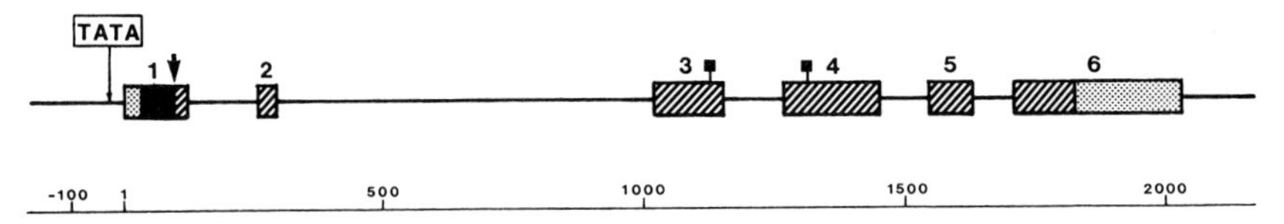


FIG. 1 Structure of the zebrafish ependymin gene, which is composed of six exons (Rinder *et al.,* 1992). 5′- and 3′-untranslated regions are dotted, whereas portions encoding the mature ependymin sequence are hatched; the signal sequence of the precursor is shown in black and the arrow indicates the cleavage site by signal peptidase. Knobs represent the two potential N-glycosylation sites.

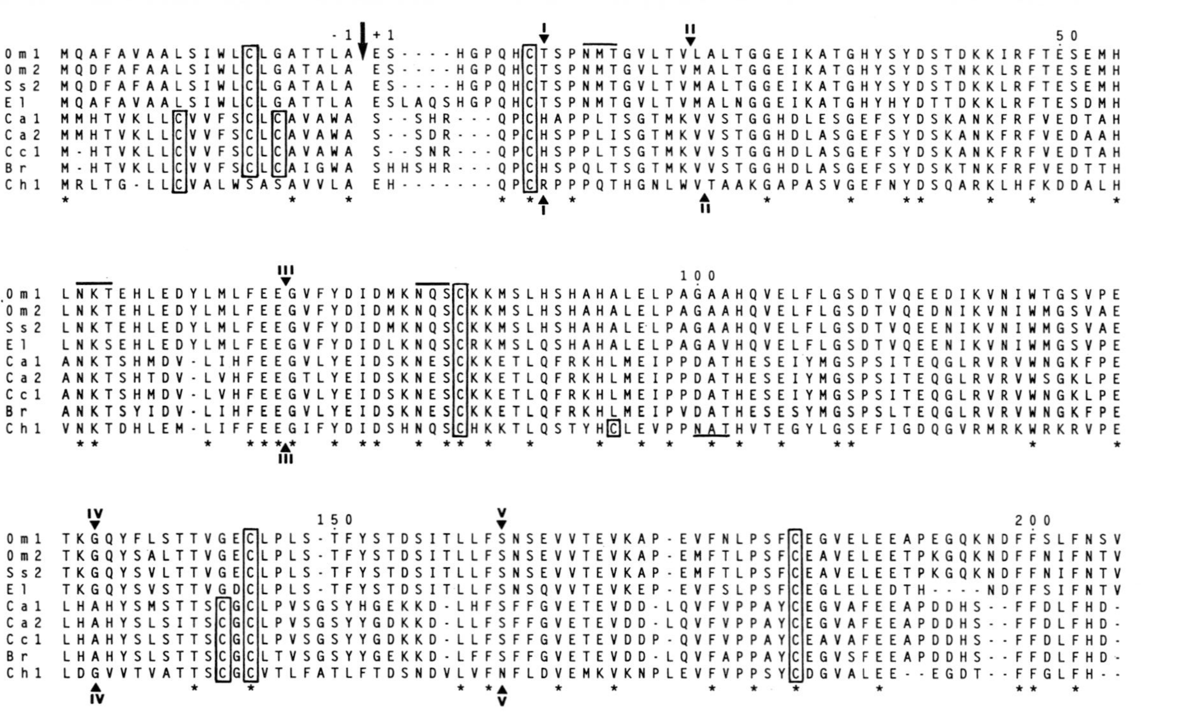


FIG. 2 Sequence comparison of all ependymin precursors characterized thus far. Om, *Oncorhynchus mykiss* (rainbow trout; Müller-Schmid *et al.,* 1992); Ss, *Salmo salar* (Atlantic salmon; Müller-Schmid *et al.,* 1992); El, *Esox lucius* (pike; Müller-Schmid *et al.,* 1993); Ca, *Carassius auratus* (goldfish; Königstorfer *et al.,* 1989a,); Cc, *Cyprinus carpio* (carp; Adams and Shashoua, 1994); Br, *Brachydanio rerio* (zebrafish; Sterrer *et al.,* 1990); Ch, *Clupea harengus* (herring; Müller-Schmid *et al.,* 1993). Gaps are introduced to maximize homologies and asterisks indicate invariant positions. The arrow represents the cleavage site for signal peptidase. Potential N-glycosylation sites are indicated by horizontal bars and cysteine residues are enclosed in boxes. The positions of the five introns are marked by triangles, either above (as determined for the salmoniform sequence Ss2; Müller-Schmid *et al.,* 1992) or below the lines (as determined for the cypriniform sequences Ca1, Cc1, and Br; Rinder *et al.,* 1992; Adams and Shashoua, 1994).

divergent (about 37–48% similarity between the different orders), with only a few strictly conserved features (Fig. 2). The most conserved region is at positions 68–77 (sequence F-E-E-G-x-x-Y-D/E-I-D) between two invariant potential N-glycosylation sites. Interestingly, this region is not encoded by a single exon but is split between exons 3 and 4. Thus far, there is no significant similarity of ependymins with any other known protein and the complete set of molecular data (Fig. 2) does not at all support a previous report speculating on similarities of ependymins with fibronectin, laminin, and so on (Shashoua *et al.*, 1990).

Mature salmoniform sequences (*Oncorhynchus mykiss, Salmo salar,* and *Esox lucius*) contain only four cysteine residues, whereas in cypriniform (*Carassius auratus, Cyprinus carpio,* and *Brachydanio rerio*) and clupeiform sequences (*Clupea harengus*) five or six residues are found, respectively. Thus, in salmoniform ependymins disulfide-linked dimerization is not observed after nonreducing SDS-PAGE, whereas cypriniform and clupeiform ependymins form dimers via an intermolecular cysteine bridge (Schmidt and Shashoua, 1981; Müller-Schmid, 1992), probably between the residues at position 142 (Fig. 2).

## C. Characterization of Mature Secretory Proteins

### 1. Glycosylation

All ependymin sequences characterized thus far contain two strictly conserved, potential N-glycosylation sites (Fig. 2). Salmoniform sequences bear one additional site. The virtual use of at least one of these sites is in agreement with the binding of ependymins to concanavalin A (Shashoua *et al.*, 1986).

Mature cypriniform (*C. auratus* and *B. rerio*) ependymins appear in two different molecular forms (termed $\beta$ and $\gamma$; apparent molecular mass on SDS-PAGE, 35 and 30 kDa; see Fig. 16) in about equal amounts. Both forms show complete immunological cross-reactivity and have nearly identical amino acid compositions (Schmidt and Shashoua, 1981, 1983). Furthermore, they contain the same amino- and carboxy-terminal sequences (Königstorfer *et al.*, 1989a,b). After digestion with glycopeptidase F, both forms can be shifted into a single band (Königstorfer *et al.*, 1989b; Sterrer *et al.*, 1990). This indicates clearly that ependymins $\beta$ and $\gamma$ represent just different glycoforms of the same protein backbone.

The existence of different glycosylation variants has also been demonstrated for salmoniform ependymins, where at least three glycoforms (termed A, B, and C) are observed (Müller-Schmid *et al.*, 1992, 1993). Here, form A represents the predominant form and microsequencing revealed the existence of additional glycoforms in the trout with a molecular mass up to 65 kDa. However, the high molecular mass forms are probably masked from immunological detection by their high sugar content (Müller-Schmid *et al.,* 1992).

Determination of the molecular mass of mature ependymins with matrix-assisted laser desorption/ionization mass spectroscopy (MALDI-MS; Stahl *et al.,* 1991) revealed a molecular mass of about 27.5 kDa for trout ependymin A, 26.6 kDa for goldfish ependymin $\beta$, and 24.4 kDa for goldfish ependymin $\gamma$ (Stahl B., Jan B. B., and Hoffmann, W., unpublished). This is considerably lower than the apparent molecular mass on SDS-PAGE (36, 35, and 30 kDa, respectively); thus, similar to a previous report (Shashoua, 1986), the sugar moiety in ependymins accounts only for about 10–20% of the molecular mass.

The composition of the N-linked sugar moiety of goldfish ependymins is complex, containing fucose, mannose, galactose, *N*-acetylglucosamine, sialic acid, and glucuronic acid (Shashoua *et al.,* 1986). The latter is responsible for the L2/HNK-1 immunoreactivity of goldfish ependymins (Shashoua *et al.,* 1986). However, this characteristic epitope is lacking in salmoniform ependymins (*O. mykiss* and *E. lucius*) and is present only in minute amounts in clupeiform ependymins (Ganß and Hoffmann, 1993). In contrast, terminal sialic acid residues seem to be one of the most conserved features of ependymins as monitored by reactivity with *Maackia amurensis* agglutinin (Ganß and Hoffmann, 1993). This lectin recognizes the structure Neu5Ac($\alpha$2–3)Gal. Interestingly, *Sambucus nigra* agglutinin, recognizing Neu5Ac($\alpha$2–6)Gal, shows only very weak reactivity. This would be an indication for "brain-type" N-glycosylation (Hoffmann *et al.,* 1994, 1995).

## 2. Microheterogeneities

Separation of goldfish ependymins by two-dimensional gel electrophoresis revealed the existence of at least seven acidic isoelectric variants (p*I* 5.0–5.6) for both the $\beta$ and $\gamma$ forms (Shashoua, 1986; Thormodsson *et al.,* 1992a, 1992b). A similar result has been obtained for the rainbow trout (Ganß, 1993). Thus far, the precise nature of these microheterogeneities is unknown. A certain degree of variability on amino acid level could originate from the existence of two highly homologous ependymin genes in goldfish (Königstorfer *et al.,* 1989b) and rainbow trout (Müller-Schmid *et al.,* 1992). This has been partially confirmed by direct microsequencing of goldfish ependymins, which revealed a series of point mutations (Königstorfer *et al.,* 1989a; Shashoua, 1991; Thormodsson *et al.,* 1992a). Since more than two variants have been identified, these may reflect additional allelic variations of the two ependymin loci. Alternatively, microheterogeneities could also arise from post-translational modifications (e.g., glycosylation). Data from Thormodsson *et al.* (1992b) would principally allow such an interpretation; here, deglycosylation resulted in the appearance of a new basic isoform. In contrast, Shashoua (1991) reported that the microheterogeneity persists after removal of the N-linked sugars. However, a report on phosphorylation as a posttranslational modification of ependymins (Nolan and Shashoua, 1989) could not be verified (Meyer L. M. G., Willy-Schmid, A., and Hoffmann, W., unpublished).

## D. Calcium-Binding and Conformational Transitions

Calcium ion-binding has been demonstrated unambiguously for goldfish and rainbow trout ependymins (Schmidt and Makiola, 1991; Ganß and Hoffmann, 1993). In the goldfish, two calcium-binding sites are believed to exist with reported dissociation constants of about $6 \times 10^{-7}$ and $2.5 \times 10^{-5}$ *M* (Schmidt and Makiola, 1991). Such values would be reasonable for extracellular proteins (extracellular $Ca^{2+}$concentration, $[Ca^{2+}]_o$, in teleost brain about $2.7 \times 10^{-3}$ *M*) but would not allow significant binding at intracellular $Ca^{2+}$ concentrations (namely, $10^{-8}$–$10^{-7}$ *M*).

Terminal sialic acids have been identified as the major $Ca^{2+}$-binding sites (Ganß and Hoffmann, 1993). Since these residues belong to the most conserved features of ependymins, one might also speculate on their importance for biological function.

This hypothesis on the functional relevance of sialic acid residues due to their $Ca^{2+}$-binding capacity is strengthened by the observation that binding of $Ca^{2+}$ induces a conformational change of ependymins. For example, circular dichroism measurements revealed that binding of $Ca^{2+}$ to ependymins alters their tertiary but not their secondary structure (Ganß and Hoffmann, 1993). Furthermore, monitoring the fluorescence of the single tryptophan residue of trout ependymins at various $Ca^{2+}$ concentrations indicated the existence of two states, namely, a holo ($Ca^{2+}$-loaded)-state and an apo ($Ca^{2+}$-free)-state. There is a relatively sharp conformational transition at about 1 m*M* $Ca^{2+}$ (Ganß, 1993), which is somewhat below $[Ca^{2+}]_o$. Thus, the holo-state can be considered as the native state of ependymins and $Ca^{2+}$ may trigger maintenance of this state similarly to SPARC/osteonectin (Maurer *et al.,* 1992). This may be essential for interaction with the extracellular matrix (ECM; see Section IV,B) or the reported aggregation to soluble multimers with a molecular mass ≥200 kDa (Thormodsson *et al.,* 1992b). Interestingly, binding of $Ca^{2+}$ to sialic acids has been shown to be important for various extracellular physiological processes (e.g., fibrin assembly) (Dang *et al.,* 1989).

Whether the apo-state is of any biological relevance is not known thus far. First, it is a question of debate if the reported fluctuations of $[Ca^{2+}]_o$ in the central nervous system (Benninger *et al.,* 1980; Nicholson, 1980; Pumain and Heinemann, 1985; Morris *et al.,* 1986) are sufficiently great as to affect ependymins. Second, the precise nature of this apo-state has not been clarified. However, in clear contrast to a previous report (Shashoua, 1988), all attempts failed to precipitate purified ependymins into insoluble aggregates solely by depletion of $Ca^{2+}$ (Ganß and Hoffmann, 1993). Consequently, the apo-state cannot be considered simply as an insoluble form of ependymins.

### E. Search for Mammalian Homologues

In the past, many reports claimed the existence of ependymin-like sequences in mammals (Schmidt *et al.*, 1986; Shashoua, 1986, 1991; Fazeli *et al.*, 1988; Shashoua *et al.*, 1990, 1992). However, the results of these publications are mainly based on immunohistochemical cross-reaction with antisera against goldfish ependymins and all these claims lack a solid molecular basis. Because ependymins are poorly conserved members of a highly divergent family of glycoproteins (Fig. 2) that only weakly cross-react immunologically (Müller-Schmid *et al.*, 1992), all these previous reports are questionable. Furthermore, reports on a neuronal expression of ependymin-like sequences in mammals (Schmidt *et al.*, 1986; Shashoua *et al.*, 1992) are in clear contrast with the nonneuronal, meningeal expression pattern observed in teleost fish (see Section III,B). However, for the future, it will certainly not be a trivial task to accumulate solid molecular evidence for the existence of ependymin-like proteins in mammals.

## III. From Synthesis to Localization: Cellular Analysis

Initial immunohistochemical studies in the goldfish brain reported on localization of these proteins typically at the ependymal zone lining the ventricles (Benowitz and Shashoua, 1977; Shashoua, 1977a). These first but incomplete investigations led to their name *ependymins* (Shashoua, 1977b). Later on, preliminary studies with cultures of zona ependyma cells (Majocha *et al.*, 1982) seemed also to indicate synthesis of ependymins in these cells. However, only molecular characterization via cDNA cloning and subsequent *in situ* hybridization studies revealed at first their real site of synthesis (i.e., meningeal fibroblasts of teleost fish) (Königstorfer *et al.*, 1990). Thus far, all attempts failed to detect expression of ependymins outside the brain or its coverings.

### A. Meninges of Teleost Fish

The microenvironment of vertebrate brain cells is formed by the cerebrospinal fluid (CSF) system, which is separated from the blood by a series of barriers (Cserr and Bundgaard, 1984). The evolution of this system is closely correlated with the development of the meninges (Hoffmann, 1993).

In fish, a basal *endomeninx* (also referred as to leptomeninx) covers the central nervous system and an *ectomeninx* is found in a subcranial position (van Gelderen, 1925). During phylogeny, the endomeninx presumably developed into the pia–arachnoid (or leptomeninges), whereas the ectomeninx probably became the precursor for the dura mater (Sagemehl, 1884; Nakao, 1979; Momose *et al.*, 1988). The space between the endomeninx and ectomeninx is filled with intermeningeal adipose tissue and perimeningeal fluid (PMF; also referred to as extradural fluid). Based on the different compositions of the PMF and CSF (Rasmussen and Rasmussen 1967; Cserr and Ostrach, 1974; Hoffmann, 1992; Müller-Schmid *et al.* 1992) and from tracer studies mainly with elasmobranchs (Zubrod and Rall, 1959; Klatzo and Steinwall, 1965; Bundgaard and Cserr, 1991), it has been concluded that the PMF is separated from the CSF compartment by a barrier in the endomeninx.

Detailed electron microscopy investigations revealed that the fish endomeninx is not as simple ("meninx primitiva"; Sterzi, 1902) as assumed in the past (Kappers, 1926). In teleost fish, three ultrastructural layers have been defined according to their proximity to the CNS (Figs. 3 and 4) (i.e., outer, intermediate, and inner layers) (Momose *et al.*, 1988; Caruncho *et al.*, 1993; Schwarz *et al.*, 1993; Wang *et al.*, 1995).

### 1. Outer Endomeningeal Layer

The outer endomeningeal layer is formed by several layers of relatively closely packed, flattened cells that are extremely electron dense. Gap junctions are formed between them and are also commonly seen at contact points with cells of the intermediate layer. Furthermore, cells of this layer are abundant in endoplasmic reticulum, indicating secretory activity (Momose *et al.*, 1988; Caruncho *et al.*, 1993; Schwarz *et al.*, 1993).

### 2. Intermediate Endomeningeal Layer

Earlier studies (Klika, 1967; Klika and Zajícová, 1975) did not describe the single-cell intermediate endomeningeal layer characteristic of teleost fish. It consists of long flat cells, which are enriched in intermediate filaments (Fig. 5); typically, these filaments confer mechanical stability and are found in cells of protective sheaths (e.g., the mammalian arachnoid) (Frank *et al.*, 1983; Achtstätter *et al.*, 1989). Generally, these cells exhibit morphological characteristics similar to endothelial cells. For example, they are nearly devoid of endoplasmic reticulum and Golgi apparatus but contain numerous mitochondria. Furthermore, these cells are abundant in pinocytotic vesicles

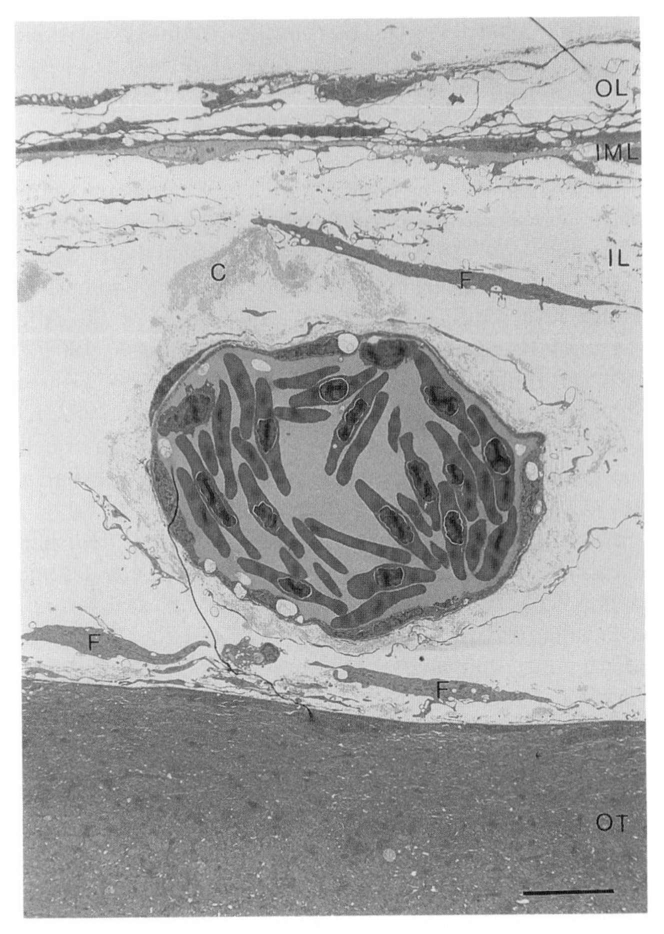


FIG. 3 Ultrastructure of the endomeninx from the rainbow trout. Visible are the electron-dense outer layer (OL), the intermediate layer (IML), and the inner layer (IL). The IL contains fibroblast-like cells (F), collagen fibrils (C), and blood vessels. Bar: 10 $\mu$m.

(Fig. 6) and vesicle openings have been characterized toward the surfaces of the adjacent layers (i.e., the outer and inner layers) (Momose *et al.,* 1988; Caruncho *et al.,* 1993; Schwarz et al., 1993). In contrast, lateral membranes between two intermediate layer cells are lacking pinocytotic vesicles. This indicates that this specialized cell may allow active transcellular transport between the outer and inner endomeningeal layer by means of transcytosis (Momose *et al.,* 1988; Caruncho *et al.,* 1993; Schwarz *et al.,* 1993).

This view is supported by the observation that cells of the intermediate layer contain well-developed junctional complexes separating the inner

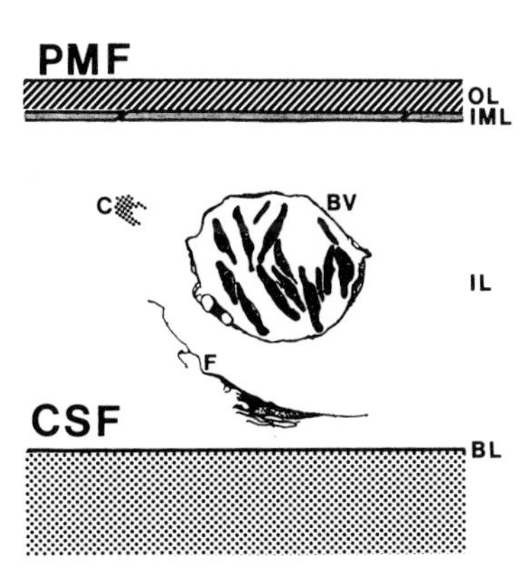


FIG. 4 Schematic representation of the endomeninx from teleost fish, consisting of the outer layer (OL), the intermediate layer (IML), and the inner layer (IL). The IL contains blood vessels (BV), fibroblast-like cells (F), and collagen fibrils (C). The barrier of the IML separates the PMF compartment from the CSF. BL, Basal lamina covering the neuropil.

layer from the outer layer. Tight junctions are found between cells of the intermediate layer whereas gap junctions occur between heterologous cells (i.e., intermediate and outer or inner layer cells, respectively) (Momose *et al.,* 1988; Caruncho *et al.,* 1993; Schwarz *et al.,* 1993). The tight junctions form the basis for the endomeningeal barrier separating the PMF from the CSF compartments (Fig. 4). In mammals, the arachnoid is the equivalent structure.

Functionally similar barrier layers of arachnoid-like cells have also been demonstrated in cyclostomes and elasmobranchs. However, in these classes of fishes the morphology of the meningeal barrier resembles a variation of the special situation described in teleosts. For example, in elasmobranchs all arachnoid-like cells are part of the barrier (i.e., an epithelium 10–15 cells thick; Bundgaard and Cserr, 1991) and in cyclostomes two sheets of arachnoid-like cells (outer part of layer II; Nakao, 1979) are held together by tight junctions. Generally, the situation is comparable with that in mammals. Thus, despite differences in the CSF system between fish and higher vertebrates, the meningeal barrier is an absolutely conserved feature during phylogeny.

Taken together, the specialized intermediate endomeningeal layer of teleost fish seems to have a crucial function for homeostasis of the brain. On the one hand, it forms a tight barrier separating the CNS from its surroundings (e.g., the PMF). On the other hand, active transcellular transport phenomena (transcytosis) may allow regulated communication of both sides controlled by meningeal cells. A similar transport phenomenon may

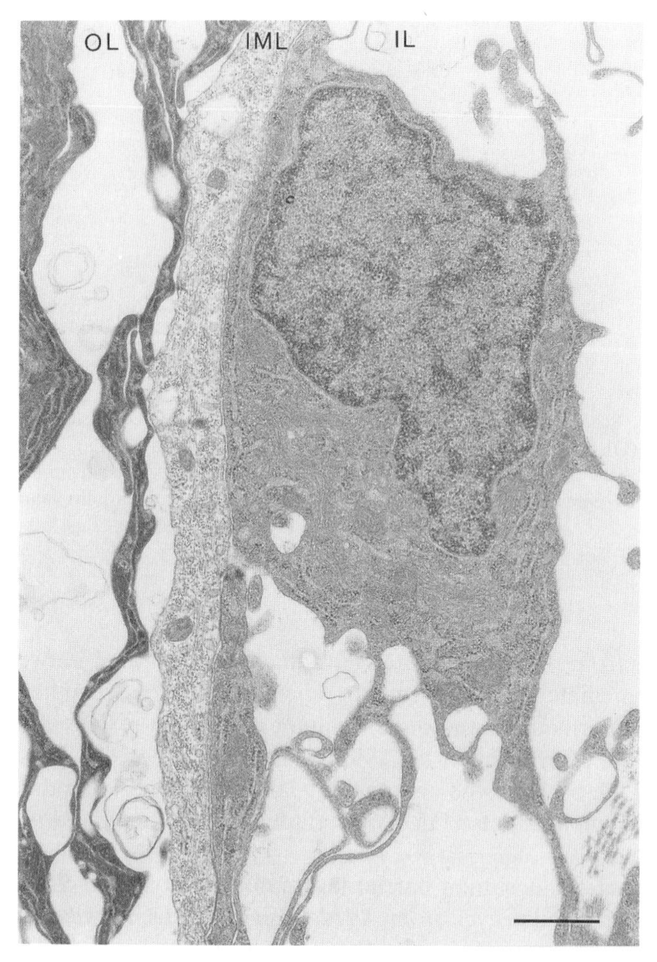


FIG. 5 Ultrastructure of the endomeningeal barrier of the rainbow trout, showing electron-dense cells of the outer layer (OL), a cell of the intermediate layer (IML), and a fibroblast of the inner layer (IL). Bar: 1 $\mu$m.

also occur in mammalian arachnoid cells, where numerous pinocytotic vesicles are present (Krisch, 1988). Interestingly, in fish barrier and transport functions seem to be subject to alterations during adaptative experiments; for example, morphological changes in junctional complexes, as well as changes in the density of plasmalemmal vesicles, have been reported to occur in the intermediate endomeningeal layer of the goldfish (Caruncho and Pinto da Silva, 1994).

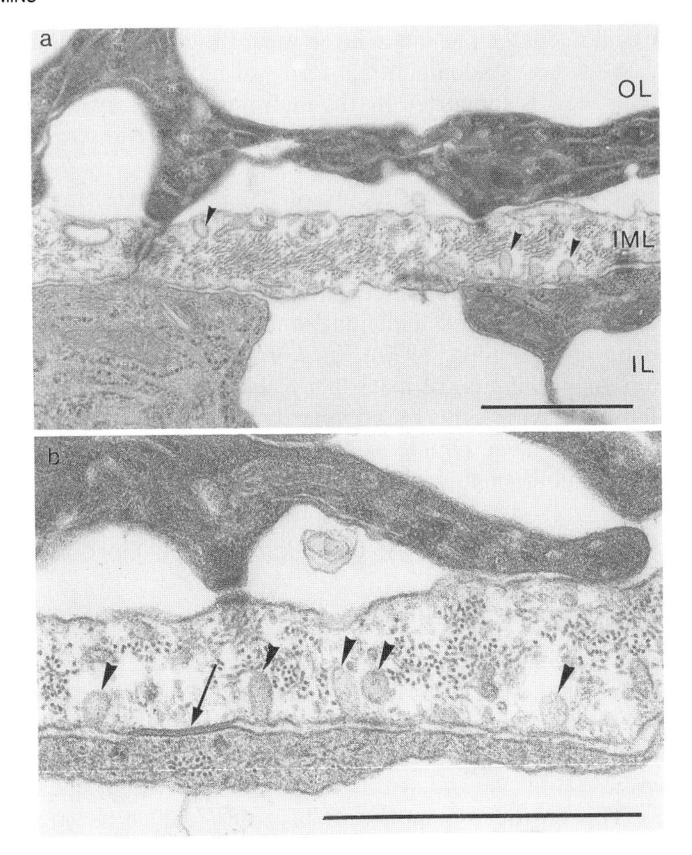


FIG. 6 Ultrastructure of the intermediate endomeningeal layer (IML) of the rainbow trout. Desmosomes are visible between cells of the IML as well as between heterologous cells of the intermediate and the outer layer (OL) and the intermediate and the inner layer (IL). (a and b) Cells of the IML contain many intermediate filaments and pinocytotic vesicles (arrowheads). (b) A typical gap junction between heterologous cells is indicated by an arrow. Bars: 1 $\mu$m.

### 3. Inner Endomeningeal Layer

Fibroblast-like elongated cells of the inner layer partly adhere to the intermediate layer via tight junctions, gap junctions, and small desmosomes (Momose *et al.,* 1988; Caruncho *et al.,* 1993; Schwarz *et al.,* 1993). Another fraction of these spindle-shaped cells is detached from both the intermediate layer and the basal lamina covering the CNS. This loose network of cells contains wide intercellular spaces, which expand with age (Wang *et al.,* 1995). Based on their abundance in rough endoplasmic reticulum, the fibroblast-like cells of the inner layer must be considered the most active cells of the teleost endomeninx in terms of biosynthesis of secretory proteins. This view is supported by the high number of nuclear pores in cells of this layer, which is correlated with the level of transcriptional activity (Caruncho *et al.,* 1993).

Other characteristic constituents of the inner layer are blood vessels tangential to the surface of the CNS (in contrast to the predominantly radial orientation of the capillaries in the neuropil). In the rainbow trout, meningeal capillaries are unfenestrated; endothelial cells contain many pinocytotic caveolae and tight junctions are present between these cells (Caruncho and Anadon, 1990). Furthermore, bundles of collagen fibrils and a minor population of macrophage-like cells containing vacuoles are typically found within this broad inner layer (Fig. 3). Generally, the inner endomeningeal layer of teleost fish is topologically homologous to the mammalian subarachnoid space.

## B. Expression of Ependymins by Endomeningeal Fibroblasts

### 1. *In Situ* Hybridization Studies

By way of *in situ* hybridization, cells of the fish endomeninx were unambiguously defined at first as the principal site of synthesis of ependymins. Generally, endomeningeal cells covering the surface of the brain as well as various invaginated parts of the endomeninx (e.g., the cavum cranii) synthesize ependymins. However, no synthesis can be observed at ependymal zones (Fig. 7). This has been demonstrated for the goldfish (Königstorfer *et al.,* 1990), the zebrafish (Sterrer *et al.,* 1990), and the rainbow trout (Müller-Schmid *et al.,* 1992).

### 2. Transient Expression Studies in Zebrafish Embryos

A typical endomeningeal expression pattern has also been obtained when a reporter gene, under the control of the zebrafish ependymin promoter, was transiently expressed in zebrafish embryos (Rinder *et al.,* 1992). This result is in perfect agreement with *in situ* hybridization histochemistry. Thus, endomeningeal cells must be considered the characteristic site of ependymin biosynthesis.

### 3. Electron Microscopy Studies

To answer the question concerning which endomeningeal layer contains ependymin-secreting cells, immunoelectron microscopy investigations wereperformed with the rainbow trout (Schwarz *et al.,* 1993). Here, the biosynthesis could be defined precisely to the characteristic fibroblast-likecells of the inner endomeningeal layer, but not at all to outer layer

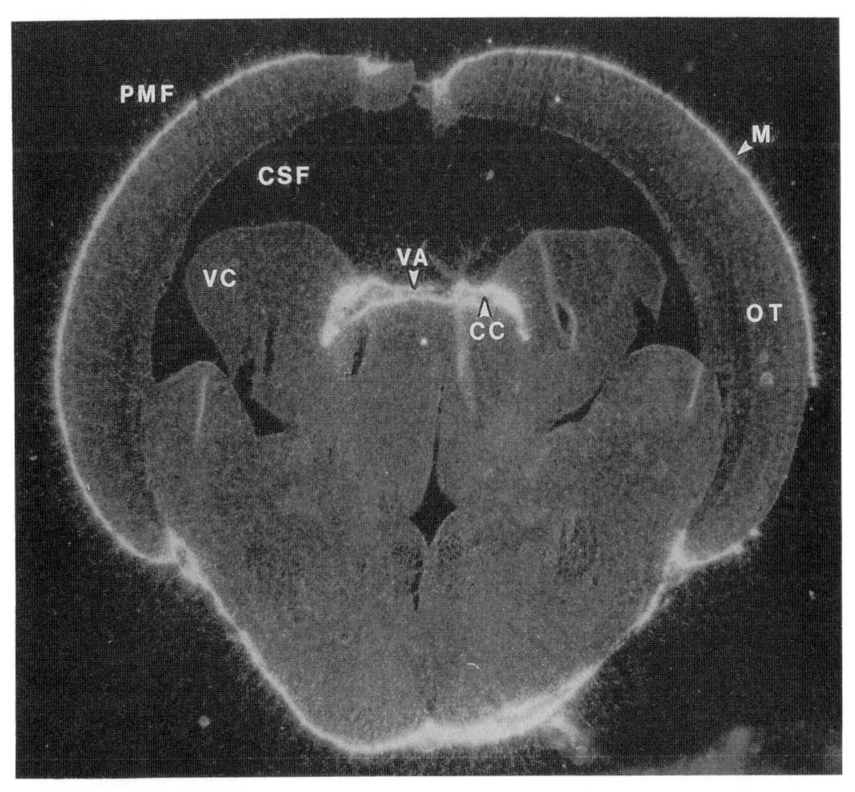


FIG. 7 *In situ* hybridization of a transverse section through the optic tectum of goldfish brain, using an ependymin-specific $^{32}$P-labeled oligonucleotide as probe. The dark-field illumination outlines ependymin expression solely in the endomeninx (M), including its invaginations, e.g., the cavum cranii (CC). CSF, Cerebrospinal fluid; OT, optic tectum; PMF, perimeningeal fluid; VC, valvula cerebelli; VA, velum anticum.

cells (Fig. 8). As is typical of secretory proteins, ependymins were clearly localized within the cisternae of the rough endoplasmic reticulum and the Golgi apparatus of fibroblasts of the inner layer (Fig. 9). A similar situation has also been reported for the goldfish (Schmidt *et al.,* 1992).

Furthermore, ependymins appear within cells of the intermediate endomeningeal layer (Fig. 10). Because cells of this layer are certainly not active in protein secretion (the cells are sparse in rough endoplasmic reticulum and contain the lowest number of nuclear pores of all endomeningeal layers; Caruncho *et al.,* 1993) but contain numerous pinocytotic vesicles, it seems likely that the presence of positive ependymin signals is not due to their active synthesis but rather to an active transport process (transcytosis). However, a definite answer cannot be given yet.

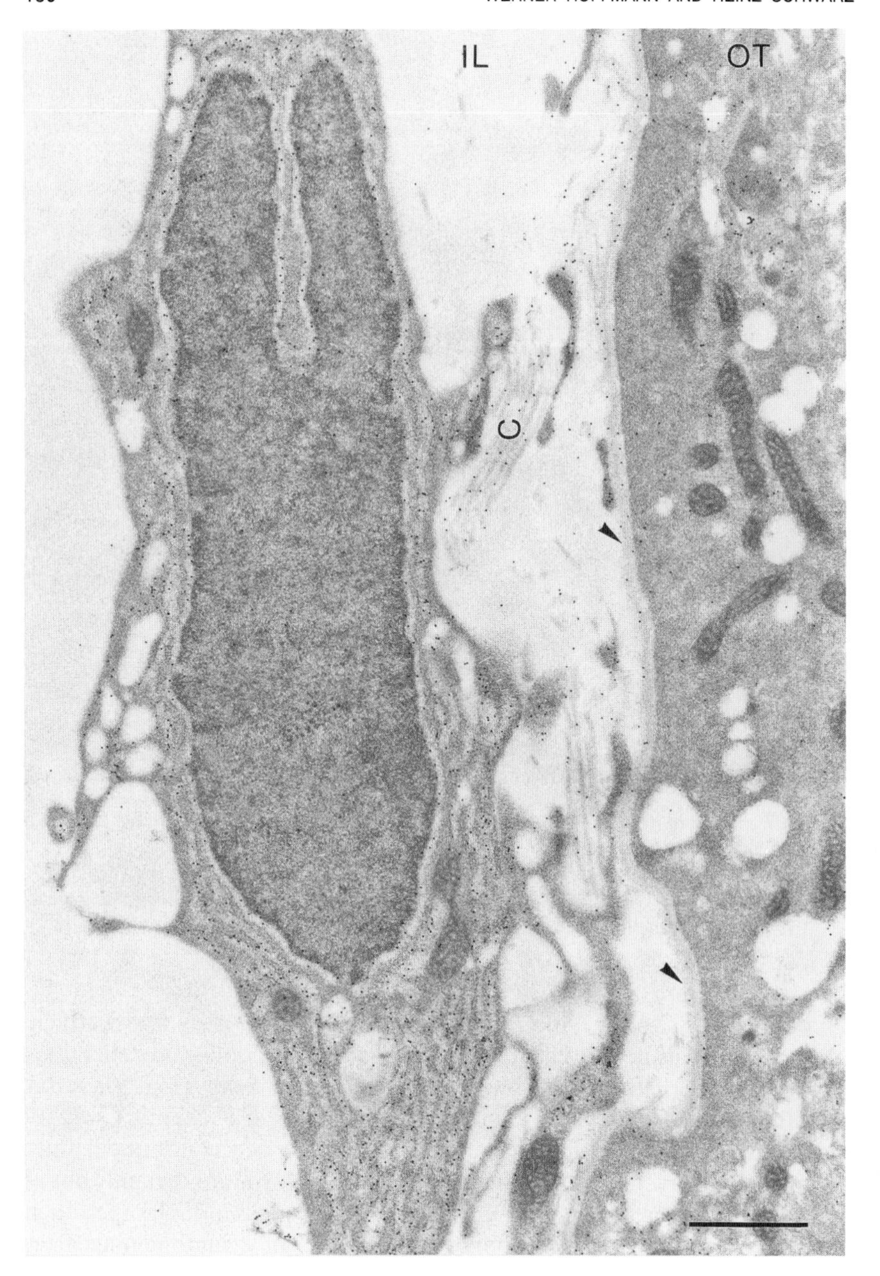


FIG. 8 Utrastructural localization of ependymins in the inner endomeningeal layer (IL) of the rainbow trout. The rough endoplasmic reticulum of a fibroblast-like cell shows pronounced immunostaining. Gold particles are also visible at collagen fibrils (C). The basal lamina covering the optic tectum (OT) is indicated by arrowheads. Bar: 1 $\mu$m.

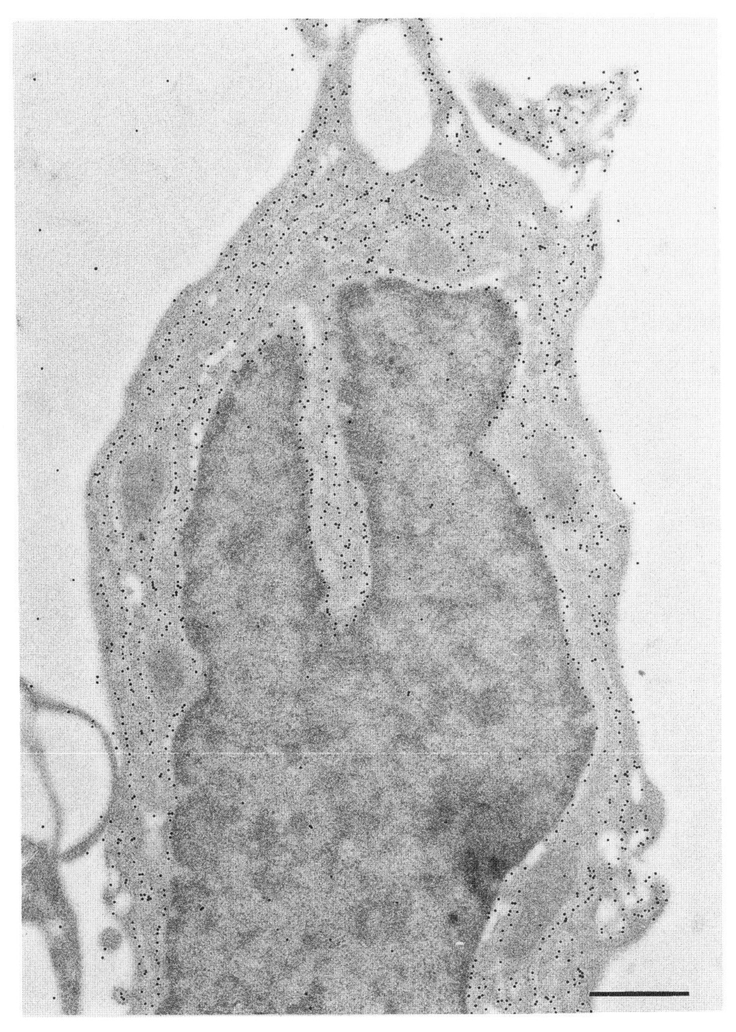

FIG. 9 Localization of ependymins in a fibroblast-like cell of the inner layer of the trout endomeninx. Gold particles at the rough endoplasmic reticulum and the Golgi apparatus indicate biosynthesis of ependymins within this cell. Bar: 1 $\mu$m.

## C. Complex Localization Pattern of Ependymins

In contrast to the relatively simple expression pattern of ependymins, their localization is more complicated. This is certainly due to the secretory nature of ependymins and their subsequent distribution by various extracellular fluids (see Section III,D).

In the past, many localization studies were performed using conventional immunohistochemistry, enzyme-linked immunosorbent assay (ELISA)

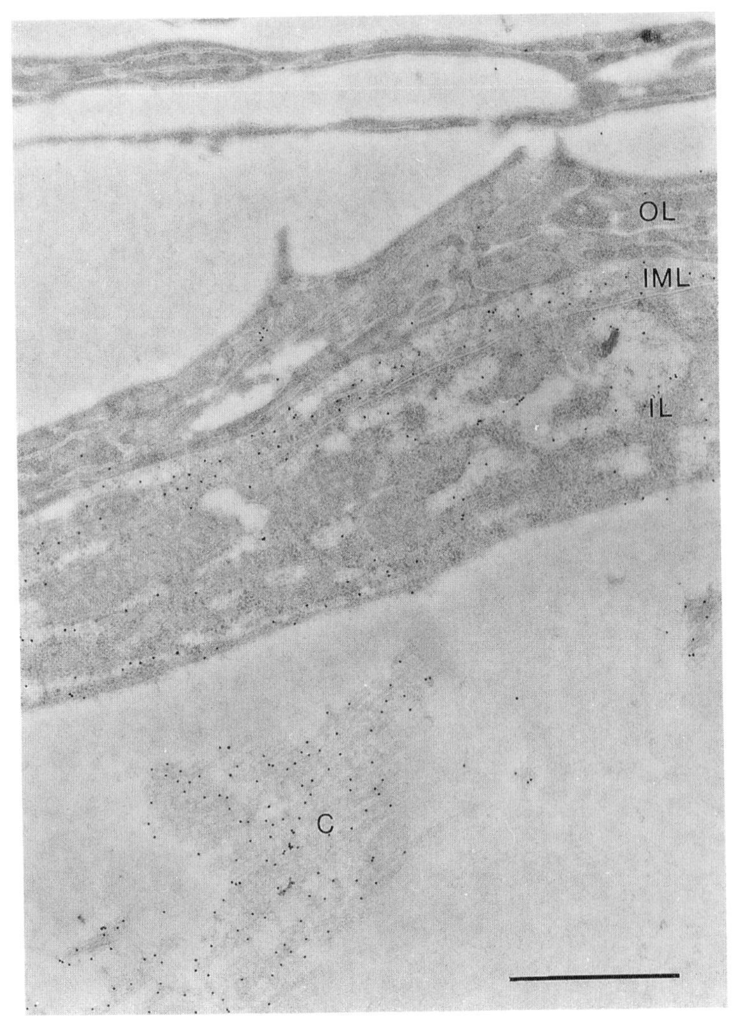


FIG. 10 Ultrastructural localization of ependymins in cells of the intermediate (IML) and inner layer (IL) of the trout endomeninx. Immunostaining is also observed at collagen fibrils (C) of the IL; cells of the outer layer (OL) are devoid of gold particles. Bar: 1 $\mu$m.

techniques, and immunoelectron microscopy (Benowitz and Shashoua, 1977; Shashoua, 1977a; Schmidt and Lapp, 1987; Schmidt and Shashoua, 1988; Schmidt, 1989; Königstorfer *et al.*, 1990; Hoffmann *et al.*, 1992; Schmidt *et al.*, 1992; Schwarz *et al.*, 1993; Lakos *et al.*, 1994). However, most publications dealt only with specific aspects of ependymin localization and real systematic studies on cellular level were rare. Thus, only recently has a conclusive picture emerged.

Figure 11 (see color insert) represents a characteristic overview of the localization of ependymins in the endomeninx and the underlaying neuropil of the rainbow trout. Clearly, different regions can be distinguished. (1) The outer layer is devoid of ependymins; (2) the intermediate and inner layers unquestionably contain the highest concentrations of ependymins; (3) in the neuropil underlaying the basal lamina the concentration of ependymins decreases continously; and (4) blood vessels are always surrounded by a sheath of ependymins, even within the neuropil.

### 1. Intermediate Layer Cells

Within the intermediate layer cells, ependymins are abundant (Fig. 10; Schwarz *et al.,* 1993). Thus far, it is not completely understood if this is due to their synthesis or to a transcellular transport phenomenon (transcytosis; see Section III,B,3). However, the latter seems to be more compatible with the morphology of these cells; furthermore, the distribution of ependymins within various brain fluids also favors the transcytosis hypothesis (see Section III,D,2).

### 2. Fibroblasts and Macrophages of Inner Endomeningeal Layer

Immunoelectron microscopy studies helped to define the precise localization within the inner endomeningeal layer; in addition to their direct place of synthesis (i.e., fibroblast-like cells; see Figs. 8 and 9 and Section III,B,3), ependymins also appear within vacuoles of flattened macrophage-like cells often associated with fibroblasts (Fig. 12; Schwarz *et al.,* 1993). This could represent a clearance pathway for ependymins and may be responsible for their reported high turnover rate (Shashoua, 1981; Schmidt and Lapp, 1987). In the past, a similar cell type has been described in the mammalian pia mater (Morse and Low, 1972).

### 3. Extracellular Matrix, Blood Vessels

A bound form of ependymins appears to be a constituent of the extracellular matrix, mainly of the inner endomeningeal layer. Here, specific association with bundles of collagen fibrils is striking (Schwarz *et al.,* 1993). These fibrils are part of the wide extracellular space of the inner endomeningeal layer (Figs. 8 and 10). Similar collagen fibrils have also been reported to be a typical part of the extracellular space of mammalian leptomeninges (Morse *et al.,* 1972; Oda and Nakanishi, 1984; Haines, 1991).

Furthermore, ependymin-decorated collagen fibrils cover the endothelial cells of the numerous blood vessels (Fig. 13; Schwarz *et al.,* 1993). This also explains the prominent staining of brain capillaries obtained by conventional immunofluorescence histochemistry (Fig. 14; Königstorfer *et al.,* 1990; Lakos *et al.,* 1994).

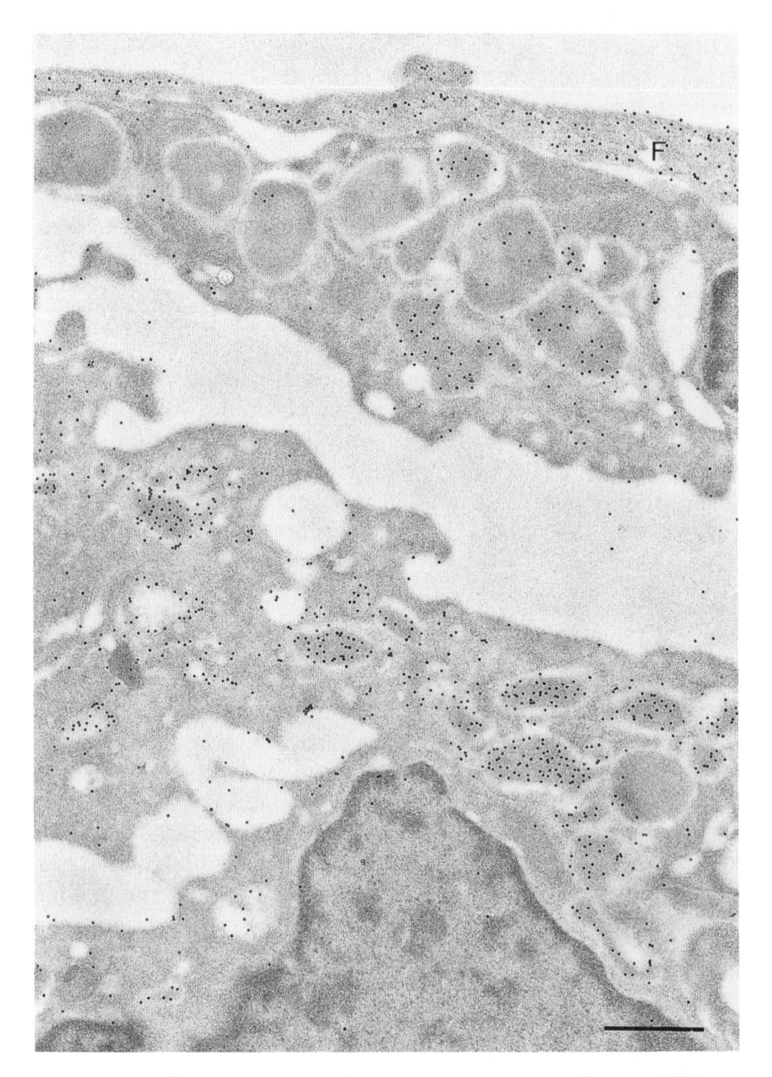


FIG. 12 Ultrastructural localization of ependymins in macrophage-like cells of the inner endomeningeal layer of the rainbow trout. The gold label is restricted to distinct vacuoles. In contrast, in the fibroblast-like cell (F) the endoplasmic reticulum is labeled. Bar: 1 $\mu$m.

### 4. Granular Cells of Periventricular Zone, Ependymal Cells, and Lymphocytes

Original localization within a "broad ependymal zone" of the optic tectum and vagal lobes (Benowitz and Shashoua, 1977; Shashoua, 1977a) obviously provided the historical basis for the name "ependymins" (Shashoua, 1977b). Later on, this area was defined as layer 2 of the periventricular gray zone

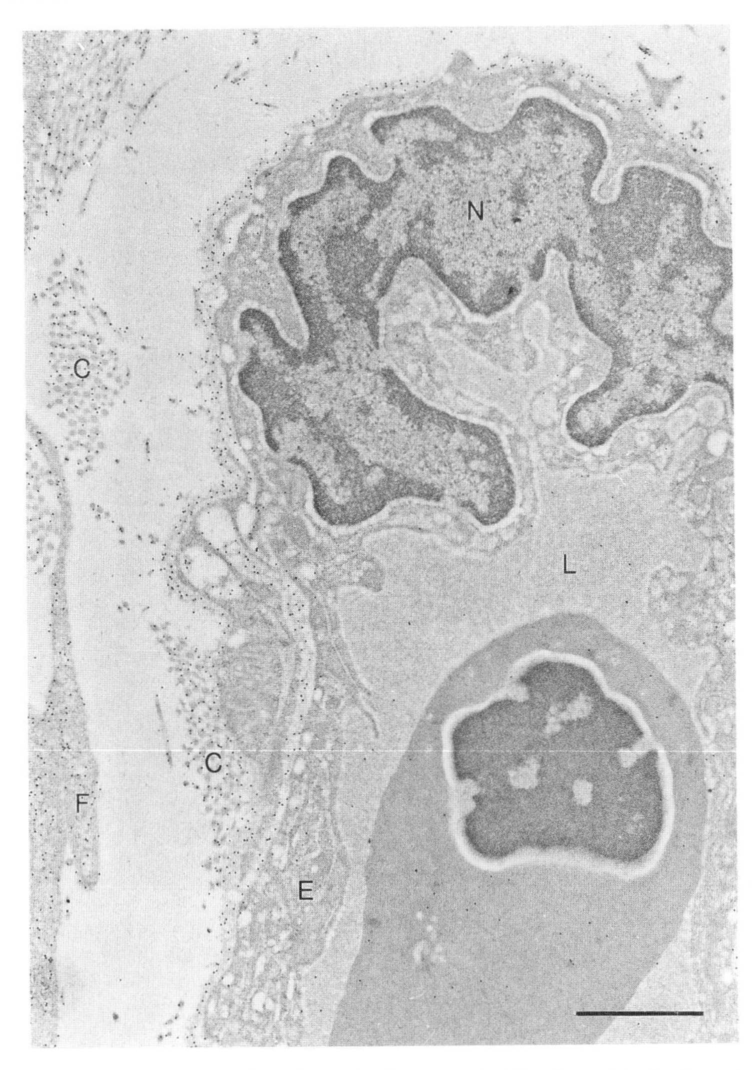


FIG. 13 Extracellular distribution of ependymins around a blood vessel in the inner layer of the trout endomeninx. Gold particles are restricted to collagen fibrils (C); endothelial cells (E) and the lumen (L) of the blood vessel are free of label. N, Nucleus of endothelial cell; F, fibroblast-like cell. Bar: 1 $\mu$m.

(terminology according to Northcutt, 1983) of the optic tectum (Königstorfer *et al.*, 1990; Lakos *et al.*, 1994). These immunopositive cells have a granular appearance and are different from the ependymal cells lining the ventricle. However, synthesis of ependymins by these cells could not be detected (Fig. 7). Occasionally, granular cells are also found near the meningeal surface of the

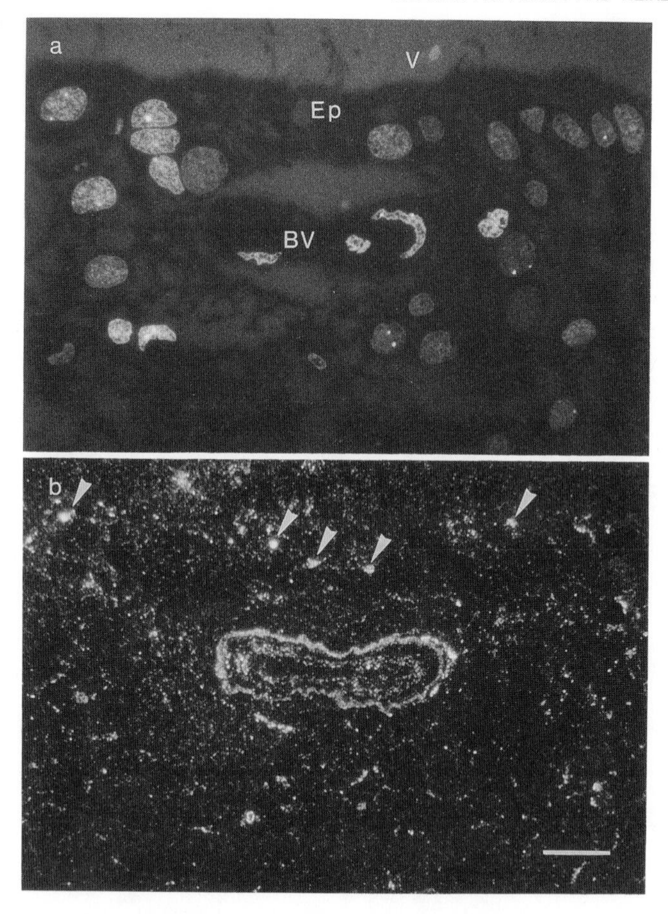


FIG. 14 Immunohistochemical localization of ependymins in the periventricular zone. (a) Counterstaining with DAPI to visualize the DNA. V, Ventricle; Ep, ependymal cell layer; BV, blood vessel. (b) Immunofluorescence indicating localization of ependymins at distinct sites in the ependyma (arrowheads) and around a blood vessel. Bar: 10 $\mu$m.

optic tectum. Generally, the biological role of these cells is not clear and a relation to macrophages or lymphocytes has been discussed (Lakos *et al.,* 1994).

Ependymal cells also contain a small amount of ependymins, probably as a result of an endocytotic process (Figs. 14 and 15). Such a pinocytotic clearance mechanism would remove ependymins from the CSF (see Section III,D,1) and would be in agreement with the general resorptive capacity of ependymal cells (e.g., as shown for ferritin) (Brightman, 1965).

More recently lymphocytes have been reported to be immunopositive for ependymins (Lakos *et al.,* 1994).

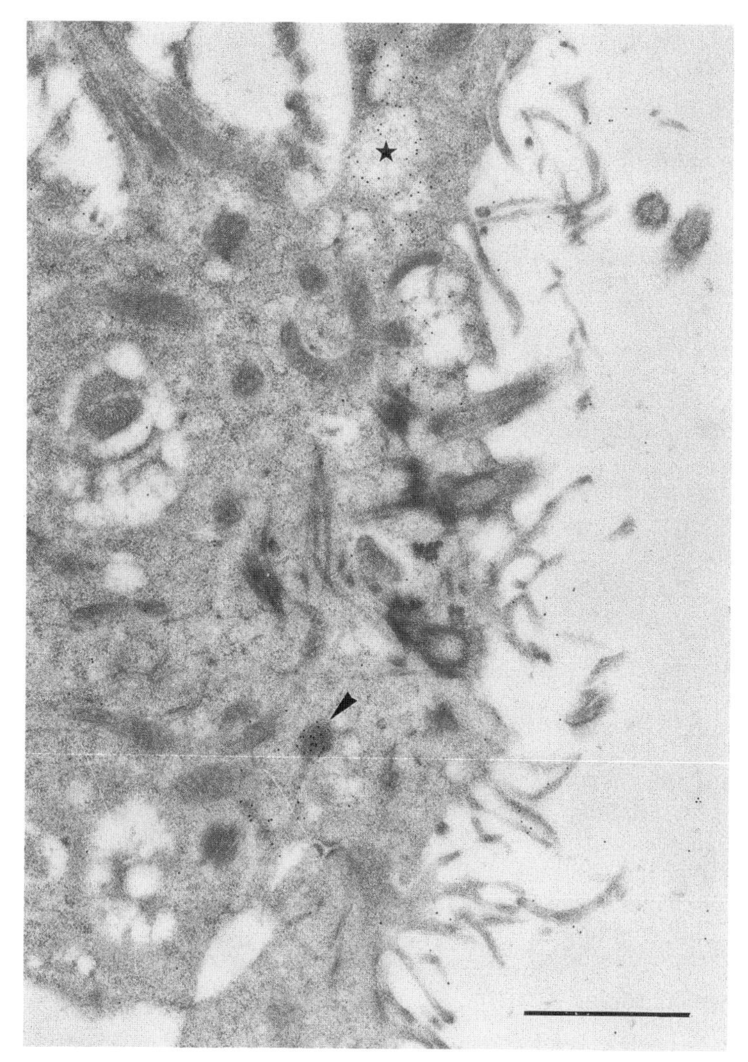

FIG. 15 Ultrastructural localization of ependymins in ciliated ependymal cells of the rainbow trout. Gold particles are present within a vesicle (star) and an electron-dense granule (arrowhead) of a cell facing the optic ventricle. Bar: 1 $\mu$m.

## 5. Choroid Plexus, Saccus Dorsalis, Saccus Vasculosus, and Reissner's Fiber

Ependymins have also been immunologically localized in the choroid plexus and the saccus dorsalis (Lakos *et al.*, 1994), as well as in the saccus vasculosus (Müller-Schmid, 1992). In the latter, coronet cells directly facing the CSF are intensely stained. However, *in situ* hybridizations revealed that there is no active biosynthesis of ependymins within these cells (Müller-Schmid, 1992). This favors the hypothesis that ependymins may be taken up by coronet cells either for active transcellular transport or for proteolytic degradation.

Intense staining of Reissner's fiber (Lakos *et al.*, 1994) is probably due to the high concentration of ependymins in the CSF and subsequent unspecific adsorption to glycoproteins of this fiber. However, molecular analysis of Reissner's fiber did not show similarity to ependymins (A. Meiniel, personal communication).

### 6. Neuronal Cells

In spite of repeated reports of ependymin immunoreactivity in piscine neuronal cells ( Schmidt, 1989; Schmidt *et al.* 1992), previous (Benowitz and Shashoua, 1977) and more recent studies (Lakos *et al.*, 1994) clearly denied any ependymin immunoreactivity of these cells. Lakos *et al.* (1994) instead described nonspecific artifactual staining of neuronal cells with ependymin antisera.

Furthermore, the report on ependymin-like proteins in mammalian neuronal cells (Schmidt *et al.*, 1986) is likely the result of unspecific immunological staining.

## D. Distribution of Ependymins within Brain Fluids

In agreement with their synthesis as typical secretory proteins, ependymins are found extracellularly. A bound form is associated with collagen fibrils of the extracellular matrix (see Section III,C,3), whereas soluble ependymins are found in the CSF, the PMF, and the blood, but to various extents (Fig. 16). They are heavily enriched in the CSF, still abundant in PMF, and nearly absent in blood.

### 1. Cerebrospinal Fluid

In many teleost fish of the orders Cypriniformes, Salmoniformes, and Clupeiformes, ependymins appear to be the predominant constituents of the CSF (Schmidt and Shashoua, 1981; Shashoua, 1981; Schmidt and Lapp, 1987; Hoffmann, 1992, 1993; Müller-Schmid *et al.*, 1992, 1993). In these orders of fish ependymins might be involved in $Ca^{2+}$ homeostasis of the brain. For example, two-thirds of $Ca^{2+}$ in the CSF of the rainbow trout is protein bound, presumably to ependymins (Ganβ and Hoffmann, 1993). In all other orders of fish investigated, ependymin-like proteins have not been detected thus far. At the moment, the reason for these enormous variations in piscine CSF is unknown. Ependymins are obviously responsible for the unusually high protein concentration in these three orders of teleost fish, where values are comparable with protein levels in the blood

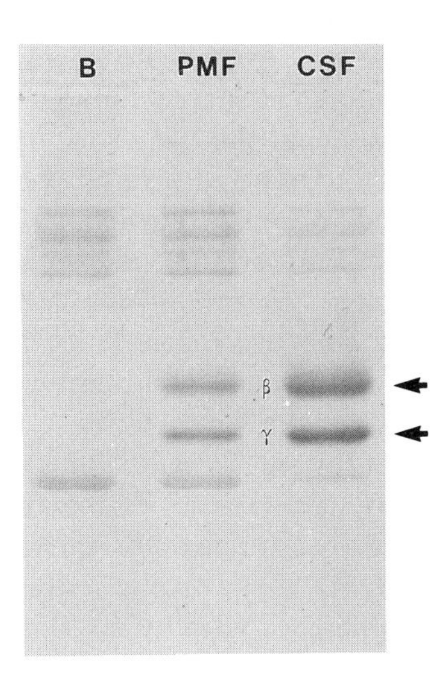


FIG. 16 Coomassie stained SDS-polyacrylamide gel of blood (B), PMF, and CSF of the goldfish. The predominant bands in the CSF represent ependymins $\beta$ and $\gamma$.

(Zucht and Rahmann, 1974; Shashoua, 1981; Müller-Schmid *et al.,* 1992; Wang *et al.,* 1995). In contrast, in most other orders of fish and in mammalian CSF, the protein concentration is diminished by at least one order of magnitude (Rasmussen and Rasmussen, 1967; Cserr *et al.,* 1972).

Ependymins in the CSF are thought to originate from their synthesis in the inner endomeningeal layer. This would strongly imply that the inner endomeningeal layer of teleost fish with its wide extracellular spaces is the equivalent to the mammalian subarachnoid space filled with external CSF. Thus far, it is unknown how internal ventricular CSF and subarachnoid fluid would communicate, because in teleost fish no macroscopic openings for drainage of ventricular CSF into the subarachnoid space have been identified (Jones, 1979). It may be that, from an invaginated part of the valvula cerebelli, the cavum cranii (Fig. 7), ependymins can be released into the CSF; here, the inner endomeningeal layer is separated from the CSF of the optic ventricles only by a delicate membrane, the velum anticum (terminology according to Franz, 1911; see Fig. 7). Furthermore, coronet cells of the saccus vasculosus have been proposed to be responsible for active transport processes between the CSF and the "endomeningeal space filled with fluid functionally similar to a liquor externus"(Jansen and Flight, 1969). Indeed, these cells stain intensely with an ependymin antiserum (Müller-Schmid, 1992). The subsequent distribution of CSF within the brain parenchyma occurs probably via the perivascular spaces (van Rijssel, 1946), similar to mammals (Rennels *et al.,* 1985; Ichimura *et al.,* 1991).

In the past, the view that the CSF system of teleost fish is comparable to the mammalian situation containing a subarachnoid fluid has been denied repeatedly. However, a subarachnoid fluid is already present in elasmobranchs (Bundgaard and Cserr, 1991) and a large pore in the posterior choroid plexus has been found to connect the subarachnoid space and ventricular system (Butt, 1987). Furthermore, also in amphibians a subarachnoid space (Cserr and Ostrach, 1974) as well as microscopic connections with the CSF have also been detected (Jones, 1978). During phylogeny, the foramina of Magendie and Luschka finally emerged as macroscopic connections between ventricular and external CSF.

The synthesis of ependymins by meningeal cells is rather unusual for major CSF proteins. Generally, constituents of mammalian CSF are of hepatic origin and enter the CSF by ultrafiltration according to their hydrodynamic radii (Felgenhauer, 1974). Alternatively, active synthesis of a variety of CSF proteins occurs in the choroid plexus (Aldred *et al.,* 1995). In spite of repeated discussions on extrachoroidal formation of CSF (Cserr, 1971; Milhorat, 1976), only recently have the leptomeninges been identified on a molecular level as a potent secretory organ for many CSF proteins (see Section IV,A). Originally, this had already been postulated by Magendie (1843).

### 2. Perimeningeal Fluid

Perimeningeal fluid fills the space between the cranium and the outer endomeningeal layer. The constitution of PMF differs clearly from that of CSF (Fig. 16). This is in perfect agreement with the view that there is no open communication between these two fluids due to the endomeningeal barrier (intermediate layer). Perimeningeal fluid seems to be derived mainly from blood plasma (Jones, 1979) but also contains ependymins (Fig. 16; Shashoua, 1981; Hoffmann, 1992, 1993; Müller-Schmid *et al.,* 1992, 1993).

The question arises as to how ependymins get into the PMF, since there is no synthesis observed in the outer endomeningeal layer (Fig. 11). Generally, possible pathways include either direct secretion or active transcellular transport or unspecific drainage from the CSF. For example, secretion by intermediate layer cells is not favored because of morphological characteristics (see Section III,B,3). However, there are indications of active transcellular transport via intermediate layer cells. In particular, the numerous pinocytotic vesicles may account for such a transcytosis process (Fig. 6). Furthermore, such a clearance mechanism (ependymins have a relatively short half-life time of about 2 hr) (Shashoua, 1981; Schmidt and Lapp, 1987) would be compatible with absorption of mammalian CSF through the Pacchionian granulations, which are abundant in pinocytotic vesicles (Krisch, 1988).

### 3. Blood

In contrast to CSF and PMF, blood serum is nearly devoid of ependymins (Fig. 16; Schmidt and Shashoua, 1981; Shashoua, 1981; Schmidt and Lapp, 1987; Hoffmann, 1992, 1993; Müller-Schmid *et al.*, 1992, 1993). This is in agreement with the observation that ependymins are not actively synthesized in the liver (Müller-Schmid, 1992). Furthermore, they are excluded from the blood by the blood–brain barrier (Lundquist, 1942; Bernstein and Streicher, 1965). However, little amounts may reach the blood stream via the PMF or by drainage via the saccus vasculosus (open connection between blood sinus and endomeningeal space; Jansen and Flight, 1969).

## IV. Functional Aspects

Based on the molecular and cellular data on biosynthesis and localization, there is certainly no reason to assume that ependymins may have a specific function for learning and memory. Consequently, their initial discovery due to enhanced turnover rates in conditioning experiments should be correlated with the fact that the piscine meninx (i.e., the place of ependymin synthesis) seems to fulfill a pivotal role for energy metabolism of the brain (Rovainen, 1970; Rovainen *et al.*, 1971); thus, fish endomeningeal cells may easily show generally enhanced metabolic activity in response to a variety of stress situations. Interestingly, alterations at least in the intermediate layer of goldfish meninges have been reported to occur during adaptative experiments (Caruncho and Pinto da Silva, 1994).

### A. Leptomeninges as Secretory Organ

Surprisingly, the importance of the meninges for the CNS was not taken into account for a long time. In fact, currently the secretory nature of the leptomeninges is hardly recognized by neurobiologists.

Synthesis of ECM molecules is a typical feature of fibroblasts. Interestingly, skin and leptomeningeal fibroblasts, in spite of their many similarities, seem to represent different cell types reflecting probably their different embryonic origin (Colombo *et al.*, 1994). Peripheral fibroblasts and dura mater seem to be mesodermal derivatives whereas the ectoderm gives rise to leptomeningeal cells of part of the mesencephalon and the entire diencephalon and telencephalon (Jacobson, 1991); interestingly, leptomeningeal fibroblasts seem to share certain characteristics with astrocytes (Colombo *et al.*, 1994). Mammalian leptomeninges and meningiomas produce numerous components of both the interstitial matrix and the basement membrane (e.g., laminin; collagen types I; III; IV; and VI; fibronectin; nidogen/entactin; and heparan sulfate proteoglycan) (Rutka *et al.,* 1986; Sievers *et al.,* 1987; Stewart and Pearlman, 1987; Azzi *et al.,* 1989; Matthiessen *et al.,* 1991; Ng and Wong, 1991; Sievers *et al.,* 1994a). Furthermore, a group of antiadhesive ECM proteins, (e.g., SPARC/osteonectin) (Mendis and Brown, 1994), SC1 (Mendis *et al.,* 1994), and maybe also tenascin/cytotactin (Grumet *et al.,* 1985) and thrombospondin (O'Shea and Dixit, 1988), are synthesized in the leptomeninges. This is in line with the view that meningeal cells influence endfoot formation of Bergmann glial cells, which organize the superficial glia limitans surrounding the central nervous system (Sievers *et al.,* 1994a). That this function is essential early in mammalian development has been demonstrated by selective destruction of meningeal cells, which results in severe developmental defects of the brain (Sievers *et al.,* 1986, 1987; Hartmann *et al.* 1992; Sievers *et al.,* 1994b).

Nonetheless, in addition to ECM molecules, the mammalian meninges also secrete many proteins, which also appear as constituents of the CSF: prostaglandin D synthase/$\beta$-trace (Urade *et al.,* 1993), transferrin and transthyretin (Blay *et al.,* 1994), amyloid precursor proteins (LeBlanc *et al.,* 1991), transforming growth factor $\beta$1 (Heine *et al.,* 1987), insulin-like growth factor (IGF) II and IGF-binding proteins (Stylianopoulou *et al.,* 1988; de Pablo and de la Rosa, 1995), various neurotrophic factors and some of their receptors (Risling *et al.,* 1994), as well as certain stimulating factors for neuronal cells (Gensburger *et al.,* 1986). A further indication for a complex secretory activity of meningeal cells is the synthesis of various processing enzymes, (e.g., dipeptidylpeptidase IV) (Haninec and Grim, 1990).

## B. Ependymins as Extracellular Matrix Proteins

Only recently electron microscopy studies revealed an association of ependymins with collagen fibrils of the ECM (Figs. 8 and 10; Schwarz *et al.,* 1993). Thus far, it is unknown whether ependymins interact directly with collagen or via another collagen associated protein. The association of ependymins with the ECM is in agreement with their synthesis by leptomeningeal fibroblasts, whose major function is certainly the production of ECM molecules.

There is no indication that both soluble and ECM-bound forms differ chemically. A similar dual existence has been observed in the past for a variety of ECM proteins, e.g., fibronectin, (Timpl and Aumailley, 1993; Vloavsky *et al.,* 1993) and SPARC/osteonectin (Lane and Sage, 1994); in particular, the antiadhesin SPARC/osteonectin resembles ependymins in some respects. For example, both proteins show calcium-induced conformational transitions (but via different mechanisms) and associate with collagen. Furthermore, SPARC is also a secretory product of leptomeningeal fibroblasts, typically synthesized at postnatal day 3 of the mouse (Mendis and Brown, 1994); however, the expression pattern of SPARC in adult animals is generally broader.

Calciumion binding and the conformational change of ependymins may be important for their association with the ECM because self-assembly of the ECM is also known to be $Ca^{2+}$ dependent (Yurchenco and Schittny, 1990; Timpl and Aumailley, 1993). Taken together, we suggest that the ECM-bound form may be the functional form of ependymins. This hypothesis is also strengthened by the fact that the soluble form in the CSF is subject to enormous variations between different orders of fish.

### 1. Adhesive versus Antiadhesive Properties

The detection of the L2/HNK-1 epitope in goldfish ependymins (Shashoua *et al.,* 1986) was the first hint that ependymins may be involved in cell contact phenomena. This epitope is characteristic of many calcium-independent cell adhesion molecules (Künemund *et al.,* 1988; Jungalwala, 1994). Since the HNK-1 epitope is not conserved in ependymins of various fish species (Ganβ and Hoffmann, 1993), it can probably be eliminated as playing an important functional role.

Second, regeneration experiments in adult goldfish also indicated a functional role of ependymins in cell–cell interactions. Here, synthesis of ependymins has been shown to increase during regeneration of the optic nerve (Schmidt and Shashoua, 1988; Thormodsson *et al.,* 1992a) and antibodies to ependymins block the sharpening of the regenerating retinotectal projection (Schmidt and Shashoua, 1988). This projection is different in the developing embryo and the regenerating adult fish (precise versus initially diffuse; Stuermer and Easter, 1984; Stuermer, 1988).

Interestingly, ependymins appear first in the zebrafish at about the time of hatching (Sterrer *et al.,* 1990; Rinder *et al.,* 1992). At this stage of development, ingrowing retinal axons have already reached the optic tectum (Stuermer, 1988) and additional axons mainly grow along preexisting tracts (Wilson *et al.,* 1990). Thus, only posthatching axons must find their path, somewhat below the pial basal lamina (Easter, 1987), in the presence of ependymins (diffuse projection), whereas in prehatching zebrafish embryos ependymins are missing (axons project precisely). Thus, hypothetically, ependymins might be responsible for a diffuse projection (e.g., due to antiadhesive properties).

A third indication for a possible function of ependymins in modulating substrate adhesion came from a report in which goldfish ependymins were described as substrates for the outgrowth of retinal axons (Schmidt *et al.,* 1991). However, in later experiments ependymins did not seem to be better substrates than untreated glass coverslips (Müller-Schmid, 1992), indicating antiadhesive rather than adhesive properties for ependymins.

Generally, it is thought that ependymins may affect hypothetical cell contact phenomena in two opposing ways (i.e., via adhesive or antiadhesive effects). On the basis of the data available, antiadhesive properties of ependymins seem to be clearly favored. Thus, ependymins may well function by preventing axon ingrowth (e.g., along blood vessels or into the meninx). Interestingly, mammalian axons show maldevelopment when forced to grow into a foreign territory (e.g., the pia) (Risling *et al.,* 1992).

The classification of ependymins as antiadhesins (similar to SPARC/osteonectin, tenascin/cytotactin, and thrombospondin) or diffusible chemorepellents (Dodd and Schuchardt, 1995) would also fit with reports on repulsive contributions of sialic acid residues to cell adhesion (Shimamura *et al.,* 1994), which indeed represent one of the most conserved features of ependymins (see Section II,C,1).

### 2. Modulation of the Blood–Brain Barrier

One striking feature is the association of ependymins with the collagen sheath of blood vessels of the endomeninx and the brain parenchyma (Figs. 11, 13, and 14). Here, endothelial cells form the blood–brain barrier at the level of tight junctions. The unique properties of endothelial cells in the CNS are induced by the specific neural environment (Risau and Wolburg, 1990). It is possible that, at this site of the blood–CSF barrier, ependymins could regulate endothelial barrier function by modulating cell–matrix interactions. Interestingly, SPARC/osteonectin has been reported to fulfill a similar function probably by influencing vascular endothelial cell shape (Goldblum *et al.,* 1994).

## V. Concluding Remarks

In contrast to the situation in mammals (Aldred *et al.,* 1995), the predominant cerebrospinal fluid proteins in many teleost fish are major secretory products of the endomeninx. Here, fibroblast-like cells of the inner endomeningeal layer synthesize ependymins. Generally, the architecture of this three-layered structure is analogous to the mammalian leptomeninges, where fibroblasts typically secrete a variety of extracellular matrix proteins.

Ependymins represent a unique family of divergent piscine glycoproteins that bind calcium via terminal sialic acid residues. Thus far, no mammalian

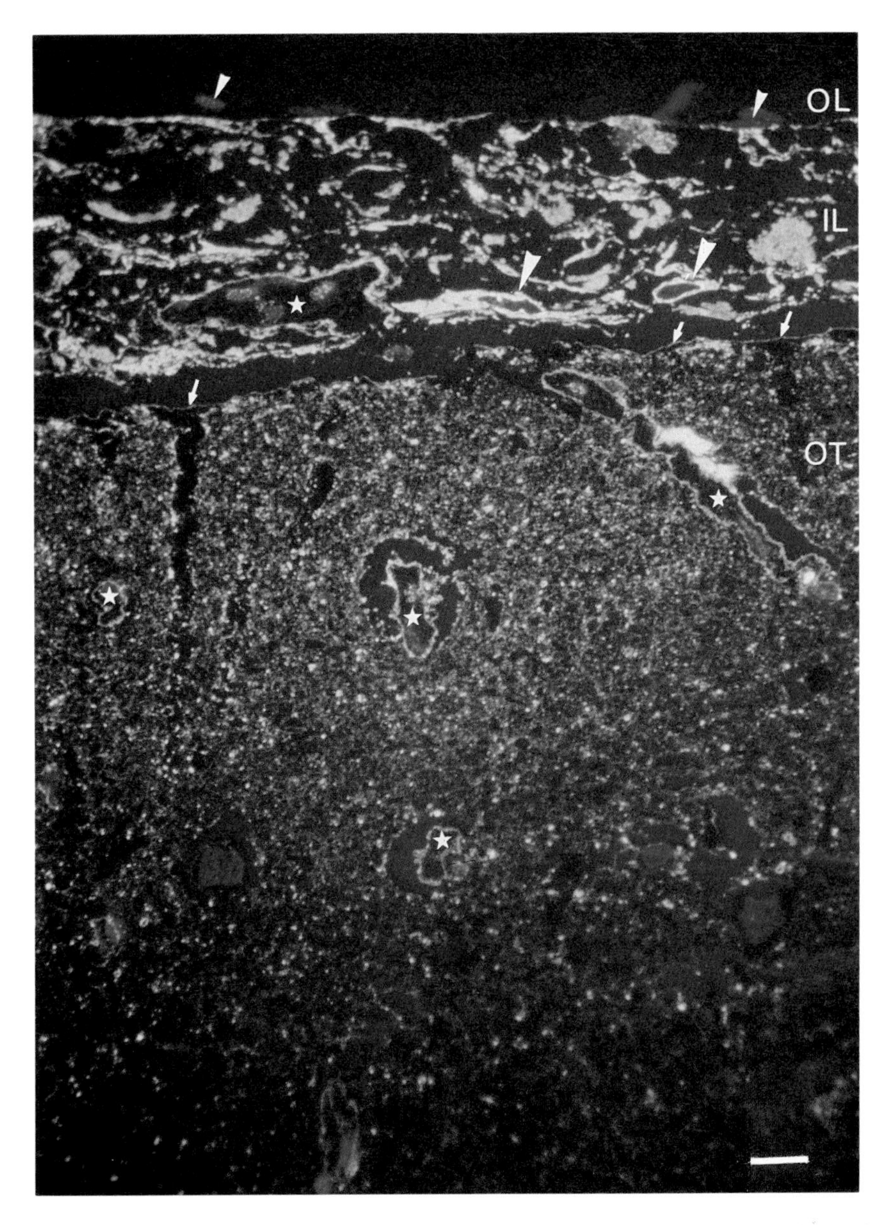


FIG. 11 Distribution of ependymins in the endomeninx and the underlaying parenchyma of the rainbow trout. Postembedding labeling on ultrathin lowicryl K4M sections with rabbit anti-ependymin serum and goat anti-rabbit Cy3 conjugates. Sections were counterstained using DAPI to visualize the DNA (appears blue). The outer layer (OL) is lacking ependymins; small arrowheads indicate nuclei of cells of the OL. The inner layer (IL) is heavily labeled, in particular the fibroblast-like cells (big arrowheads). In the underlaying optic tectum (OT) a decreasing gradient of ependymins is observed from the basal lamina (arrows) toward the ventricle. Stars indicate blood vessels covered by a sheath of ependymins. Bar: 10 $\mu$m.

homologs have been characterized. However, ependymins appear in a soluble as well as a bound form, associated with collagen fibrils of the extracellular matrix. Here, they might have antiadhesive properties due to sialic acid residues preventing axon ingrowth into "false territories" (i.e., the meninx or along blood vessels). This would also be in context with the observed increased synthesis during regeneration of the optic nerve (Schmidt and Shashoua, 1988; Thormodsson *et al.,* 1992a). Antiadhesive properties might also be important for the sharpening process of the regenerating retinotectal projection observed in adult goldfish; here, ependymins could weaken the multiple interactions of regenerating axons and their environment (keeping them in a "plastic state") similar to a selection barrier allowing corrections of non properly ingrown axons. The observed high turnover rate of ependymins would be in agreement with this model, as well as the observation that the sharpening process can be blocked by infusion of ependymin antibodies (Schmidt and Shashoua, 1988).

Also, modulation of the endothelial barrier function by ependymins similar to SPARC/osteonectin is within the limits of expectation. However, there is no evidence thus far that ependymins are directly involved in memory processes. In addition, ependymins are not synthesized by neuronal cells.

In the future it will be interesting—but certainly not easy—to determine if ependymin-homologous molecules also exist in mammals. Hypothetically, it could be that ependymins are expressed only in those species that retain the capacity to regenerate axons of the CNS until adulthood (e.g., fish and amphibia).

Taken together, the molecular function of ependymins is far from being understood and the name "ependymins," which had been erroneously introduced in the past, reflects neither the correct place of synthesis nor the typical localization nor the biological relevance of these still enigmatic glycoproteins.

## Acknowledgments

We thank I. Zimmermann and B. Lattemann for skillful assistance in immunolabeling experiments and E.-M. Gertzen for valuable technical help. We are also indebted to Dr. B. Grafstein (New York) for making results available prior to publication, to Dr. B. Krisch (Kiel) for discussions, and to Dr. H. Wolburg (Tübingen) for critically reading the manuscript.

## References

Achtstätter, T., Fouquet, B., Rungger-Brändle, E., and Franke, W. W. (1989). Cytokeratin filaments and desmosomes in the epithelioid cells of the perineural and arachnoidal sheaths of some vertebrate species. *Differentiation* **40,** 129–149.

Adams, D. S., and Shashoua, V. E. (1994). Cloning and sequencing the genes encoding goldfish and carp ependymin. *Gene* **141,** 237–241.

Aldred, A. R., Brack, C. M., and Schreiber, G. (1995). The cerebral expression of plasma protein genes in different species. *Comp. Biochem. Physiol.* **111B,** 1–15.

Azzi, G., Jouis, V., Godeau, G., Groult, N., and Robert, A. M. (1989). Immunolocalisation of extracellular matrix macromolecules in the rat spinal cord. *Matrix* **9**, 479-485.

Bailey, G. S., Cocks, G. T., and Wilson, A. C. (1969). Gene duplication in fishes: Malate dehydrogenases of salmon and trout. *Biochem. Biophys. Res. Commun.* **34,** 605–612.

Benninger, C., Kadis, J., and Prince, D. A. (1980). Extracellular calcium and potassium changes in hippocampal slices. *Brain Res.* **187,** 165–182.

Benowitz, L. I., and Shashoua, V. E. (1977). Localization of a brain protein metabolically linked with behavioral plasticity in the goldfish. *Brain Res.* **136,** 227–242

Bernstein, J. L., and Streicher, E. (1965). The blood–brain barrier of fish. *Exp. Neurol.* **11,** 464–473.

Blay, P., Nilsson, C., Hansson, S., Owman, C., Aldred, A., and Schreiber, G. (1994). An *in vivo* study of the effect of 5-HT and sympathetic nerves on transferrin and transthyretin mRNA expression in rat choroid plexus and meninges. *Brain Res.* **662,** 148-154.

Brightman, M. W. (1965). The distribution within the brain of ferritin injected into cerebrospinal fluid compartments. I. Ependymal distribution. *J. Cell Biol.* **26,** 99–123.

Bundgaard, M., and Cserr, H. F. (1991). Barrier membranes at the outer surface of the brain of an elasmobranch, *Raja erinacea. Cell Tissue Res.* **265,** 113–120.

Butt, A. M. (1987). A subarachnoid space in the elasmobranch brain–macro- and microscopic evidence using large molecular weight fluorescent markers. *Biol. Bull.* **173,** 421.

Caruncho, H. J., and Anadon, R. (1990). Endothelial cells of the meningeal capillaries in the rainbow trout: A freeze-fracture study. *J. Morphol.* **206,** 327–332.

Caruncho, H. J., Pinto da Silva, P., and Anadon, R. (1993). The morphology of teleost meningocytes as revealed by freeze-fracture. *J. Submicrosc. Cytol. Pathol.* **25,** 397–406.

Caruncho, H. J., and Pinto da Silva, P. (1994). Alterations in the intermediate layer of goldfish meninges during adaptation to darkness. *J. Anat.* **184,** 355–362.

Colombo, J. A., Napp, M. I., and Puissant, V. (1994). Leptomeningeal and skin fibroblasts: Two different cell types? *Int. J. Dev. Neurosci.* **12,** 57–61.

Cserr, H. F. (1971). Physiology of the choroid plexus. *Phsysiol. Rev.* **51,** 273–311.

Cserr, H. F., and Bundgaard, M. (1984). Blood–brain interfaces in vertebrates: A comparative approach. *Am. J. Physiol.* **246,** R277–R288.

Cserr, H. F., Fenstermacher, J. D., and Rall, D. P. (1972). Brain-barrier systems in sharks. *Comp. Biochem. Physiol.* **42A,** 73–78.

Cserr, H. F., and Ostrach, L. H. (1974). On the presence of subarachnoid fluid in the mudpuppy, *Necturus maculosus. Comp. Biochem. Physiol.* **48A,** 145–151.

Dang, C. V., Shin, C. K., Bell, W. R., Nagaswami, C., and Weisel, J. W. (1989). Fibrinogen sialic acid residues are low affinity calcium-binding sites that influence fibrin assembly. *J. Biol. Chem.* **264,** 15104–15108.

de Pablo, F., and de la Rosa, E. J. (1995). The developing CNS: A scenario for the action of proinsulin, insulin and insulin-like growth factors. *Trends Neurosci.* **18,** 143–150.

Dodd, J., and Schuchardt, A. (1995). Axon guidance: A compelling case for repelling growth cones. *Cell* **81,** 471–474.

Easter, S. S. (1987). Retinal axons and the basal lamina. *In* "Mesenchymal-Epithelial Interactions in Neural Development" (J. R. Wolff, J. Sievers, and M. Berry, eds.), pp. 385–396. Springer-Verlag, Berlin.

Fazeli, M. S., Errington, M. L., Dolphin, A. C., and Bliss, T. V. P. (1988). Proteins of the extracellular fluid of dentate gyrus: Presence of proteins S-100 and ependymins but absence of NGF *in vivo. Eur. J. Neurosci.* **Suppl. 1,** 90.

Felgenhauer, K. (1974). Protein seize and cerebrospinal fluid composition. *Klin. Wschr.* **52,** 1158–1164.

Frank, E. H., Burge, B. W., Liwnicz, B. H., Lotspeich, L. J., White, J. C., Wechsler, S. L., Mayfield, F. H., and Keller, J. T. (1983). Cytokeratin provides a specific marker for human arachnoid cells grown *in vitro. Exp. Cell Res.* **146,** 371–376.

Franz, V. (1911). Das Kleinhirn der Knochenfische. *Zool. Jahrb. Abt. Anat.* **32,** 401–464.

Ganβ, B. (1993). "Molekulare Analyse der Ependymine: Calcium-bindende Glykoproteine der Cerebrospinalflüssigkeit von Fischen." Ph.D. Thesis, Universität Regensburg (Naturwissenschaftliche Fakultät III), Regensburg, Germany.

Ganβ, B., and Hoffmann, W. (1993). Calcium binding to sialic acids and its effect on the conformation of ependymins. *Eur. J. Biochem.* **217,** 275–280.

Gensburger, C., Labourdette, G., and Sensenbrenner, M. (1986). Influence of meningeal cells on the proliferation and maturation of rat neuroblasts in culture. *Exp. Brain. Res.* **63,** 321–330.

Goldblum, S. E., Ding, X., Funk, S. E., and Sage, E. H. (1994). SPARC (secreted protein acidic and rich in cysteine) regulates endothelial cell shape and barrier function. *Proc. Natl. Acad. Sci. U.S.A.* **91,** 3448–3452.

Grumet, M., Hoffman, S., Crossin, K. L., and Edelman, G. M. (1985). Cytotactin, an extracellular matrix protein of neural and non-neural tissues that mediates glia-neuron interaction. *Proc. Natl. Acad. Sci. U.S.A.* **82,** 8075–8079.

Haines, D. E. (1991). On the question of a subdural space. *Anat. Rec.* **230,** 3–21.

Haninec, P., and Grim, M. (1990). Localization of dipeptidylpeptidase IV and alkaline phosphatase in developing spinal cord meninges and peripheral nerve coverings of the rat. *Int. J. Dev. Neurosci.* **8,** 175–185.

Hartmann, D., Sievers, J., Pehlemann, F. W., and Berry, M. (1992). Destruction of meningeal cells over the cerebral hemisphere of newborn hamsters prevents the formation of the infrapyramidal blade of the dentate gyrus. *J. Comp. Neurol.* **320,** 33–61.

Heine, U. I., Munoz, E. F., Flanders, K. C., Ellingsworth, L. R., Lam, H.-Y., Thompson, N. L., Roberts, A. B., and Sporn, M. B. (1987). Role of transforming growth factor-β in the development of the mouse embryo. *J. Cell Biol.* **105,** 2861–2876.

Hoffmann, W. (1992). Goldfish ependymins: Cerebrospinal fluid proteins of meningeal origin. *Progr. Brain Res.* **91,** 13–17.

Hoffmann, W. (1993). Biosynthesis of ependymins: Implications for barrier concepts in fish brain. *In* "CNS Barriers and Modern CSF Diagnostics" (K. Felgenhauer, M. Holzgraefe, and H. W. Prange, eds.), pp. 247–256. VCH-Verlag, Weinheim.

Hoffmann, W., Königstorfer, A., and Sterrer, S. (1992). Biosynthesis and expression of goldfish ependymins: Potential candidates in neural plasticity and regeneration? *In* "Development and Regeneration of the Nervous System" (S. Nona, J. Cronly-Dillon, C. Stafford, and M. Ferguson, eds.), pp. 255–265. Chapman and Hall, London.

Hoffmann, A., Nimtz, M., Getzlaff, R., and Conradt, H. S. (1995). "Brain-type" N-glycosylation of asialo-transferrin from human cerebrospinal fluid. *FEBS Lett.* **359,** 164–168.

Hoffmann, A., Nimtz, M., Wurster, U., and Conradt, H. S. (1994). Carbohydrate structures of β-trace protein from human cerebrospinal fluid: Evidence for "brain-type" N-glycosylation. *J. Neurochem.* **63,** 2185–2196.

Ichimura, T., Fraser, P. A., and Cserr, H. F. (1991). Distribution of extracellular tracers in perivascular spaces of the rat brain. *Brain Res.* **545,** 103–113.

Jacobson, M. (1991). "Developmental Neurobiology," 3rd Ed. Plenum Press, New York.

Jansen, W. F., and Flight, W. F. G. (1969). Light- and electronmicroscopical observations on the saccus vasculosus of the rainbow trout. *Z. Zellforsch.* **100,** 439–465.

Jones, H. C. (1978). Continuity between the ventricular and subarachnoid cerebrospinal fluid in an amphibian. *Cell Tissue Res.* **195,** 153–167.

Jones, H. C. (1979). Comparative aspects of the cerebrospinal fluid systems in vertebrates. *Sci. Prog. Oxford.* **66,** 171–190.

Jungalwala, F. B. (1994). Expression and biological functions of sulfoglucuronyl glycolipids (SGGLs) in the nervous system—a review. *Neurochem. Res.* **19,** 945–957.

Kappers, C. U. A. (1926). The meninges in lower vertebrates compared with those in mammals. *Arch. Neurol. Psych.* (*Chicago*) **15,** 281–296.

Klatzo, I., and Steinwall, O. (1965). Observation on cerebrospinal fluid pathways and behaviour of the blood–brain barrier in sharks. *Acta Neuropathol.* **5,** 161–175.

Klika, E. (1967). The ultrastructure of meninges in vertebrates. *Acta Univ. Carol. Med.* (*Praha*) **13,** 53–71.

Klika, E., and Zajíková, A. (1975). The ultrastructure of leptomeninx in fish. *Folia Morphol.* (*Praha*) **23,** 380–389.

Königstorfer, A., Sterrer, S., Eckerskorn, C., Lottspeich, F., Schmidt, R., and Hoffmann, W. (1989a). Molecular characterization of an ependymin precursor from goldfish brain. *J. Neurochem.* **52,** 310–312.

Königstorfer, A., Sterrer, S., and Hoffmann, W. (1989b). Biosynthesis of ependymins from goldfish brain. *J. Biol.Chem.* **264,** 13689–13692.

Königstorfer, A., Sterrer, S., and Hoffmann, W. (1990). Ependymins are expressed in the meninx of goldfish brain. *Cell Tissue Res.* **261,** 59–64.

Krisch, B. (1988). Ultrastructure of the meninges at the site of penetration of veins through the dura mater, with particular reference to Pacchionian granulations. *Cell Tissue Res.* **251,** 621–631.

Künemund, V., Jungalwala, F. B., Fischer, G., Chou, D. K. H., Keilhauer, G., and Schachner, M. (1988). The L2/HNK-1 carbohydrate of neural cell adhesion molecules is involved in cell interactions. *J. Cell Biol.* **106,** 213–223.

Lakos, S. F., Thormodsson, F. R., and Grafstein B. (1994). Immunolocalization of exoglycoproteins ("ependymins") in the goldfish brain. *Neurochem. Res.* **19,** 1401–1412.

Lane, T. F., and Sage, E. H. (1994). The biology of SPARC, a protein that modulates cell–matrix interactions. *FASEB J.* **8,** 163–173.

LeBlanc, A. C., Chen, H. Y., Autilio-Gambetti, L., and Gambetti, P. (1991). Differential APP gene expression in rat cerebral cortex, meninges, and primary astroglial, microglial and neuronal cultures. *FEBS Lett.* **292,** 171–178.

Lundquist, F. (1942). The blood–brain barrier in some freshwater teleosts. *Acta Physiol. Scand.* **4,** 201–206.

Magendie, F. (1843). "Physiologische und klinische Untersuchungen über die Hirn- und Rückenmarksflüssigkeit." Kollmann, Leipzig.

Majocha, R. E., Schmidt, R., and Shashoua, V. E. (1982). Cultures of zona ependyma cells of goldfish brain: An immunological study of the synthesis and release of ependymins. *J. Neurosci. Res.* **8,**, 331–342.

Matthiessen, H. P., Schmalenbach, C., and Müller, H. W. (1991). Identification of meningeal cell released neurite promoting activities for embryonic hippocampal neurons. *J.Neurochem.* **56,** 759–768.

Maurer, P., Mayer, U., Bruch, M., Jenö, P., Mann, K., Landwehr, R., Engel, J., and Timpl, R. (1992). High-affinity and low-affinity calcium binding and stability of the multidomain extracellular 40-kDa basement membrane glycoprotein (BM-40/SPARC/osteonectin). *Eur. J. Biochem.* **205,** 233-240.

Mendis, D. B., and Brown, I. R. (1994). Expression of the gene encoding the extracellular matrix glycoprotein SPARC in the developing and adult mouse brain. *Mol. Brain Res.* **24,** 11–19.

Mendis, D. B., Shahin, S., Gurd, J. W., and Brown, I. R. (1994). Developmental expression in the rat cerebellum of SC1, a putative brain extracellular matrix glycoprotein related to SPARC. *Brain Res.* **633,** 197–205.

Milhorat, T. H. (1976). Structure and function of the choroid plexus and other sites of cerebrospinal fluid formation. *Int. Rev. Cytol.* **47,** 225–288.

Momose, Y., Kohno K., and Ito, R. (1988). Ultrastructural study on the meninx of the goldfish brain. *J. Comp. Neurol.* **270,** 327–336.

Morris, M. E., Ropert, N., and Shashoua, V. E. (1986). Stimulus-evoked changes in extracellular calcium in optic tectum of the goldfish: Possible role in neuroplasticity. *Ann. N.Y. Acad. Sci.* **481,** 375–377.

Morse, D. E., and Low, F. N. (1972). The fine structure of the pia mater of the rat. *Am. J. Anat.* **133,** 349–368.

Müller-Schmid, A. (1992). "Die Ependymine der Regenbogenforelle und des Herings: Studien zur molekularen Struktur, Lokalisation, Phylogenie und Funktion von sekretorischen Produkten der Endomeninx." Ph.D. Thesis, Universität Tübingen (Fakultät für Biologie), Tübingen, Germany.

Müller-Schmid, A., Ganß, B., Gorr, T., and Hoffmann, W. (1993). Molecular analysis of ependymins from the cerebrospinal fluid of the orders Clupeiformes and Salmoniformes: No indication for the existence of an euteleost infradivision. *J. Mol. Evol.* **36,** 578–585.

Müller-Schmid, A., Rinder, H., Lottspeich, F., Gertzen, E.-M., and Hoffmann, W. (1992). Ependymins from the cerebrospinal fluid of salmonid fish: Gene structure and molecular characterization. *Gene* **118,** 189–196.

Nakao, T. (1979). Electron microscopic studies on the lamprey meninges. *J. Comp. Neurol.* **183,** 429–454.

Nelson, T. J., and Alkon, D. L. (1989). Specific protein changes during memory acquisition and storage. *BioEssays* **10,** 75–79.

Ng, H. K., and Wong, A.T. C. (1991). Immunohistochemical expression of extracellular matrix proteins in meningiomas. *Lab. Invest.* **64,** 102A.

Nicholson, C. (1980). Modulation of extracellular calcium and its functional implications. *Fed. Proc.* **39,** 1519–1523.

Nolan, P. M., and Shashoua, V. E. (1989). Mechanisms of ependymin phosphorylation in goldfish brain ECF. *Soc. Neurosci. Abstr.* **15,** 961.

Northcutt, R. G. (1983). Evolution of the optic tectum in ray-finned fishes. *In* "Fish Neurobiology" (R. G. Northcutt, and R. E. Davis, eds.), Vol. 2: Higher Brain Areas and Functions, pp. 1–41. University of Michigan Press, Ann Arbor.

Oda, Y., and Nakanishi, I. (1984). Ultrastructure of the mouse leptomeninx. *J. Comp. Neurol.* **225,** 448–457.

Ohno, S., Muramoto, J., and Christian, L. (1967). Diploid–tetraploid relationship among old-world members of the fish family *Cyprinidae. Chromosoma* (*Berl.*) **23,** 1–9.

O'Shea, K. S., and Dixit, V. M. (1988). Unique distribution of the extracellular component thrombospondin in the developing mouse embryo. *J. Cell Biol.* **107,** 2737–2748.

Piront, M.-L., and Schmidt, R. (1988). Inhibition of long-term memory formation by anti-ependymin antisera after active shock-avoidance learning in goldfish. *Brain Res.* **442,** 53–62.

Pumain, R., and Heinemann, U. (1985). Stimulus- and amino acid-induced calcium and potassium changes in rat neocortex. *J. Neurophysiol.* **53,** 1–15.

Rasmussen, L., and Rasmussen, R. (1967). Comparative protein and enzyme profiles of the cerebrospinal fluid, extradural fluid, nervous tissue, and sera of elasmobranchs. *In* "Sharks, Skates and Rays" (P. W. Gilbert, R. F. Mattewson, and D. P. Rall, eds.), pp. 361–379. Johns Hopkins University Press, Baltimore.

Rennels, M. L., Gregory, T. F., Blaumanis, O. R., Fujimoto, K., and Grady, P. A. (1985). Evidence for a "paravascular" fluid circulation in the mammalian central nervous system, provided by the rapid distribution of tracer protein throughout the brain from the subarachnoid space. *Brain Res.* **326,** 47–63.

Rinder, H., Bayer, T. A., Gertzen, E.-M., and Hoffmann, W. (1992). Molecular analysis of the ependymin gene and functional test of its promoter region by transient expression in *Brachydanio rerio. DNA Cell Biol.* **11,** 425–432.

Risau, W., and Wolburg, H. (1990). Development of the blood–brain barrier. *Trends Neurosci.* **13,** 174–178.

Risling, M., Dalsgaard, C.-J., Frisén, J., Sjögren, A.-M., and Fried, K. (1994). Substance-P, calcitonin gene-related peptide, growth-associated protein-43, and neurotrophin receptor-like immunoreactivity associated with unmyelinated axons in feline ventral roots and pia mater. *J. Comp. Neurol.* **339,** 365–386.

Risling, M., Sörbye, K., and Cullheim, S. (1992). Aberrant regeneration of motor axons into the pia mater after ventral root neuroma formation. *Brain Res.* **570,** 27–34.

Rovainen, C. M. (1970). Glucose production by lamprey meninges. *Science* **167,** 889–890.

Rovainen, C. M., Lemcoe, G. E., and Peterson, A. (1971). Structure and chemistry of glucose-producing cells in meningeal tissue of the lamprey. *Brain Res.* **30,** 99–118.

Rutka, J. T., Giblin, J., Dougherty, D. V., McCulloch, J. R., DeArmond, S. J., and Rosenblum, M. L. (1986). An ultrastructural and immunochemical analysis of leptomeningeal and meningioma cultures. *J. Neuropathol. Exp. Neurol.* **45,** 285–303.

Sagemehl, M. (1884). Beiträge zur vergleichenden Anatomie der Fische. II. Einige Bemerkungen über die Gehirnhäute der Knochenfische. *Morphol. Jahrb.* **9,** 457-475.

Schmidt, J. T., Schmidt, R., Lin, W., Jian, X., and Stuermer, C. A. O. (1991). Ependymin as a substrate for outgrowth of axons from cultured explants of goldfish retina. *J. Neurobiol.* **22,** 40–54.

Schmidt, J. T., and Shashoua, V. E. (1988). Antibodies to ependymin block the sharpening of the regenerating retinotectal projection in goldfish. *Brain Res.* **446,** 269–284.

Schmidt, R. (1986). Biochemical participation of glycoproteins in memory consolidation after two different training paradigms in goldfish. *Adv. Biosci.* **59,** 213–222.

Schmidt, R. (1987). Changes in subcellular distribution of ependymins in goldfish brain induced by learning. *J. Neurochem.* **48,** 1870–1878.

Schmidt, R. (1989). Glycoproteins involved in long-lasting plasticity in the teleost brain. *In* "Fundamentals of Memory Formation: Neuronal Plasticity and Brain Function" (H. Rahmann, ed.), *Progr. Zool.* **37,** 327–339. Gustav Fischer Verlag, Stuttgart.

Schmidt, R., and Lapp, H. (1987). Regional distribution of ependymins in goldfish brain measured by radioimmunoassay. *Neurochem. Int.* **10,** 383–390.

Schmidt, R., and Makiola, E. (1991). Calcium and zinc ion binding properties of goldfish brain ependymin. *Neurol. Chem. (Life Sci. Adv.)* **10,** 161–171.

Schmidt, R., and Shashoua, V. E. (1981). A radioimmunoassay for ependymins $\beta$ and $\gamma$: Two goldfish brain proteins involved in behavioral plasticity. *J. Neurochem.* **36,** 1368–1377.

Schmidt, R., and Shashoua, V. E. (1983). Structural and metabolic relationships between goldfish brain glycoproteins participating in functional plasticity of the central nervous system. *J. Neurochem.* **40,** 652–660

Schmidt, R., Löffler, F., Müller, H. W., and Seifert, W. (1986). Immunological cross-reactivity of cultured rat hippocampal neurons with goldfish brain proteins synthesized during memory consolidation. *Brain Res.* **386,** 245–257.

Schmidt, R., Rother, S., Schlingensiepen, K.-H., and Brysch, W. (1992). Neuronal plasticity depending on a glycoprotein synthesized in goldfish leptomeninx. *Progr. Brain Res.* **91,** 7–12.

Schwarz, H., Müller-Schmid, A., and Hoffmann, W. (1993). Ultrastructural localization of ependymins in the endomeninx of the brain of the rainbow trout: Possible association with collagen fibrils of the extracellular matrix. *Cell Tissue Res.* **273,** 417–425.

Shashoua, V. E. (1976a). Identification of specific changes in the pattern of brain protein synthesis after training. *Science* **193,** 1264–1266.

Shashoua, V. E. (1976b). Brain metabolism and the acquisition of new behaviors. I. Evidence for specific changes in the pattern of protein synthesis. *Brain Res.* **111,** 347–364.

Shashoua, V. E. (1977a). Brain protein metabolism and the acquisition of new patterns of behavior. *Proc. Natl. Acad. Sci. U.S.A.* **74,** 1743–1747.

Shashoua, V. E. (1977b). Ependymin $\beta$: A brain protein metabolically linked with behavioral plasticity in the goldfish. *In* "Mechanisms, Regulation and Special Functions of Protein Synthesis in the Brain" (S. Roberts, A. Lajtha, and W. H. Gispen, eds.), pp. 331–342. Elsevier, Amsterdam.

Shashoua, V. E. (1981). Extracellular fluid proteins of goldfish brain: Studies of concentration and labeling patterns. *Neurochem. Res.* **6,** 1129–1147.

Shashoua, V. E. (1986). The role of brain extracellular proteins in learning and memory. *In* "Neural Mechanisms of Conditioning" (D. L. Alkon, and C. D. Woody), pp. 459–490. Plenum Press, New York.

Shashoua, V. E. (1988). Monomeric and polymeric forms of ependymin: A brain extracellular glycoprotein implicated in memory consolidation processes. *Neurochem. Res.* **13,** 649–655.

Shashoua, V. E. (1991). Ependymin, a brain extracellular glycoprotein, and CNS plasticity. *Ann. N.Y. Acad. Sci.* **627,** 94–114.

Shashoua, V. E., Daniel, P. F., Moore, M. E., and Jungalwala, F. B. (1986). Demonstration of glucuronic acid on brain glycoproteins which react with HNK-1 antibody. *Biochem. Biophys. Res. Commun.* **138,** 902–909.

Shashoua, V. E., and Hesse, G. W. (1989). Classical conditioning leads to changes in extracellular concentrations of ependymin in goldfish brain. *Brain Res.* **484,** 333–339.

Shashoua, V. E., Hesse, G. W., and Milinazzo, B. (1990). Evidence for the in vivo polymerization of ependymin: A brain extracellular glycoprotein. *Brain Res.* **522,** 181–190.

Shashoua, V. E., and Moore, M. E. (1978). Effect of antisera to $\beta$ and $\gamma$ goldfish brain proteins on the retention of a newly acquired behaviour. *Brain Res.* **148,** 441–449.

Shashoua, V. E., Nolan, P. M., Shea, T. B., and Milinazzo, B. (1992). Dibutyryl cyclic AMP stimulates expression of ependymin mRNA and the synthesis and release of the protein into the culture medium by neuroblastoma cells (NB2a/d1). *J. Neurosci. Res.* **32,**, 239–244.

Shimamura, M., Shibuya, N., Ito, M., and Yamagata, T. (1994). Repulsive contribution of surface sialic acid residues to cell adhesion to substratum. *Biochem. Mol. Biol. Int.* **33,** 871–878.

Sievers, J., Hartmann, D., Gude, S., Pehlemann, F. W., and Berry, M. (1987). Influences of meningeal cells on the development of the brain. *In* "Mesenchymal–Epithelial Interactions in Neural Development" (J. R. Wolff, J. Sievers, and M. Berry, eds.), pp. 171–188. Springer-Verlag, Berlin.

Sievers, J., Pehlemann, F. W., Gude, S., and Berry, M. (1994a). Meningeal cells organize the superficial glia limitans of the cerebellum and produce components of both the interstitial matrix and the basement membrane. *J. Neurocytol.* **23,** 135–149.

Sievers, J., Pehlemann, F. W., Gude, S., and Berry, M. (1994b). A time course study of the alterations in the development of the hamster cerebellar cortex after destruction of the overlying meningeal cells with 6-hydroxydopamine on the day of birth. *J. Neurocytol.* **23,** 117–134.

Stahl, B., Steup, M., Karas, M., and Hillenkamp, F. (1991). Analysis of neutral oligosaccharides by matrix-assisted laser desorption/ionization mass spectroscopy. *Anal. Chem.* **63,** 1463–1466.

Sterrer, S., Königstorfer, A., and Hoffmann, W. (1990). Biosynthesis and expression of ependymin homologous sequences in zebrafish brain. *Neuroscience* **37,** 277–284.

Sterzi, G. (1902). Recherches sur l'anatomie comparée et sur l'ontogenèse des méninges. *Arch. Ital. Biol.* **37,** 257–269.

Stewart, G. R., and Pearlman, A. L. (1987). Fibronectin-like immunoreactivity in the developing cerebral cortex. *J. Neurosci.* **7,** 3325–3333.

Stuermer, C. A. O. (1988). Retinotopic organization of the developing retinotectal projection in the zebrafish embryo. *J. Neurosci.* **8,** 4513–4530.

Stuermer, C. A. O., and Easter, S. S. (1984). A comparison of the normal and regenerated retinotectal pathways of goldfish. *J. Comp. Neurol.* **223,** 57–76.

Stylianopoulou, F., Herbert, J., Soares, M. B., and Efstratiadis, A. (1988). Expression of the insulin-like growth factor II gene in the choroid plexus and the leptomeninges of the adult central nervous system. *Proc. Natl. Acad. Sci. U.S.A.* **85,** 141–145.

Thormodsson, F. R., Antonian, E., and Grafstein, B. (1988). Extracellular glycoproteins of the goldfish optic tectum are labelled by intraocular injection of $^{3}$H-proline. *Soc. Neurosci. Abstr.* **14,** 805.

Thormodsson, F. R., Antonian, E., and Grafstein, B. (1992a). Extracellular proteins of goldfish optic tectum labeled by intraocular injection of $^{3}$H-proline. *Exp. Neurol.* **117,** 260–268.

Thormodsson, F. R., Parker, T. S., and Grafstein, B. (1992b). Immunochemical studies of extracellular glycoproteins (X-GPs) of goldfish brain. *Exper. Neurol.* **118,** 275–283.

Timpl, R., and Aumailley, M. (1993). Other basement membrane proteins and their calcium-binding potential. *In* "Molecular and Cellular Aspects of Basement Membranes" (D. H. Rohrbach, and R. Timpl, eds.), pp. 211–235. Academic Press, San Diego.

Urade, Y., Kitahama, K., Ohishi, H., Kaneko, T., Mizuno, N., and Hayaishi, O. (1993). Dominant expression of mRNA for prostaglandin D synthase in leptomeninges, choroid plexus, and oligodendrocytes of the adult brain. *Proc. Natl. Acad. Sci. U.S.A.* **90,** 9070–9074.

Uyeno, T., and Smith, G. R. (1972). Tetraploid origin of the karyotype of catostomid fishes. *Science* **175,** 644–646.

van Gelderen, C. (1925). Über die Entwicklung der Hirnhäute bei Teleostiern. *Anat. Anz.* **60,** 48–57.

van Rijssel, T. G. (1946). Circulation of cerebrospinal fluid in *Carassius gibelio. Arch. Neurol. Psychiatr.* **56,** 522–543.

Vloavsky, I., Bar-Shavit, R., Korner, G., and Fuks, Z. (1993). Extracellular matrix-bound growth factors, enzymes, and plasma membranes. *In* "Molecular and Cellular Aspects of Basement Membranes" (D. H. Rohrbach, and R. Timpl, eds.), pp. 327–343. Academic Press, San Diego.

Wang, J., Murray, M., and Grafstein, B. (1995). Cranial meninges of goldfish: Age-related changes in morphology of meningeal cells and accumulation of surfactant-like multilamellar bodies. *Cell Tissue Res.* **281,** 349–358.

Wilson, S. W., Ross, L. S., Parrett, T., and Easter, S. S.(1990). The development of a simple scaffold of axon tracts in the brain of the embryonic zebrafish, *Brachydanio rerio. Development* **108,** 121–145.

Yurchenco, P. D., and Schittny, J. C. (1990). Molecular architecture of basement membranes. *FASEB J.* **4,** 1577–1590.

Zubrod, C. G., and Rall, D. P. (1959). Distribution of drugs between blood and cerebrospinal fluid in the various vertebrate classes. *J. Pharmacol.* **125,** 194–197.

Zucht, B., and Rahmann, H. (1974). Protein concentration in serum and cerebrospinal fluid of different vertebrates (fish, frog, bird and man). *J. Int. Res. Commun.* **2,** 1471.

# Cadherin Cell Adhesion Molecules in Differentiation and Embryogenesis

James A. Marrs* and W. James Nelson†
*Departments of Medicine, Physiology, and Biophysics, Indiana University Medical Center, Indianapolis, Indiana 46202-5116
†Department of Molecular and Cellular Physiology, Stanford University School of Medicine, Beckman Center for Molecular and Genetic Medicine, Stanford, California 94305-5426

The cadherin gene superfamily of calcium-dependent cell–cell adhesion molecules contains more than 40 members. We summarize functions attributed to these proteins, especially their roles in cellular differentiation and embryogenesis. We also describe hierarchies of protein–protein interactions between cadherins and cadherin-associated proteins (catenins). Several signal transduction pathways converge on, and diverge from, the cadherin/catenin complex to regulate its function; we speculate on roles of these signaling processes for cell structure and function. This review provides a framework for interpretation of developmental functions of cadherin cell adhesion molecules.

**KEY WORDS:** Cadherin, Catenin, Adherens junction, Desmosome, Epithelial cell polarity, Membrane–cytoskeleton, Differentiation, Embryogenesis.

## I. Introduction

Multicellular organisms are composed of heterogeneous cell types that are organized during development into distinct patterns to form tissues and organs. One of the most important primary processes involved in regulating the establishment and maintenance of these patterns is cell–cell adhesion. Pioneering studies by Holtfreter and colleagues, Moscona and colleagues, and Steinberg (Holtfreter, 1944; Moscona, 1952; Moscona and Moscona, 1952; Townes and Holtfreter, 1955; Steinberg, 1964) (reviewed in Trinkaus, 1965; Grunwald, 1991) established as a central principle that the interaction among cells within heterogeneous cell populations was based on the specificity and extent of adhesion among cells; cells of one type aggregate together and sort out from other cells. More recently, the molecular basis for cell adhesion has been shown to be due to the cell surface expression of a family of glycoproteins that bind with high specificity to each other on adjacent cells (Edelman, 1985, 1986; Takeichi, 1988, 1990; Edelman and Crossin, 1991; Kemler, 1992). Significantly, these proteins are expressed in distinct patterns during tissue and organ morphogenesis, suggesting that they play important and direct roles in the temporal and spatial regulation of cell interactions and cell sorting during tissue formation (Takeichi, 1988). Furthermore, loss of expression of these proteins correlates with the loss of intercellular adhesion, which is an early event in metastatic disease (Behrens *et al.,* 1989; Hashimoto *et al.,* 1989; Vleminckx *et al.,* 1991; Takeichi, 1993).

Two functionally distinct mechanisms of cell–cell adhesion have been described, $Ca^{2+}$ dependent and $Ca^{2+}$ independent, and the proteins involved have been classified into two major classes of cell–cell adhesion proteins based on protein structure and functional requirement for $Ca^{2+}$: the immunoglobulin (Ig) superfamily and the cadherins (Edelman, 1985, 1986; Takeichi, 1988, 1990; Edelman and Crossin, 1991). Neural–cell adhesion molecule (N-CAM) is one of the most thoroughly characterized members of the Ig superfamily of CAMs. N-CAM interactions on adjacent cells are homotypic and $Ca^{2+}$-independent. $Ca^{2+}$-independent cell adhesion mechanisms are not discussed further in this article, and the reader is directed to reviews of the proteins involved (Buck, 1992; Goridis and Brunet, 1992; Walsh and Doherty, 1993). This article focuses on cadherins, which regulate $Ca^{2+}$-dependent cell–cell adhesion during embryogenesis and in the adult body, and discusses how these molecules play important roles in the structural and functional organization of membrane domains in polarized cells.

## II. Structural Similarities and Functional Diversity in the Cadherin Family of Cell Adhesion Proteins

Cadherins share several common structural motifs that characterize this family of proteins (Fig. 1). They are type I integral membrane proteins. The extracellular domain contains "cadherin repeats," each of which has two calcium-binding sites. Repeats usually form head-to-tail homotypic interactions with repeats in cadherins on neighboring cells. Determinations of nuclear magnetic resonance (NMR) solution structure (Overduin *et al.,* 1995) and X-ray crystal structure (Shapiro *et al.,* 1995) of the first cadherin repeat of E- and N-cadherin, respectively, have provided molecular details for homotypic interactions and calcium requirements of cadherins. Al-

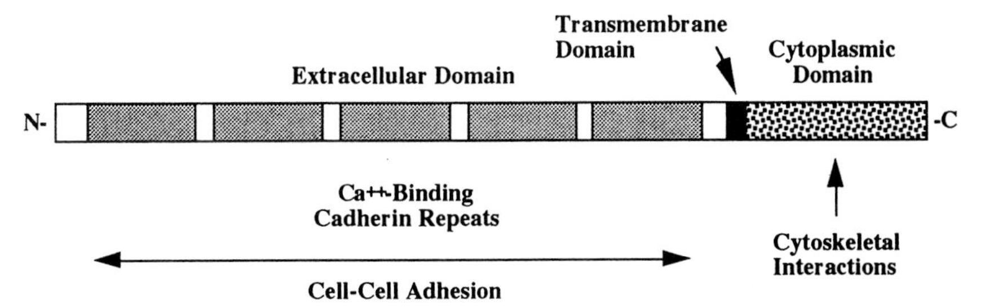


FIG. 1 Generalized cadherin gene superfamily member. This schematic diagram illustrates the features that are shared by nearly all cadherin gene superfamily members. Cadherins are type I membrane proteins. Extracellular domains contain "cadherin repeats" that bind calcium and interact with the repeats of cadherins on neighboring cells. The cytoplasmic domains of most cadherin molecules have been shown to bind cytoskeletal proteins. There are, of course, exceptions to the generalizations presented here. For example, the *fat* gene product contains 34, rather than 5, "cadherin repeats"; T-cadherin is GPI anchored and, thus, does not contain transmembrane or cytoplasmic domain sequences. For details, see text.

though there are some discrepancies in the two studies (e.g., the oligomeric state of the cadherin repeats in solution; see Weis, 1995), these detailed structures provide a useful framework for studies of intra- and intermolecular interactions. With the exception of the glycosylphosphatidylinositol (GPI)-anchored T-cadherin (Ranscht and Dours-Zimmermann, 1991; Vestal and Ranscht, 1992 Sacristán *et al.,* 1993), all cadherins contain a single membrane-spanning domain and a cytoplasmic domain. Detailed structural comparisons of cadherins have been published previously (Buxton and Magee, 1992; Geigerand Ayalon, 1992; Kemler, 1992, 1993).

The genes encoding cadherins comprise a large superfamily (Buxton and Magee, 1992; Geiger and Ayalon, 1992; Kemler, 1992). On the basis of functional considerations and sequence comparisons, the cadherin superfamily can be grouped into three classes: (1) *fat*-like cadherins or protocadherins, (2) desmosomal cadherins, and (3) classical cadherins. Although based on common structural motifs (see above), these classes have significant structural differences, and they have distinct functions and subcellular localizations.

### A. *fat*-Like or Protocadherins

There is considerable variation in the structure of the *fat*-like or protocadherin family (Mahoney *et al.,* 1991; Sano *et al.,* 1993), indicating that the criteria for classification of these proteins into one class may be weak. They are defined as members of the cadherin family because they contain "cadherin repeats" in the extracellular domain, and vertebrate members of the protocadherin family have cell adhesion activity (Sano *et al.,* 1993). The *fat* gene of *Drosophila* was identified in mutants that lacked growth control of imaginal disc epithelium, indicating that this gene functions as a tumor suppressor during embryogenesis (Bryant *et al.,* 1988; Bryant and Schmidt, 1990). The cytoplasmic domain of *fat*-like or protocadherin family does not share sequence similarity with the other two cadherin classes, and interactions with cytoplasmic proteins have not been reported (Mahoney *et al.,* 1991; Sano *et al.,* 1993).

## B. Desmosomal Cadherins

Desmosomal cadherins represent the $Ca^{2+}$-dependent cell adhesion molecules of desmosomes (discrete adhesion plaques at cell–cell contacts) (Schwarz *et al.,* 1990; Buxton and Magee, 1992). Desmosomal cadherins consist of two subclasses of proteins termed desmogleins and desmocollins (Buxton *et al.,* 1993). Desmogleins and desmocollins are structurally different, which may reflect different functions within the desmosome; they also comprise several isoforms (Kowalczyk *et al.,* 1994) that are expressed in tissue-specific and developmentally regulated patterns. Desmocollins have two alternatively spliced variants within cytoplasmic domain sequences; the longer desmocollin splice variant, but not the shorter variant, contains sequences similar to those in classical cadherins (Parker *et al.,* 1991). The cytoplasmic domain of desmoglein also contains sequence identical to that of classical cadherin cytoplasmic domains (Goodwin *et al.,* 1990; Koch *et al.,* 1990). On the basis of their similarity to classical cadherins, desmosomal cadherins are thought to form homophilic interactions, but definitive evidence for these interactions is lacking (Kowalczyk *et al.,* 1995). However, when specific antibodies to desmosomal cadherins, or autoantibodies from patients with certain blistering diseases, are exposed to cells, cell–cell adhesion is lost (Cowin *et al.,* 1984; Stanley, 1993). Sequence similarities between desmosomal and classical cadherins in the cytoplasmic domain are also reflected by the fact that plakoglobin (also termed $\gamma$-catenin) interacts with both desmosomal and classical cadherins (see Sections III,A and III,D) (Cowin *et al.,* 1985, 1986; Franke *et al.,* 1987a,b, 1989; Korman *et al.,* 1989; Knudsen and Wheelock, 1992; Peifer *et al.,* 1992; Piepenhagen and Nelson, 1993; Mathur *et al.,* 1994; Troyanovsky *et al.,* 1994; Kowalczyk *et al.,* 1995).

Desmosomes are localized in discrete spotlike, trilaminar plaque structures at the lateral plasma membrane of adjacent epithelial cells (Farquhar and Palade, 1963; Schwarz *et al.,* 1990). The plaque is composed of membrane and peripheral membrane proteins that link cytoplasmic keratin intermediate filaments to the plasma membrane (Schwarz *et al.,* 1990; Kowalczyk *et al.*, 1994). This linkage generates a structural continuum throughout the epithelium that may be required for the maintenance of tissue integrity. Mutations in keratin genes that disrupt keratin filaments in the epidermis also disrupt the structural organization of the epidermis, leading to epidermal blistering (Fuchs, 1994). Desmosome assembly requires initial cell–cell interactions mediated by classical cadherins (Gumbiner *et al.*, 1988; Watabe *et al.*, 1994; Marrs *et al.*, 1995), and may represent a differentiation step that stabilizes epithelial cell adhesion and serves to remodel the intermediate filament cytoskeleton.

### C. Classical Cadherins

Classical cadherins comprise a large family of genes (Edelman and Crossin, 1991; Suzuki *et al.*, 1991; Kemler, 1992), of which there are now more than 40, and include the first members of the cadherin superfamily that were identified (E-, N-, and P-cadherin) (Kemler *et al.*, 1977; Takeichi, 1977; Hyafil *et al.*, 1981; Hatta *et al.*, 1985; Peyrieras *et al.*, 1985; Nose and Takeichi, 1986; Schuh *et al.*, 1986; Shirayoshi *et al.*, 1986; Nose *et al.*, 1987; Ringwald *et al.*, 1987). Classical cadherins share significant structural conservation (Kemler, 1992, 1993). The extracellular domain contains five "cadherin repeats" with 30–60% sequence conservation between different cadherins. Homology in the cytoplasmic domain can reach 90% within the same species, and up to 60% between cadherins from mammals to *Drosophila* (Oda *et al.*, 1994). Classical cadherins regulate $Ca^{2+}$-dependent cell adhesion at the adherens junction, and also at cell–cell contacts where there is not a discrete ultrastructural organization (Boller *et al.*, 1985). The adherens junction is a site for localization of signal transduction molecules (e.g., receptor and nonreceptor tyrosine kinases, Sevenless/Bride of Sevenless, Notch/Delta) (Tsukita *et al.*, 1991, 1992; Woods and Bryant, 1993) and for the attachment of the actin-based membrane–cytoskeleton (Geiger *et al.*, 1985; Geiger, 1989). The juxtaposition of signal transduction systems and the membrane–cytoskeleton indicates a functional linkage between cadherin-mediated cell adhesion (adherens junction) and cellular machinery that may regulate differentiation (signaling), which may in turn generate morphogenetic movements (cytoskeletal rearrangements) during embryogenesis.

## III. Interactions between Cadherins and Cytoplasmic Proteins (Catenins)

Studies have established that members of the cadherin superfamily bind directly to cytoplasmic proteins. The identification and characterization of cadherin-associated proteins have provided important clues to mechanisms of linkage of cadherins to the cytoskeleton and to intracellular signaling pathways.

### A. Hierarchy of Cadherin/Catenin Protein–Protein Interactions

Initial studies showed that antibodies to classical cadherins coimmunoprecipitated a complex that contained three prominent cytoplasmic proteins (Vestweber and Kemler, 1984a; Peyrieras *et al.,* 1985; Vestweber *et al.,* 1987; Nagafuchi and Takeichi, 1989; Ozawa *et al.,* 1989). The coimmunoprecipitated proteins have molecular weights of 102,000, 94,000, and 86,000, and were later termed $\alpha$-, $\beta$-, and $\gamma$-catenin, respectively ($\gamma$-catenin was later identified as plakoglobin). Deletions in the cadherin cytoplasmic domain showed that catenins bind within the carboxy-terminal 70 amino acids of cadherins (Nagafuchi and Takeichi, 1989; Ozawa *et al.,* 1989). More recently, Stappert and Kemler (1994) showed that point mutations in a short, serine-rich, phosphorylated sequence in the center of the cytoplasmic domain is required for cadherin/catenin interactions. In addition, catenins have been shown to bind proteins other than cadherins, including the ademomatosis polyposis coli (APC) tumor suppressor gene product (Rubinfeld *et al.,* 1993, 1995; Su *et al.,* 1993) and the epidermal growth factor (EGF) receptor (Hoschuetzky *et al.,* 1994).

The hierarchy of protein–protein interactions between E-cadherin and different catenins was determined using the yeast two-hybrid system and *in vitro* binding assays with purified, bacterially expressed proteins (Aberle *et al.,* 1994; Jou *et al.,* 1995). These studies have shown that the hierarchy of protein–protein interactions is as follows: cadherin $\leftrightarrow$ {$\beta$-catenin or $\gamma$-catenin} $\leftrightarrow$ $\alpha$-catenin (Fig. 2). This is supported by studies in whole cells that revealed two, mutually exclusive cadherin/catenin complexes: one containing $\beta$-catenin and $\alpha$-catenin, and another containing $\gamma$-catenin and $\alpha$-catenin (see Section III,B) (Hinck *et al.,* 1994a,b; Näthke *et al.,* 1994). Another protein, p120, which shares sequence homology with $\beta$-catenin and $\gamma$-catenin (see Section III,C) (Reynolds *et al.,* 1992; Peifer *et al.,* 1994a), binds to the cadherin cytoplasmic domain (Jou *et al.,* 1995; Shibamoto *et al.,* 1995; Staddon *et al.,* 1995). However, unlike $\beta$-catenin and $\gamma$-catenin, p120 does not bind $\alpha$-catenin in an *in vitro* binding assay (Jou *et al.,* 1995), but other studies showed that p120 associates with $\alpha$-catenin in extracts from whole cells (Jou *et al.,* 1995; Shibamoto *et al.,* 1995); the basis for this difference is not understood.

Yeast two-hybrid analysis, biochemical experiments, and transfection studies have defined sequences required for cadherin $\leftrightarrow$ $\beta$-catenin and $\beta$-

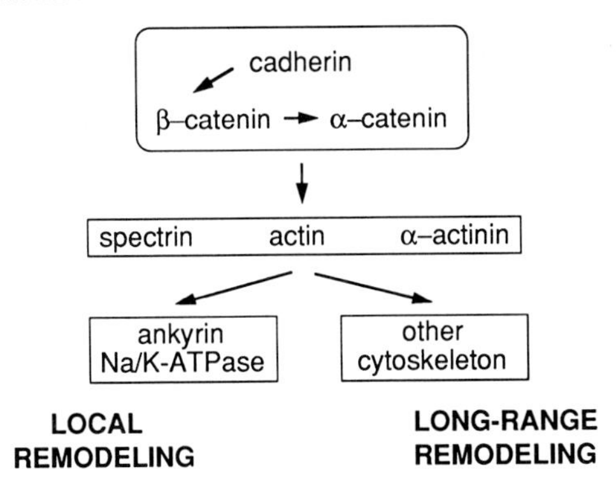


FIG. 2 Hierarchy of protein–protein interactions between cadherins, the cytoskeleton, and other membrane proteins. Biochemical studies have demonstrated protein interactions from cadherins to other membrane proteins through cytoskeletal proteins. These interactions lead to both local remodeling of membrane protein distributions (e.g., $Na^+$,$K^+$-ATPase) and long-range remodeling by changes in cytoskeletal architecture throughout the cell. In this view, cadherin is at the top of the molecular cascade of protein–protein interactions leading to these remodeling events. For details, see text.

catenin ↔ $\alpha$-catenin interactions (Fig. 2). A large number of deletion mutations in the cadherin cytoplasmic domain have been tested for their ability to coimmunoprecipitate the catenin complex, and for cell adhesion (Nagafuchi and Takeichi, 1989; Ozawa *et al.,* 1989; Stappert and Kemler, 1994). Maintenance of catenin binding correlated precisely with the function of cadherin in cell adhesion. The catenin-binding site on the cytoplasmic domain of cadherin has been narrowed down to a short stretch of serine residues in the center of the domain (Stappert and Kemler, 1994). Twenty-five amino acids surrounding this site were also deleted from the cytoplasmic domain of E-cadherin, and protein–protein interactions were tested using the yeast two-hybrid system; the short stretch of serines was shown to be required for binding directly to $\beta$-catenin (Jou *et al.,* 1995). Neither the complete E-cadherin cytoplasmic domain nor the deletion mutant interacted with $\alpha$-catenin (Jou *et al.,* 1995). In transfection studies, $\gamma$-catenin was also not coimmunoprecipitated with mutant cadherins that had the serine-rich domain deleted, indicating that $\beta$- and $\gamma$-catenin bind directly and independently to the same site on the cytoplasmic domain of cadherin (Stappert and Kemler, 1994).

Mapping binding sites on $\beta$-catenin protein by deletion analysis showed that there are two distinct domains required for binding cadherin cytoplasmic domain (Hülsken *et al.,* 1994). Beta-catenin has 13 imperfect repeats, each containing approximately 42 amino acids flanked by short, nonrepeat N- and C-terminal domains (McCrea *et al.,* 1991; Butz *et al.,* 1992). Two subdomains within the core repeating domain are required, cooperatively, to bind cadherin (Hülsken *et al.,* 1994). Different regions of the repeat domain from those that bind cadherin are required for β-catenin binding to APC (Hülsken *et al.,* 1994). The authors of this study also show evidence for competition between cadherins and APC for β-catenin binding.

Using the yeast two-hybrid system, deletion analysis of α-catenin showed that an amino-terminal domain binds to β-catenin (Jou *et al.,* 1995). In another study, Tsukita and colleagues (Nagafuchi *et al.,* 1994) showed that a cadherin/α-catenin chimeric protein functioned in cell–cell adhesion; the chimeric protein contained the extracellular, transmembrane, and amino-terminal half of the cytoplasmic domains of E-cadherin fused to either full-length sequence, or carboxy-terminal half of α-catenin. A cadherin/α-catenin chimera containing only the amino-terminal half of α-catenin did not exhibit cell adhesion function. Considering these two studies together, it appears that α-catenin is an adapter protein containing separate binding sites for different proteins: the amino-terminal half of α-catenin binds β-catenin, and the carboxy-terminal half confers $Ca^{2+}$-dependent cell adhesion function to the cadherin/catenin complex. The role of α-catenin in conferring $Ca^{2+}$-dependent cell adhesion function to the cadherin/catenin complex may be to bind to the actin cytoskeleton (Rimm *et al.,* 1995) (Fig. 2); such interactions might increase the valency of cadherin–cadherin interactions, decrease cadherin diffusion away from adhesive sites, and restrict cadherin molecules to a subdomain, producing a highly adhesive zone on the cell surface.

## B. Dynamics of Cadherin/Catenin Complex Assembly

Early studies of cadherin/catenin complex assembly and organization relied on the coimmunoprecipitation of the complex with cadherin antibodies, and were limited to the analysis of the nonionic detergent Triton X-100 (TX-100)-soluble fraction of these proteins (Nagafuchi and Takeichi, 1989; Ozawa *et al.,* 1989; Shore and Nelson, 1991). These studies concluded that only one complex exists, one that contains cadherin and all of the catenins.

However, more recent studies in Madin-Darby canine kidney (MDCK) epithelial cells have shed light on the hierarchy of assembly and interactions in the cadherin/catenin complex *in vivo* (Hinck *et al.,* 1994b; Näthke *et al.,* 1994). Complexes containing E-cadherin and catenins were separated in sucrose density gradients following protein extraction from cells in buffers containing TX-100. Immediately following synthesis, E-cadherin, β-catenin, and plakoglobin cosedimented as complexes, but α-catenin was not associated with these complexes. Later, a subpopulation of α-catenin joined the complex at a time coincident with the arrival of the E-cadherin–β-catenin (or $\gamma$-catenin) complex at the plasma membrane. Significantly, arrival of E-cadherin at the plasma membrane and addition of $\alpha$-catenin to the complex coincided with an increase in its TX-100 insolubility, indicating that $\alpha$-catenin may link the complex to the actin-based membrane–cytoskeleton.

Previous studies were not able to analyze the TX-100-insoluble complex of cadherin/catenins because extraction in 1% sodium dodecyl sulfate (SDS), required for solubilizing the TX-100-insoluble complex, disrupts the complex. Therefore, reversible chemical cross-linking was used prior to cell extraction with TX-100 and solubilization of the insoluble material with SDS (Hinck *et al.,* 1994b; Näthke *et al.,* 1994). Analysis of cross-linked complexes from cells labeled with [$^{35}$S] methionine/cysteine to steady state indicated that, in addition to cadherin/catenin complexes, there were cadherin-independent pools of catenins present in both the TX-100-soluble and -insoluble fractions. In addition, pulse–chase analysis, combined with chemical cross-linking, showed that immediately following synthesis, cadherin/$\beta$-catenin and cadherin/plakoglobin complexes were present in the TX-100-soluble fraction. Approximately 50% of complexes were titrated into the TX-100-insoluble fraction coincident with the arrival of the complexes at the plasma membrane and the incorporation of $\alpha$-catenin. Subsequently, >90% of labeled cadherin, but no additional labeled catenin complexes, entered the TX-100-insoluble fraction. These results have been interpreted as suggesting that catenins either interact with cadherin complexes synthesized at different times, or there is exchange of catenins between labeled and unlabeled cadherin complexes. In addition, cross-linking revealed additional proteins associated with catenin complexes (Hinck *et al.,* 1994b; Näthke *et al.,* 1994). The identity of these proteins is unknown at present.

The spatial distributions of E-cadherin, $\alpha$-catenin, $\beta$-catenin, and plakoglobin have been determined in detail (Näthke *et al.,* 1994). Protein distributions were visualized by wide-field, optical sectioning, double-immunofluorescence microscopy, followed by reconstruction of three-dimensional images. In cells that were extracted with TX-100 and then fixed (TX-100-insoluble fraction), more E-cadherin is concentrated at the apicolateral junction relative to other areas of the lateral membrane. Alpha-Catenin and $\beta$-catenin colocalize with E-cadherin at the apical junctional complex. There is some overlap in the distribution of these proteins in the lateral membrane, but there are also areas where the distributions are distinct. Plakoglobin is excluded from the apicolateral junction, and its distribution on the lateral membrane is different from that of E-cadherin, consistent with its preferential localization in desmosomes. Cells were also fixed and then permeabilized to reveal the total cellular pool of each protein (TX-100-soluble and -insoluble fractions). This analysis showed more lateral membrane localization of α-catenin, β-catenin, and plakoglobin, and also revealed that they are distributed throughout the cytoplasm.

Chemical cross-linking of proteins and analysis with specific antibodies confirmed the presence at steady state of E-cadherin/catenin complexes containing either β-catenin or plakoglobin, and catenin complexes devoid of E-cadherin. Complexes containing E-cadherin/β-catenin and E-cadherin/α-catenin are present in both the TX-100-soluble and -insoluble fractions, but E-cadherin/plakoglobin complexes are not detected in the TX-100-insoluble fraction. Taken together these results show that different complexes of cadherin and catenins accumulate in fully polarized epithelial cells and that they distribute to different sites. Localization of cadherin/catenin and catenin complexes to different subcellular sites may be important for establishing and maintaining the structural and functional organization of polarized epithelial cells.

Studies on the dynamics of catenin assembly in cultured vascular endothelial cells (Lampugnani *et al.,* 1995) showed low amounts of plakoglobin in the cadherin/catenin complex in sparse cultures and in cells proximal to experimentally produced wounds in endothelial cell monolayers. However, plakoglobin became associated with cadherins after the cells formed mature, confluent monolayers. Levels of cadherin-associated α- and β-catenin were found to be similar under either culture condition. Significantly, the steady state amount of plakoglobin mRNA was selectively upregulated in mature cultures. These data indicate that assembly of the cadherin/catenin complex is also regulated at the level of transcript abundance.

Understanding the hierarchy of interactions in the cadherin/catenin complex provides a foundation for understanding the mechanism for regulating cadherin/catenin complex assembly and cadherin adhesion (Aberle *et al.,* 1994; Jou *et al.,* 1995). Analysis of catenin dynamics has shown that assembly of catenins with other proteins is highly regulated, and that distinct cadherin-associated and cadherin-independent pools of catenins appear to exchange with one another (Hinck *et al.,* 1994b; Näthke *et al.,* 1994). A detailed molecular understanding of how catenin assembly is regulated during specific events (cellular differentiation, metastasis, etc.), and how changes in catenin interactions result in changes in adhesion or propagation of intracellular signals, will be important questions for future studies.

### C. Molecular Characterization of Catenins

cDNAs encoding each of the three catenins have been cloned. The deduced amino acid sequences suggest roles for the cadherin/catenin complex in development (Franke *et al.,* 1989; Herrenknecht *et al.,* 1991; McCrea *et al.,* 1991; Nagafuchi *et al.,* 1991; Butz *et al.,* 1992; Hirano *et al.,* 1992; Reynolds *et al.,* 1992; Claverie *et al.,* 1993; Oda *et al.,* 1993; Furukawa *et al.,* 1994; Uchida *et al.,* 1994). Two genes encoding α-catenin have been identified. One gene is expressed in epithelial and most other cell types (αE-catenin) (Herrenknecht *et al.,* 1991; Nagafuchi *et al.,* 1991), and the other is expressed in neuronal tissues (αN-catenin) (Hirano *et al.,* 1992). Both genes have 25–30% sequence homology with vinculin, a membrane–cytoskeletal protein in focal adhesion plaques that is involved in linkage of integrin adhesion molecules to the actin cytoskeleton. The sequence homology to vinculin indicates that α-catenin may link the cadherin/catenin complex to the actin cytoskeleton (Rimm *et al.,* 1995). The αN-catenin gene also generates two protein isoforms by alternative splicing of a 144-nucleotide insert that encodes a 48-amino acid peptide near the carboxy-terminus (Uchida *et al.,* 1994). This insert sequence shows similarity to a portion of the neurofibromatosis type 1 gene product, a member of the GTPase-activating protein family (Claverie *et al.,* 1993; Uchida *et al.,* 1994). The significance of these alternative splice forms for αN-cadherin function(s) is not understood.

Beta-Catenin, γ-catenin, and p120 share sequence and structural similarity with one another (McCrea *et al.,* 1991; Butz *et al.,* 1992; Reynolds *et al.,* 1992), and sequence analysis and antibody cross-reactivity show that γ-catenin and plakoglobin are identical (Butz *et al.,* 1992; Knudsen and Wheelock, 1992; Peifer *et al.,* 1992; Piepenhagen and Nelson, 1993). Beta-Catenin, γ-catenin, and p120 have up to 65% sequence homology with Armadillo, the product of a *Drosophila* segment polarity gene and component of the *wingless* signaling pathway in early embryogenesis (see Section X) (Peifer and Wieschaus, 1990; McCrea *et al.,* 1991; Butz *et al.,* 1992; Reynolds *et al.,* 1992; Peifer *et al.,* 1994a). In *Drosophila,* the homolog of E-cadherin, termed DE-cadherin, forms a complex with Armadillo and the homolog of α-catenin (Oda *et al.,* 1994).

Analysis of the nucleotide sequence of *armadillo* revealed the presence of 12 imperfect repeats of 42 amino acids that form a "central rod domain," flanked by amino- and carboxy-terminal nonrepeat domains. The function(s) of these Armadillo repeats is unknown (see Section III,A). Both β-catenin and plakoglobin contain 12 Armadillo repeats. Interestingly, several other proteins also contain Armadillo repeats, including SRP1, adenomatosis polyposis coli (APC) protein, p120, and smgGDS. It has been suggested that the Armadillo repeat motif is important in protein–protein interactions (Peifer *et al.,* 1994a).

### D. Desmosomal Catenins

*armadillo* gene family members are also localized to the desmosome plaque. Desmoglein binds directly to plakoglobin (γ-catenin) (Mathur *et al.,* 1994; Troyanovsky *et al.,* 1994). However, $\alpha$-catenin is not incorporated into the desmoglein/plakoglobin complex, in contrast to the classical cadherin/catenin complex of the adherens junction (Peifer *et al.,* 1992). Because the desmosome plaque associates with keratin and not actin filaments, the lack of $\alpha$-catenin in the desmoglein/plakoglobin complex may help specify the interaction with keratin, rather than actin, filaments.

Another component of the desmosome plaque, band 6, shares sequence identity with members of the *armadillo* gene family (Hatzfeld *et al.,* 1994), suggesting an interaction of band 6 with desmosomal cadherins. Finally, it remains unclear how "desmosomal catenins" link to other proteins such as desmoplakins, a major desmosomal plaque component that directly associates with the keratin filament cytoskeleton (Kouklis *et al.,* 1994). Future studies should define other protein–protein interactions within the desmosome that link desmosomal cadherins and catenins to desmoplakins and the keratin filament cytoskeleton.

## IV. Catenin Function: Requirement for Cadherin-Mediated Cell–Cell Adhesion and Linkage to the Actin-Based Cytoskeleton

Formation of the cadherin/catenin complex is required for the adhesive function of classical cadherins (Fig. 3). Deletion of the catenin-binding site(s) in the cadherin cytoplasmic domain results in loss of both cadherin-

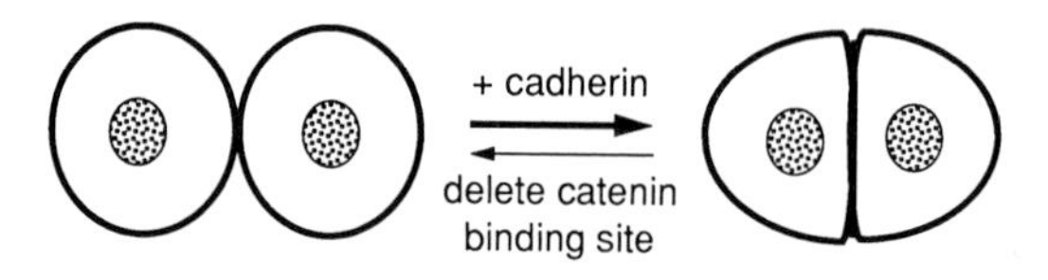


•strengthen cell-cell adhesion (compaction)
•cell sorting
•permissive for assembly of intercellular junctions
•redistribution of membrane & cytoskeletal proteins
•change cell phenotype

FIG. 3 Consequence of cadherin-mediated cell–cell adhesion. Cadherin expression leads to cell–cell adhesion. Strong cell adhesion is manifested by close attachment of adjacent cells (compaction). Loss of interactions between cadherins and catenins results in loss of cell–cell adhesion. Cadherin-mediated adhesion leads to dramatic changes in cell phenotype that are characterized by assembly of other junctional complexes (tight junction, desmosomes, gap junctions), and the redistribution of membrane and cytoskeletal proteins. For details, see text.

mediated adhesion, and cadherin association with catenins and the actin cytoskeleton (Nagafuchi and Takeichi, 1988, 1989; Ozawa *et al.*, 1989). A consequence of the loss of cell–cell adhesion is that cells exhibit increased invasiveness and metastatic potential (see Section V,D).

Loss of $\alpha$-catenin expression also results in loss of cadherin function and $Ca^{2+}$-dependent cell–cell adhesion (Hirano *et al.*, 1992; Shimoyama *et al.*, 1992; Breen *et al.*, 1993). Cell–cell adhesion is rescued by expression of exogenous $\alpha$-catenin, demonstrating that cadherin-mediated cell–cell adhesion requires the association of $\alpha$-catenin with the cadherin/catenin complex (Hirano *et al.*, 1992; Watabe *et al.*, 1994). A cadherin/$\alpha$-catenin chimera, comprising the extracellular and transmembrane domains, and the amino-terminal half of the cytoplasmic domain of E-cadherin fused to $\alpha$-catenin, was found to associate with a detergent insoluble cell fraction, indicating binding to the cytoskeleton, and confers strong $Ca^{2+}$-dependent cell–cell adhesion (Nagafuchi *et al.*, 1994).

Studies indicate that $\beta$-catenin acts as a linker protein between cadherin and $\alpha$-catenin. Studies have identified cells that express truncated $\beta$-catenin, and have shown that this defect coincides with a loss of cell–cell adhesion (Oyama *et al.*, 1994). This mutant $\beta$-catenin forms a complex with E-cadherin but not with $\alpha$-catenin; deletions in $\beta$-catenin were identified only in the amino-terminus, indicating that this region contains the binding site for $\alpha$-catenin. This has been confirmed independently by deletion analysis of $\beta$-catenin and *in vitro* binding experiments, showing that the $\alpha$-catenin-binding site is in the amino-terminus of $\beta$-catenin (Aberle *et al.*, 1994). Also, mosaic analysis of *armadillo* in *Drosophila* embryos revealed that deletions leading to protein truncation in the carboxy-terminus result in loss of cell–cell adhesion and alterations in the actin cytoskeleton (Peifer *et al.*, 1993b). Because the only binding site for $\alpha$-catenin on $\beta$-catenin is at the amino-terminus (Aberle *et al.*, 1994), it is unclear at present why deletions in the carboxy-terminus of Armadillo lead to defects in cellular organization. It is possible that Armadillo/$\beta$-catenin bind to proteins, in addition to cadherin and $\alpha$-catenin, that are important in cellular organization.

The cadherin/catenin complex may coordinate assembly of other membrane–cytoskeletal proteins at sites of cell–cell adhesion (Geiger, 1989; Tsukita *et al.*, 1992). For example, cadherins form a complex, either directly or indirectly, with the membrane–cytoskeletal proteins ankyrin and fodrin (Nelson *et al.*, 1990). Linkage between the cadherin/catenin complex and the ankyrin/fodrin membrane–cytoskeleton plays an important role in the generation of epithelial cell surface polarity (see Sections V,A ad V,B) (McNeill *et al.*, 1990; Nelson *et al.*, 1990; Marrs *et al.*, 1993, 1995). Preliminary evidence suggests that the cadherin/catenin complex also associates with other cytoskeletal proteins (Fig. 2), including association with $\alpha$-actinin (Knudsen *et al.*, 1995) and direct interaction of $\alpha$-catenin with spectrin and actin (Lombardo *et al.,* 1994; Rimm *et al.,* 1995). Defining the molecular nature of these protein–protein interactions, and their importance in cellular organization, will be important future goals.

## V. Cadherin Function in Regulating Changes in Cell Structure and Function

Classical cadherins, in addition to regulating the physical recognition and adhesion between cells, perform specialized functions that indicate that they are morphoregulatory molecules during embryogenesis (see Section VIII) and differentiation. Some functions (e.g., remodeling protein distributions and the establishment of cell surface polarity) can be attributed to downstream consequences of the adhesive function of cadherins and subsequent assembly of the membrane–cytoskeleton (Fig. 2). Other functions (e.g., changes in gene expression) may be initiated by cadherin-mediated cell–cell contact and involve immediate and early intracellular signaling events (McNeill *et al.,* 1993).

### A. Effects of Cadherin Expression in Fibroblasts

Expression of classical cadherins in fibroblasts results inphenotypic transformation of cells toward an epithelial cell type (Edelman *et al.,* 1987; Nagafuchi *et al.,* 1987; Ozawa and Kemler,1990). Cadherin expression induces $Ca^{2+}$-dependent cell–cell adhesion, resulting in the growth of cells in colonies, similar to epithelial cells, rather than single cells (Fig. 3). As a consequence of cadherin-mediated cell–cell adhesion, the actin cytoskeleton reorganizes into a prominent circumferential ring of membrane-associated actin filaments (Hirano *et al.,* 1987), which is also characteristic of epithelial cells in colonies. It is noteworthy, however, that cadherin-expressing fibroblast and epithelial cell lines differ in the way that the actin cytoskeleton is reorganized during cell contact formation. In fibroblast cell lines, adherens junction formation occurs in small patches where actin filaments attach perpendicularly; in epithelial cell lines, initial stages of cadherin-mediated cell contacts resemble that of fibroblasts, but a circumferential belt of actin eventually develops (Yonemura *et al.,* 1995). The difference in actin organization between endogenous cadherin-expressing fibroblasts and cadherin transfected fibroblasts may reflect the level of cadherin expression.

These changes in cellular organization require not only the expression of cadherin, but also the formation of the cadherin/catenin complex. Deletion of the catenin-binding site in the cytoplasmic domain of cadherin inhibits cell–cell adhesion, changes in cell shape and actin organization, and the development of cell surface polarity in transfected fibroblasts (Fig. 3) (Nagafuchi and Takeichi, 1988, 1989; Ozawa *et al.*, 1989; McNeill *et al.*, 1990).

The epithelial transformation of cadherin-expressing fibroblasts is partial. For example, neither desmosomes nor gap junctions are formed in L cells expressing cadherin. This suggests that additional cues are required for further epithelial differentiation of these cells (McNeill *et al.*, 1990). S180 cells constitutively express gap junction subunits (connexins), but only assemble functional gap junctions following expression of exogenous cadherin (Musil *et al.*, 1990). Therefore, cadherin expression appears to be permissive for assembly of junctional complexes that are constitutively expressed in these cell types (Musil *et al.*, 1990; Jongen *et al.*, 1991).

Studies in other cells indicate that expression of catenins is also required for protein complex assembly at sites of cell–cell adhesion. In some tumor cell lines such as PC9 cells, E-cadherin is expressed at normal levels, but $\alpha$-catenin is not expressed, resulting in little or no cell–cell adhesion (Hirano *et al.*, 1992; Shimoyama *et al.*, 1992; Breen *et al.*, 1993) and lack of tight junctions and desmosomes (Watabe *et al.*, 1994). Transfection of PC9 cells with $\alpha$-catenin cDNA restores E-cadherin function (Hirano *et al.*, 1992; Watabe *et al.*, 1994), and results in assembly of tight junctions and desmosomes (Watabe *et al.*, 1994).

In addition to formation of different intercellular junctional complexes, cadherin expression in polarized cells may establish and maintain distributions of other membrane proteins in distinct membrane subdomains (Fig. 2). A characteristic of epithelial cells is the nonrandom distribution of membrane proteins between structurally and functionally distinct membrane domains, termed apical and basal-lateral (Rodriguez-Boulan and Nelson, 1989). For example, $Na^+$,$K^+$-ATPase distribution is restricted to the basal-lateral membrane in most epithelial cells. In fibroblasts, $Na^+$,$K^+$-ATPase is randomly distributed in the plasma membrane. Significantly, expression of E-cadherin L cells results in the redistribution of $Na^+$,$K^+$-ATPase to membrane only at sites of cadherin-mediated cell–cell contact; this is similar to its localization to the lateral membrane domain of renal epithelial cells (McNeill *et al.*, 1990). Examination of the distributions of other proteins detected with the lectin wheat germ agglutinin revealed that cadherin-induced redistribution of $Na^+$,$K^+$-ATPase is specific.

E-cadherin-induced redistribution of $Na^+$,$K^+$-ATPase coincides with a similar redistribution of the membrane–cytoskeletal protein, fodrin (McNeill *et al.*, 1990). In fibroblasts, fodrin is diffusely distributed throughout the cytoplasm and at the plasma membrane. Expression of E-cadherin results in redistribution of fodrin to sites of cell–cell contact similar to that of E-cadherin and $Na^+$,$K^+$-ATPase, indicating that these proteins are linked to one another. Significantly, both the cadherin/catenin complex (Nelson *et al.,* 1990) and $Na^+,K^+$-ATPase associate with fodrin (Nelson and Veshnock, 1987; Koob *et al.,* 1988; Morrow *et al.,* 1989; Nelson and Hammerton, 1989; Davis and Bennett, 1990; Devarajan *et al.,* 1994); fodrin may interact directly with $\alpha$-catenin (see Section IV) (Lombardo *et al.,* 1994), and $Na^+,K^+$-ATPase associates with fodrin through their mutual binding to ankyrin (Nelson and Veshnock, 1987; Koob *et al.,* 1988; Morrow *et al.*). These results support a hypothesis that the redistribution of $Na^+,K^+$-ATPase to sites of cell–cell contact is a consequence of its assembly into a supramolecular cytoskeletal complex comprising $Na^+,K^+$-ATPase $\leftrightarrow$ ankyrin $\leftrightarrow$ fodrin $\leftrightarrow$ $\alpha$-catenin $\leftrightarrow$ $\beta$-catenin $\leftrightarrow$ cadherin. Significantly, when the link between cadherin and catenins is broken (by deletion of the catenin-binding site on cadherin), then the distributions of neither $Na^+,K^+$-ATPase nor fodrin become polarized in transfected fibroblasts (McNeill *et al.,* 1990). Together, these observations indicate that E-cadherin provides positional information on the cell surface that leads to assembly of the membrane–cytoskeleton and recruitment of $Na^+,K^+$-ATPase to sites of cell–cell contact (Nelson, 1992).

Results from the L cell system indicate that E-cadherin-induced assembly of the membrane–cytoskeleton and redistribution of $Na^+,K^+$-ATPase may be a general mechanism for restricting $Na^+,K^+$-ATPase distribution to the basal-lateral membrane in epithelial cells. Indeed, studies in polarized MDCK epithelial cells demonstrate that E-cadherin-mediated cell–cell adhesion results in the assembly of the membrane–cytoskeleton and redistribution of $Na^+,K^+$-ATPase to sites of cell–cell adhesion (Nelson and Veshnock, 1986; Nelson and Hammerton, 1989; Wang *et al.,* 1990). An important consequence of the incorporation of $Na^+,K^+$-ATPase into the membrane–cytoskeleton is that it is retained in the membrane for longer times. Studies of MDCK cells show that the half-life of $Na^+,K^+$-ATPase on the basal-lateral membrane, where membrane–cytoskeleton assembly is induced, is 30 times longer than that in the apical membrane, where the membrane–cytoskeleton is not induced (Hammerton *et al.,* 1991; Seimers *et al.,* 1993). Incorporation of $Na^+,K^+$-ATPase into the membrane–cytoskeletal complex may restrict the rate of $Na^+,K^+$-ATPase diffusion in the plane of the lipid bilayer, and sequester $Na^+,K^+$-ATPase from the endocytosis machinery. Consequently, $Na^+,K^+$-ATPase is retained and accumulates at those membrane sites (Nelson, 1992).

### B. Cadherin Expression and $Na^+,K^+$-ATPase Polarity in Choroid Plexus and Retinal Pigmented Epithelia

In contrast to its distribution in the basal-lateral membrane domain in reabsorptive epithelia such as renal cells, $Na^+,K^+$-ATPase is localized to the apical membrane domain in choroid plexus and retinal pigment epithelia (Wright, 1972; Quinton *et al.,* 1973; Steinberg and Miller, 1979; Bok, 1982). Similar to renal epithelial cells, $Na^+,K^+$-ATPase is complexed and codistributed with ankyrin and fodrin in the apical membrane of the choroid plexus (Marrs *et al.,* 1993) and retinal pigmented epithelia (Gundersen *et al.,* 1991). In contrast to other epithelia, however, the choroid plexus and retinal pigment epithelium do not express E-cadherin (Gundersen *et al.,* 1991; Marrs *et al.,* 1993). Instead, these cells express B-cadherin (in the chicken) (Murphy-Erdosh *et al.,* 1994), a cadherin closely related to but distinct from E-cadherin (Napolitano *et al.,* 1991; Sorkin *et al.,* 1991). Significantly, expression of B-cadherin in L cells induces $Ca^{2+}$-dependent cell–cell adhesion, but it does not result in a redistribution of either $Na^+,K^+$-ATPase or fodrin to sites of cell–cell adhesion (Marrs *et al.,* 1993). Because choroid plexus epithelial cells do not form cell–cell contacts at the apical cell surface, localization of $Na^+,K^+$-ATPase to the apical membrane domain in choroid plexus epithelium indicates that a mechanism other than differential stabilization by cadherin-mediated membrane–cytoskeleton assembly is responsible for organizing $Na^+,K^+$-ATPase distribution in these cells. For example, $Na^+,K^+$-ATPase may be delivered directly from the Golgi complex to the apical membrane and subsequently retained there by assembly with the membrane–cytoskeleton. Because the choroid plexus epithelium has a free apical cell surface (does not form cell–cell contacts), this implies that membrane–cytoskeleton assembly and retention of $Na^+,K^+$-ATPase at the apical plasma membrane is cadherin independent in this cell type. This is in contrast to that of E-cadherin-expressing epithelia (e.g., kidney or intestine).

Differences in the capacity of cadherins to induce the redistribution of $Na^+,K^+$-ATPase to different membrane domains have been further examined in a cell line derived from retinal pigmented epithelium, termed RPE-J (Nabi *et al.,* 1993). RPE-J cells form a structurally polarized monolayer in culture, and display several phenotypic features of retinal pigmented epithelial cells *in situ,* including apical membrane phagocytosis, expression of an RPE-specific antigen, and a high transepithelial electrical resistance (Nabi *et al.,* 1993). In dissociated primary cell culture (Rizzolo, 1990, 1991) and in RPE-J cells (Nabi *et al.,* 1993), the membrane distribution of $Na^+,K^+$-ATPase is nonpolarized. Therefore, interactions with the neural retina that abuts the apical membrane of the retinal pigmented epithelium *in situ* may be required for establishing and maintaining the apical $Na^+,K^+$-ATPase distribution, either directly by cell–cell contact-mediated assembly of the membrane–cytoskeleton, or indirectly by regulating protein-trafficking pathways between the Golgi complex and the cell surface (Gundersen *et al.,* 1993).

RPE-J cells express a cadherin related to P- or B-cadherin, but not E-cadherin (Marrs *et al.,* 1995). Therefore, these cells provide an interesting paradigm to test the hypothesis that E-cadherin-mediated cell–cell adhesion induces $Na^+$,$K^+$-ATPase redistribution in a bona fide epithelial cell. Stable expression of E-cadherin in RPE-J cells (RPE-J + EC cells) results in restriction of $Na^+$,$K^+$-ATPase distribution to sites of cell–cell adhesion (Table I) (Marrs *et al.,* 1995), thus confirming results obtained in MDCK cells and by transfection of E-cadherin cDNAs into fibroblasts. Because RPE-J + EC cells form a tight cell monolayer, it was possible to investigate mechanisms involved in the redistribution of $Na^+$,$K^+$-ATPase. Results showed that newly synthesized $Na^+$,$K^+$-ATPase is delivered from the Golgi complex to both apical and basal-lateral membrane domains. However, $Na^+$,$K^+$-ATPase was retained in the basal-lateral membrane and rapidly removed from the apical membrane (Marrs *et al.,* 1995). This result indicated that $Na^+$,$K^+$-ATPase redistribution in RPE-J + EC cells was the result of localized assembly of the membrane–cytoskeleton.

Studies in L cells and MDCK cells showed that selective retention of $Na^+$,$K^+$-ATPase in the basal-lateral membrane coincides with the recruitment and assembly of the membrane–cytoskeleton (Nelson and Hammerton, 1989; McNeill *et al.,* 1990). Significantly, expression of E-cadherin in RPE-J cells had profound effects on both the expression and distribution of the membrane–cytoskeleton (Marrs *et al.,* 1995). In RPE-J cells, the ankyrin-3 isoform (also termed $ankyrin_G$) is constitutively expressed. However, E-cadherin expression induced the synthesis and accumulation of a different ankyrin isoform, either ankyrin-1 and/or -2 (also termed $ankyrin_R$ and $ankyrin_B$, respectively), which coincided with the redistribution of $Na^+$,$K^+$-ATPase to the basal-lateral membrane in RPE-J + EC cells (Table I). In RPE-J + EC cells, amounts of both fodrin and ankyrins increased significantly due to increased protein stability. These data support the im-

TABLE I
Comparison of RPE-J and RPE-J + EC Cells

| Characteristic | RPE-J | RPE-J + EC |
|---|---|---|
| Tight junction | Yes | Yes |
| Desmosomes | No | Yes |
| Desmoglein synthesis | No | Yes |
| Keratin synthesis | Yes | Yes |
| Keratin filaments | No | Yes |
| $Na^+$,$K^+$-ATPase | Apical/lateral | Lateral |
| Ankyrin synthesis | Ankyrin-3 | Ankyrin-3, -2, -1 |
| Ankyrin distribution | (Lateral) | Lateral |
| Fodrin distribution | (Lateral) | Lateral |

portant role of E-cadherin expression in the induction of cell surface polarity of $Na^+$,$K^+$-ATPase, and the role of the membrane–cytoskeleton in generating and maintaining $Na^+$,$K^+$-ATPase polarity. In addition, because expression of E-cadherin induced changes in the patterns of ankyrin isoform synthesis, it is likely that cadherin expression also affects gene transcription, mRNA splicing, or mRNA translation.

E-cadherin expression also affected the assembly of other intercellular junctions in RPE cells. Although tight junctions are expressed constitutively in RPE-J cells, their location along the lateral membrane was variable and occasionally coincided with the apicolateral membrane junction as in most other epithelia (Nabi *et al.,* 1993). Expression of E-cadherin resulted in a refinement of the location of tight junctions to the apex of the lateral membrane, at the boundary with the apical membrane (Marrs *et al.,* 1995).

E-cadherin expression in RPE-J + EC cells also resulted in *de novo* assembly of desmosomes (Table I) (Marrs *et al.,* 1995). Desmosomes are not expressed in either rat retinal pigmented epithelial cells *in situ* (Owaribe *et al.,* 1988), or in RPE-J cells (Marrs *et al.,* 1995). The peripheral desmosomal proteins, desmoplakins and plakoglobin, are constitutively expressed in RPE-J and RPE-J + EC cells. In contrast, a desmosomal cadherin, desmoglein, was not detected in RPE-J cells at either the mRNA or protein levels. However, E-cadherin expression in RPE-J + EC cells induced accumulation of both desmoglein mRNA and protein (Table I).

At present, the mechanism(s) involved in E-cadherin induction of desmoglein transcript accumulation are unknown, but data from RPE-J cells strongly support the concept that E-cadherin-specific intracellular signals influence epithelial phenotype. Interestingly, cadherin-mediated cell adhesion has been shown to regulate intracellular signaling processes that lead to neurite outgrowth (Neugebauer *et al.,* 1988; Tomaselli *et al.,* 1988; Bixby and Zhang, 1990; Drazba and Lemmon, 1990; Doherty *et al.,* 1991a,b; Walsh and Doherty, 1992; Bixby *et al.,* 1994). Similar signaling mechanisms may also occur in retinal pigmented epithelial cells, leading to observed changes in epithelial phenotype.

### C. Cadherin Expression in Keratinocytes

Cadherins play important roles in keratinocyte differentiation (Hirai *et al.,* 1989; Wheelock and Jensen, 1992; Hodivala and Watt, 1994). During stratification, suprabasal cells downregulate expression of involucin, peanut lectin-binding glycoproteins, and integrins (Watt and Green, 1982; Watt, 1983; Morrison *et al.,* 1988; Hodivala and Watt, 1994). Inhibition of cadherin function, either with function-blocking antibodies or low concentrations of extracellular calcium, blocked stratification and maintained expression of involucin, peanut lectin-binding glycoprotein, and integrin (Hodivala and Watt, 1994). Integrin expression is controlled at the level of transcript accumulation. These data suggest that cadherin function regulates integrin gene expression. However, these processes may also be controlled by stratification and the loss of basal substrate adhesion, which have not been uncoupled from inhibition of cadherin function.

High integrin expression has been shown to be a marker of stem cells in the epidermis (Jones *et al.,* 1995). Keratinocytes reduce integrin expression after transition from the stem cell population to transit-amplifying cells, which subsequently undergo terminal differentiation and lose integrin expression. Watt and colleagues (Jones *et al.,* 1995) suggest that cadherin molecules may regulate these transitions, establish pattern in the epidermis, and determine keratinocyte fate.

### D. Cadherin Expression in Tumor Cells

In general, expression of cadherins is downregulated in many types of tumor cells, and correlates with increased metastatic potential and invasiveness (Takeichi, 1991, 1993; Birchmeier *et al.,* 1993). Cell adhesion and normal growth in these tumor cells were rescued by expression of high levels of cadherins (Fig. 3) (Vleminckx *et al.,* 1991). In some tumor cells, however, cadherin expression is high but cadherin function is low, resulting in high metastatic potential (Matsuyoshi *et al.,* 1992); in some of these cells, $\alpha$-catenin expression is absent (Shimoyama *et al.,* 1992; Breen *et al.,* 1993).

In other tumor cell lines, $\beta$-catenin is expressed with an amino-terminal truncation (Oyama *et al.,* 1994). This aberrant $\beta$-catenin interacts with E-cadherin but does not interact with $\alpha$-catenin, thereby disrupting linkage of $\alpha$-catenin with the cadherin/$\beta$-catenin complex. Surprisingly, complexes of cadherin/plakoglobin are formed in these cells, but $\alpha$-catenin did not coimmunoprecipitate with E-cadherin. Note that plakoglobin interacts with $\alpha$-catenin (Aberle *et al.,* 1994; Jou *et al.,* 1995). Therefore, it is unclear why in these cell lines the cadherin/plakoglobin complex does not incorporate $\alpha$-catenin into the cadherin/catenin complex which may allow cadherin function. It would be interesting to show whether transfection of normal $\beta$-catenin cDNA rescues cadherin-mediated cell–cell adhesion in these cells, as was shown for the $\alpha$-catenin-deficient cell line PC9 (Hirano *et al.,* 1992; Watabe *et al.,* 1994), or whether truncated $\beta$-catenin is acting as a dominant negative mutant subunit or otherwise disrupting the function of the cadherin/catenin complex.

Cadherin-mediated cell–cell adhesion is decreased in cells that have high levels of tyrosine kinase activity, such as that induced by expression of v-*src* (Volberg *et al.,* 1991; Matsuyoshi *et al.,* 1992; Behrens *et al.,* 1993;

Hamaguchi *et al.,* 1993). Loss of cell–cell adhesion coincides with increased phosphorylation of tyrosine residues in all proteins in the cadherin/catenin complex, although the levels are highest in $\beta$-catenin. Ligand-induced activation of some membrane receptor tyrosine kinases (e.g., EGF $\rightarrow$ EGF-R, HGF (hepatocyte growth factor) $\rightarrow$ *c*-Met) also results in increased levels of phosphotyrosine in the cadherin/catenin complex, especially that of $\beta$-catenin, decreased $Ca^{2+}$-dependent cell–cell adhesion, and increased cell migration (Watabe *et al.,* 1993; Hoschuetzky *et al.,* 1994; Shibamoto *et al.,* 1994). However, the mechanism involved in disruption of cell–cell adhesion following tyrosine phosphorylation of $\beta$-catenin is not understood. Increased levels of phosphotyrosine do not result in the disassembly of the cadherin/catenin complex. It is possible that target tyrosine residues for specific kinases alter the conformation of these proteins which then disrupts protein interactions between the cadherin/catenin complex, the cytoskeleton, and other proteins.

## VI. Desmosomal Cadherins and Effects on Cell Function and Phenotype

The function of desmosomal cadherins in cell–cell adhesion and cytoskeleton organization has been tested by expression of chimeric proteins comprising the extracellular and transmembrane domains of gap junction proteins (connexin) fused to the cytoplasmic domain of either desmoglein 1, or each of the two splice variants of desmocollin 1 (Troyanovsky *et al.,* 1993). Constructs were expressed in A-431 cells, which constitutively express desmosomes. Connexin/desmoglein 1 chimeric proteins localized to the cell surface, but desmosomes were absent, indicating that the chimera disrupted endogenous desmosome organization, perhaps by competing for binding of cytoplasmic proteins important for desmosome organization (i.e., the chimera behaved as a dominant negative mutation) (Troyanovsky *et al.,* 1993).

Expression of a chimeric protein comprising connexin and the longer splice variant of desmocollin, which contains sequences homologous to the cytoplasmic domain of classical cadherins, resulted in the formation of electron-dense plaques on the plasma membrane that morphologically resembled desmosome plaques attached to gap junctions. These plaque structures contained desmoplakins and plakoglobin, but endogenous desmoglein was excluded (Troyanovsky *et al.,* 1993). A chimeric protein containing the short splice variant of desmocollin, which does not have sequences homologous to classical cadherins, also formed gap junction-like structures, but these structures did not colocalize desmoplakins or plakoglobin (Troyanovsky *et al.,* 1993). These results demonstrate the importance of desmosomal cadherin cytoplasmic sequences for assembly of desmosome plaque structures, and the data illustrate how desmoglein and desmocollin sequences affect plaque assembly differently.

In addition to mediating cell–cell adhesion, desmosomes are also membrane attachment sites for the keratin cytoskeleton. The maintenance of the structural integrity of the tissues, such as the epidermis, may require the presence of both desmosomes and associated keratin filaments. Mutations in keratins that disrupt keratin filament organization do not disrupt keratinocyte growth in tissue culture (Fuchs, 1994). However, expression of these mutant proteins in epidermal tissues *in vivo* caused cytolysis following mechanical stress, resulting in severe blistering of the epidermis (Coulombe *et al.,* 1991; Vassar *et al.,* 1991).

Assembly of desmosomal plaques is mediated by desmosomal cadherins. Binding of keratin filaments to desmosomes appeared to be mediated by the peripheral plaque component, desmoplakins. Expression of mutant desmoplakins in cultured epithelial cells results in disruption of the keratin filament network, similar to that induced by keratin mutations (Stappenbeck and Green, 1992; Stappenbeck *et al.,* 1993, 1994; Kouklis *et al.,* 1994). Direct interactions between desmoplakins and keratin proteins were demonstrated *in vitro* (Kouklis *et al.,* 1994). In fact, the domain of keratin protein that interacts with desmoplakins is the same domain that is most frequently mutated in human blistering diseases (Kouklis *et al.,* 1994). These data indicate that epithelial structures that experience mechanical stress require a three-dimensional scaffold of keratin filaments linked to the desmosome cell–cell contact sites, resulting in cell-to-cell integration of tissue structure.

## VII. Do Desmosomal Cadherins Have Morphoregulatory Roles during Embryogenesis?

Desmosome formation is regulated during early mouse development (Fleming and Johnson, 1988; Fleming *et al.,* 1991, 1993). However, expression of desmosomal plaque components and desmosomal glycoproteins is not coincident. In late morula stage embryos, plakoglobin appears first and is localized at the cell surface, which may reflect its association with classical cadherin. Membrane staining of desmoplakins is detected at the early blastocyst stage. However, the desmosomal cadherins, desmoglein and desmocollin, do not appear at the cell surface until the late blastocyst stage. Expression of desmocollins appears to be controlled at the level of transcription (Collins *et al.,* 1995). Expression of desmosomal cadherins correlates with the morphological appearance of desmosomes and blastocoel expansion. Studies in which desmosomal components are mutated or deleted during development have not been reported, and there are no mutants in genes encoding desmosomal components known in any organism. Thus, the function of desmosomes during blastocyst formation and subsequent embryogenesis remains an open question. However, the demonstration that dominant negative mutants of desmoglein (Troyanovsky *et al.,* 1993) and desmoplakins (Stappenbeck and Green, 1992; Stappenbeck *et al.,* 1993, 1994; Kouklis *et al.,* 1994) disrupt desmosome assembly and attachment to keratin filaments may provide impetus to address experimentally these questions using transgenic mouse technology.

## VIII. Cadherin Function in Embryogenesis

### A. Selective Cell Adhesion is Cadherin Mediated

Classic studies showed that mixed suspensions of single cells, which had been dissociated from different embyronic tissues, sort out and reassociate into topologically correct layers of tissues (Holtfreter, 1944; Moscona, 1952; Moscona and Moscona, 1952; Townes and Holtfreter, 1955; Steinberg, 1964). For example, mixtures of cells from mesoderm and epidermis sorted into specific cell types and formed an aggregate in which epidermal cells adhered on the outside and mesodermal cells adhered in the center. This cell-sorting phenomenon is regulated by selective adhesion between cells derived from the same tissue type (i.e., homophilic adhesion), and is $Ca^{2+}$dependent. Sorting out of embryonic cell suspensions *in vitro* is reminiscent of normal processes in development in which cells migrate through and associate with new cell layers, even of different embryological origin (Trinkaus, 1965; Takeichi, 1988).

With the identification of cadherins as the family of cell adhesion proteins responsible for $Ca^{2+}$-dependent cell interactions, it became possible to test directly whether selective adhesion between cells was responsible for cell sorting (Takeichi *et al.,* 1981; Nose *et al.,* 1988; Takeichi, 1988; Friedlander *et al.,* 1989; Miyatani *et al.,* 1989; Steinberg and Takeichi, 1994). Fibroblasts, expressing either E- or P-cadherin, were mixed and found to sort into two separate groups of cells, each of which expressed the same cadherin (Fig. 4); similar results were obtained by pairwise combinations of cells expressing N-, P-, and E-cadherin. Furthermore, mixing single-cell suspensions of E-cadherin-transfected L cells and dissociated embyronic lung epithelia resulted in the incorporation of transfected L cells into lung epithelial tissue-like structures (Nose *et al.,* 1988). In control experiments, untransfected L

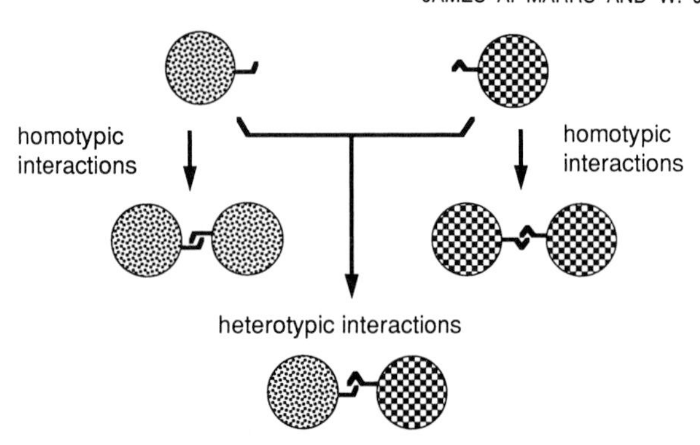


FIG. 4 Cell sorting by homotypic adhesion. In general, cadherins form homotypic interactions with the same cadherin type on adjacent cells. When two populations of cells expressing different cadherins are mixed, this binding property specifies interactions between cells expressing the same cadherin, resulting in sorting of cell populations into two distinct groups. However, heterotypic interactions also exist. For details, see text.

cells segregated and associated with mesenchymal regions of the aggregates (Nose *et al.,* 1988).

Although cell sorting involves predominantly homophilic cell interactions, heterophilic interactions also exist (Fig. 4). B- and E-cadherin (L-CAM) engage in heterophilic interactions (Murphy-Erdosh *et al.,* 1995); B- and E-cadherin are often coexpressed in chick tissues (Murphy-Erdosh *et al.,* 1994), suggesting that heterophilic interactions may be utilized *in vivo.* R- and N-cadherin also form heterophilic combinations (Inuzuka *et al.,* 1991). This is consistent with the spatial organization of neurites in motor and sensory fiber neurons in the spinal cord in which mixed and separate neurite fascicles exist that express both N- and R-cadherin (Redies *et al.,* 1992; Redies and Takeichi, 1993b). In addition, in the optic nerve, R-cadherin expressing glial cells may promote outgrowth of N-cadherin expressing retinal neurites during their projection back to the brain (Redies and Takeichi, 1993b).

### B. Developmentally Regulated Expression of Cadherins

The developmental expression patterns of cadherins also suggest a role for cadherins in cell sorting of specific cell types during morphogenetic cell movements (Takeichi, 1988, 1990). During neurulation in mammals, ectodermal cells change cadherin expression from E- to N-cadherin (Thiery *et al.,* 1984; Hatta and Takeichi, 1986; Hatta *et al.,* 1987; Duband *et al.,* 1988).

This may allow segregation of neural precursor cells from other cells derived from the ectoderm. Subsequently, neural crest cells migrate from the dorsal ectoderm to specific locations in different germ layers. During migration, the cells down-regulate N-cadherin expression (Aoyama *et al.,* 1985; Hatta and Takeichi, 1986; Hatta *et al.,* 1987; Duband *et al.,* 1988; Ranscht and Bronner-Fraser, 1991). Migrating neural crest cells may initiate expression of different cadherin types during migration in order to navigate their pathway through various tissues (Aoyama *et al.,* 1985), or to orchestrate *en masse* migration of neural crest cell populations due to coordinate expression of a particular cadherin. Expression analysis suggests that cadherin-11 is expressed in migrating neural crest cells in mouse (Hoffmann and Balling, 1995; Kimura *et al.,* 1995; Nakagawa and Takeichi, 1995). On arrival at a target location, neural crest cells express cadherins present on the target tissue. Hence, morphogenetic cell movements appear to be regulated by an initial downregulation of cadherin expression that allows dissociation of cells from the tissue of origin, and then reinitiation of expression at target tissues to allow cell recognition and adhesion.

Posttranslational mechanisms may also regulate cadherin function during morphogenetic cell movements. Using activin-induced *Xenopus* blastula animal caps as a model system for studying gastrulation-like morphogenetic movements, Brieher and Gumbiner showed that C-cadherin-mediated cell adhesion activity of blastomeres is downregulated by activin treatment (Brieher and Gumbiner, 1994). Reduction of cadherin-mediated cell adhesion occurred without reduction in the amount of C-cadherin on blastomere cell surfaces, suggesting that cadherin activity was regulated posttranslationally (e.g., $\beta$-catenin phosphorylation; see Section V,D). It was suggested that the activin-regulated cadherin-mediated cell–cell adhesion allows cells to orchestrate morphogenetic movements by using cell–cell interactions as a cell migration substrate (Brieher and Gumbiner, 1994). The idea that cells use cadherin contacts as a migration substrate is supported by experiments using transfected fibroblasts. L cells transfected with full-length E-cadherin migrate within a monolayer of E-cadherin expressing cells, but L cells expressing a cadherin/$\alpha$-catenin chimera comprising the extracellular, transmembrane, and amino-terminal half of the cytoplasmic domain of E-cadherin fused to $\alpha$-catenin, conferred strong $Ca^{2+}$-dependent cell–cell adhesion, but do not permit migration within the monolayer (Nagafuchi *et al.,* 1994). Mechanisms for regulating cadherin function are unknown, but may involve $\beta$-catenin (see Section V,D) (Hinck, 1994). The ability to regulate cadherin activity may be lost in the cadherin/$\alpha$-catenin chimera because cadherin association of $\alpha$-catenin is no longer regulated by its association with $\beta$-catenin (Nagafuchi *et al.,* 1994).

## C. The Requirement for Cadherin-Mediated Cell–Cell Adhesion During Embryogenesis

What are the consequences of cadherin-mediated cell–cell adhesion on the organization and fate of cells in the developing embryo? This question has been examined principally during early mammalian and amphibian development. Compaction of preimplantation embryos in eutherians, which is the first sign of overt cellular differentiation, absolutely requires changes in $Ca^{2+}$-dependent cell–cell adhesion mediated by E-cadherin (Hyafil *et al.*, 1981; Vestweber and Kemler, 1984a, 1985; Vestweber *et al.*, 1987; Johnson *et al.*, 1988). During compaction of the early morula, blastomeres on the outer surface of the early morula increase cell–cell adhesion and flatten on one another. Compaction eventually gives rise to an outer layer of polarized transporting epithelial cells, termed the trophectoderm, while the blastomeres in the center of the embryo remain loosely aggregated (the inner cell mass). E-cadherin is expressed even in the loosely adhering blastomeres before compaction (Johnson *et al.*, 1988); then, at compaction, E-cadherin function is somehow activated. Although the mechanism(s) involved in the increase in adhesivity is poorly understood, there is evidence that activation of protein kinase C and increased serine/threonine phosphorylation of E-cadherin are involved (Bloom, 1989).

The requirement for E-cadherin in early embryogenesis has been demonstrated unequivocally in mice in which the gene has been ablated (Larue *et al.*, 1994). The genetic knockout results in embyronic death prior to implantation. Although compaction is initiated in the early morula, because E-cadherin is expressed from maternal mRNAs, the blastomeres rapidly lose intercellular adhesion and disaggregate. The embryos fail to hatch from the zona pellucida and do not implant. Interestingly, embryonic stem cells derived from E-cadherin −/− mice are viable, grow in tissue culture, and express the epithelial marker protein cytokeratin, indicating that some differentation occurs perhaps as a consequence of the presence of maternal E-cadherin (Larue *et al.*, 1994).

The functions of cadherins in later stages of development in multicellular organisms have been examined in *Xenopus* embryos by injecting synthetic mRNAs encoding normal and mutant versions of different classical cadherins. In the first studies, the effect of ectopic, overexpression of full-length N-cadherin was investigated (Detrick *et al.*, 1990; Fujimori *et al.*, 1990). Embyros displayed an abnormal phenotype that appeared to be due to disorganization of the ectoderm and abnormal histogenesis. A mutant N-cadherin, composed of only the transmembrane and cytoplasmic domains, was expressed in order to create a dominant negative phenotype by competing with endogenous cadherin for binding to catenins (Kintner, 1992). This mutant also resulted in disruption of cell–cell adhesion and dissaggregation/

disorganization of tissues in the early embryo. Kintner (1992) presented evidence that the mutant protein competes for catenin binding with endogenous cadherin. Interestingly, a mutant N-cadherin, which contained only the membrane-proximal 27 amino acids of the cytoplasmic domain (which does not contain the catenin-binding site, and so would not compete for catenin binding), also disrupted cell–cell adhesion and tissue integrity (Kintner, 1992). This suggests that a region outside the catenin-binding site on the cadherin cytoplasmic domain must play an as yet unidentified role in cadherin function. Similar truncated, cadherin cytoplasmic domain constructs have been expressed in tissue culture cells (Fujimori and Takeichi, 1993). Examination of the mutant phenotype in this study suggested that competition for catenin binding was not occurring, but there may be competition for some other, unknown cytoplasmic proteins leading to defects in endogenous cadherin function (Fujimori and Takeichi, 1993).

Specific disruption of tissues was achieved by injection of mutant N- or XP-cadherin constructs, similar to those described above, into the four animal blastomeres of 32-cell embryos. In this way, anterior neural structures could be targeted (Dufour *et al.,* 1994). Expression of these proteins disrupted neural morphogenesis: the emergence of cranial nerves was delayed; eye development was abnormal due to fusion to the diencephalon floor; and cell layers in the retina were disorganized (Dufour *et al.,*1994). A similar phenotype was observed by inhibition of cadherin function using antibodies during retinal histogenesis (Matsunaga *et al.,* 1988). Together, these results show that the maintenance of normal cell–cell adhesion and tissue organization requires specific levels of cadherin function (either too little or too much cadherin activity induces abnormalities), and that normal interactions between cadherins and catenins or other unidentified proteins are required.

Disruption of enterocyte cadherin function was evaluated by expressing a dominant negative, cadherin transmembrane and cytoplasmic domain construct at high levels specifically in enterocytes in transgenic mice (Hermiston and Gordon, 1995). Enterocyte cell–cell adhesion, cell-substrate contacts, the actin cytoskeleton, and cell polarity were disrupted. Also, enterocytes expressing mutant cadherin displayed increased rates of migration in the crypt–villus axis. Increased apoptosis in enterocytes expressing the dominant negative cadherin was also observed (Hermiston and Gordon, 1995). Interestingly, loss of cell–substratum contact can cause apoptosis (programmed cell death) in epithelial cells (Frisch and Francis, 1994; Montgomery *et al.,* 1994). These and other data (Watabe *et al.,* 1994) support the concept that cell–cell and cell–substratum contacts converge on common signaling pathways, or have feedback signaling pathways in which these two types of cell adhesion systems influence one another.

Studies have also examined roles of cadherin-mediated cell–cell adhesion in development by use of mutant cadherins encoding only the extracellular domain (Levine *et al.,* 1994). Such mutant E- or N-cadherins resulted in developmental abnormalities that were restricted to cells that either expressed E-cadherin (e.g., ectodermal lesions during gastrulation) or N-cadherin (e.g., neural tube), respectively. Normal development was rescued by coinjecting full-length E-cadherin and truncated E-cadherin constructs, indicating that the truncated protein acted as a dominant negative mutation by competing for extracellular interactions with endogenous E-cadherin. Coinjection of full-length C-cadherin with truncated E-cadherin did not rescue embryos from ectodermal defects, suggesting that homophilic interactions with E-cadherin are responsible for rescue. However, injection of C-cadherin alone produced ectodermal lesions like those produced by truncated E-cadherin (Levine *et al.,* 1994), demonstrating an overexpression phenotype. Again, these data show the requirement for specific levels of cadherin function, but the data also show the importance of the cadherin extracellular domain for tissue organization and maintenance.

## IX. Role of Catenins in Development

Developmentally regulated patterns of expression of the two $\alpha$-catenin isoforms have been described in the mouse (Nagafuchi and Tsukita, 1994; Uchida *et al.,* 1994). The more general isoform, $\alpha$E-catenin, was found to be expressed in all cadherin-expressing tissues at the neurula stage, even in the neural tube and heart, which eventually express only N-cadherin (Nagafuchi and Tsukita, 1994). Later, in 12.5-day-old embryos, expression of $\alpha$E-catenin was progressively lost from the central nervous system tissues (Nagafuchi and Tsukita, 1994). This was concomitant with an increase in the expression of $\alpha$N-catenin in those tissues (Uchida *et al.,* 1994). Similar $\alpha$-catenin isoform expression patterns occur during early development of newt and zebrafish (Nagafuchi and Tsukita, 1994); evolutionary conservation suggests these are important developmental events. Although it is not known whether the $\alpha$E- and $\alpha$N-isoforms of catenin are functionally different, it has been suggested that they play differential roles in neural and nonneural cell–cell interactions, and the switch in isoform expression may be essential for CNS development. Future experiments will be necessary to test these hypotheses, but these observations demonstrate the potential importance of cell-type-specific and developmentally regulated expression of different $\alpha$-catenin isoforms.

## X. β-Catenin and *Wnt/wingless* Signaling Pathway

Because β-catenin and plakoglobin are mammalian homologs of the segment polarity gene *armadillo* in *Drosophila* (McCrea *et al.,* 1991; Butz *et al.,* 1992), insights into the function of this gene family in development are provided by analysis of effects of mutations in segment polarity genes on *Drosophila* development (Peifer *et al.,* 1993a). After cellularization of the syncytial blastula to form the cellular blastula, cell–cell interactions are induced that establish segment identity by localized expression of segment polarity genes, which in turn results in further elaboration of patterning and specification of new cellular identities. A subset of the segment polarity genes, which can be grouped together by genetic criteria, forms a signal transduction pathway that includes *armadillo.* At the top of the hierarchy of this subset of genes is *wingless.* There are several homologs of *wingless* in both invertebrate and vertebrate organisms that constitute the *Wnt* gene family (Nusse and Varmus, 1992).

Epistasis analysis of genes in the *wingless* signaling pathway has established a pathway of interactions that leads to the induction of *engrailed* gene expression in neighboring cells (Fig. 5) (Noordermeer *et al.,* 1994; Peifer *et al.,* 1994b; Siegfried *et al.,* 1994). Wingless is a secreted glycoprotein that provides a graded paracrine signal within the embyro from a discrete subset of *wingless*-expressing cells. The receptor for Wingless on cells expressing *engrailed* has not yet been identified genetically or biochemically. The putative wingless receptor is thought to activate the *disheveled* gene product, which inhibits a serine/threonine kinase encoded by the *shaggy* or *zeste-white 3* gene, which in turn inhibits further downstream signaling through *armadillo;* the function of *armadillo* in the *wingless* pathway is not

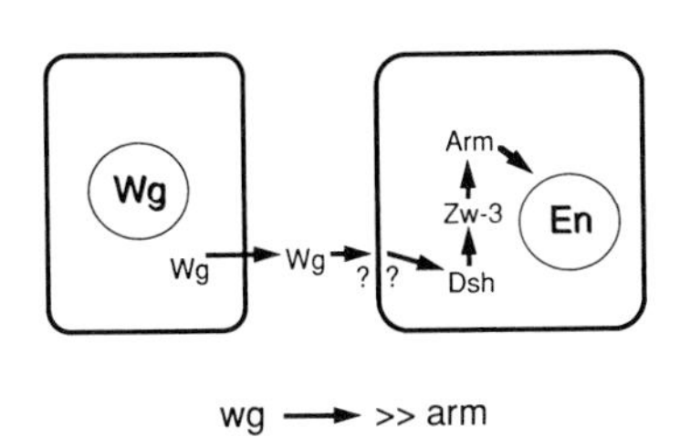


FIG. 5 Simplified *wingless* signaling pathway in *Drosophila.* In *Drosophila* embryos, *wingless* is secreted and binds to an unknown receptor (?) on the adjacent cell, which precipitates a signal transduction cascade in that cell. This signal transduction cascade involves a hierarchy of genetic interactions between *disheveled* (*dsh*), *zest-white 3* (*zw-3*), and *armadillo* (*arm*), and results in the activation of *engrailed* (*en*) transcription. For details, see text.

known at present. Analysis has shown that mutations in *wingless* and *zeste-white 3* increase and decrease, respectively, the levels of phosphoserine/threonine and phosphotyrosine in Armadillo (Peifer *et al.*, 1995). Activation of the *wingless* signaling pathway results in the accumulation of Armadillo in a narrow strip of *engrailed*-expressing cells that are immediately adjacent to cells that express *wingless*. Significantly, *armadillo* mRNA is expressed throughout the embryo (Riggleman *et al.*, 1990). Accumulation of Armadillo leads to the activation of *engrailed* transcription through an unknown mechanism. Other pathways exist for the maintenance of *engrailed* and *wingless* transcription by feedback loops (Hooper, 1994), but the one outlined above describes the general features of a signal transduction pathway initiated by transcription of *wingless* and ending in *engrailed* gene activation (Fig. 5).

Analysis of mutations in *armadillo* has identified domains of the protein that are required for the *wingless* signal transduction pathway. Increasingly large truncations of Armadillo from the carboxy-terminal resulted in a corresponding increase in severity of segment polarity defects (Peifer and Wieschaus, 1990). Note that binding sites for cadherin and $\alpha$-catenin are located in the core domain and amino-terminal domain, respectively, of mammalian $\beta$-catenin (see Sections III,A and IV). Although Armadillo binds to *Drosophila* (DE-) cadherin (Oda *et al.*, 1994), the fact that the domains of Armadillo involved in the *wingless* signaling pathway and binding to cadherin or $\alpha$-catenin are different may indicate that the cadherin/catenin complex is not directly involved in the *wingless* signaling pathway.

Several studies have sought to determine the role of the vertebrate homologs of *wingless* in development. One of the first assays was to examine the developmental consequences of injecting synthetic *Wnt-1* mRNA into *Xenopus* oocytes (McMahon and Moon, 1989). Ectopic *Wnt-1* expression increased gap junction communication between bastomeres (Olson *et al.*, 1991), and subsequently induced duplication of the embryonic axis (McMahon and Moon, 1989). Injection of *Wnt-8* transcripts had an effect similar to that of *Wnt-1*, that is, increased gap junctional communication, but *Wnt-5A* had no effect (Olson *et al.*, 1991). The outcome of these experiments suggests that *Wnt* genes may function to determine dorsal–ventral polarity in the early *Xenopus* embryo.

Other vertebrate homologs of the *wingless* signaling pathway (Fig. 5) have been shown to effect dorsal–ventral polarity in *Xenopus* embryos, giving more support to a role for *Wnt* signaling in Spemann's organizer. In the *wingless* signaling pathway, *dishevelled* is activated by *wingless;* ecoptic overexpression by injecting mRNA encoding the *Xenopus dishevelled* homolog into prospective ventral mesodermal cells induces dorsal axis duplication (Sokol *et al.*, 1995). The vertebrate homolog of the *shaggy* or *zeste-white 3* gene is glycogen synthase-3 (GSK-3) (Siegfried *et al.*, 1992). In the *wingless* signaling pathway, *dishevelled* inhibits the activity of the *zeste-*

*white 3* gene (Noordermeer *et al.,* 1994; Peifer *et al.,* 1994; Siegfried *et al.,* 1994). Injection of dominant negative GSK-3 induced dorsalized embryos (duplication of the embryonic axis) and wild-type GSK-3 induced ventralization (He *et al.,* 1995; Pierce and Kimelman, 1995). Beta-Catenin is the vertebrate homolog of the *armadillo* gene (McCrea *et al.,* 1991). Duplication of the embryonic axis was produced by ectopic overexpression of $\beta$-catenin (Funayama *et al.,* 1995). In contrast, however, injection of antibodies to $\beta$-catenin also induced duplication of the embryonic axis, although it was suggested that the antibodies activated, rather than inhibited, $\beta$-catenin function (McCrea *et al.,* 1993). Interestingly, overexpression of $\beta$-catenin constructs that contain only the Armadillo repeats appears to be sufficient to induce duplication of the embryonic axis (Funayama *et al.,* 1995). Although it has been inferred from these results that the functions of $\beta$-catenin in *Xenopus* development and those of *armadillo* in *Drosophila* development are similar, note that domains of Armadillo that are important in development are different.

Studies have sought to examine the function of *Wnt-1* in mammalian cells. Using *Wnt-1*-responsive cells, these studies showed that *armadillo* gene family members, $\beta$-catenin and plakoglobin, accumulate following expression of Wnt-1 (Bradley *et al.,* 1993; Hinck *et al.,* 1994c). The amounts of cadherin and $\alpha$-catenin did not change on expression of *Wnt-1.* Accumulation of $\beta$-catenin and plakoglobin was regulated at the posttranslational level by increased metabolic stability of protein bound to cadherin, which in turn increased the strength of cell–cell adhesion (Hinck *et al.,* 1994c).

Increasing cadherin function may be an important role of*Wnt* genes during embryogenesis, perhaps in the formation of cellular boundaries within the embryo (Hinck *et al.,* 1994c): tightening adhesion and decreasing migration in *Wnt/wingless*-expressing regions, and loosening adhesion and increasing migration outside the *Wnt/wingless*-expressing regions. These effects may be translated into specific morphogenetic movements that are controlled by *Wnt/wingless* gene expression. Although intriguing, these ideas need to be tested.

*Wnt* signaling may also affect desmosome cell–cell adhesion. Given the presence of *armadillo* gene family members in the desmosome (plakoglobin and band 6), one can speculate about possible developmental signaling roles (*Wnt/wingless* signaling roles) for these molecules by analogy to the $\beta$-catenin and *armadillo* functions.

The role of *Wnt* gene expression in embryonic development has been demonstrated by using targeted gene disruption in transgenic mice. The developmental defects observed may reflect alterations in the cadherin/catenin complex; where this was tested, however, the effects differed from that of the *wingless* signaling pathway in *Drosophila* embryogenesis (Shimamura *et al.,* 1994). *Wnt-1* homozygous null mutant leads to embryonic death due to the complete lack of cerebellum development (McMahon and Bradley, 1990; Thomas and Capecchi, 1990). Using the *Wnt-1* knockout mouse and a more mild, spontaneous, recessive mutation in *Wnt-1,* the expression patterns of components of the cadherin/catenin complex were compared to their normal expression patterns (Shimamura *et al.,* 1994). There was no change in the expression of *armadillo* gene family members, $\beta$-catenin and plakoglobin. However, the distribution of E-cadherin, which is expressed in limited regions of the CNS, and $\alpha$N-catenin was altered in the *Wnt-1* mutant mice. Normally, expression of E-cadherin and $\alpha$N-catenin coincides temporally and spatially with *Wnt-1* in the brain (Shimamura and Takeichi, 1992). In *Wnt-1* mutant embryos, however, the onset of E-cadherin expression is earlier, and expression is found in a broader region of the developing brain (Shimamura *et al.,* 1994). In contrast, expression of $\alpha$N-catenin was suppressed in response to *Wnt-1* mutations. Changes in expression of both E-cadherin and $\alpha$N-catenin as a consequence of *Wnt-1* mutations occurred at the transcriptional level (Shimamura *et al.,* 1994). This is in contrast to the accumulation of Armadillo protein in *Drosophila* which was regulated at the posttranslational level by *wingless* signaling (Riggleman *et al.,* 1990).

In Table II, we outline the differences between the *Wnt* signaling responses in the cadherin/complex observed in different experimental systems. These changes include differences in the target proteins and the mechanisms in which the changes occur (transcriptional versus posttransla-

TABLE II
*Wnt-1* Signaling Effects on Cadherin Cell Adhesion System

| Potential effects | *Drosophila* development and mammalian tissue culture systems | Mouse brain development |
|---|---|---|
| Effects on *armadillo* gene family members | Increase in protein levels by posttranslational mechanism | No effect |
| Effects on cadherin levels and distribution | Little or no increase in levels, no change in distribution | Maintains restricted distribution, bifurcating pattern by transcriptional mechanism |
| Effects on cadherin activity | Increased cell adhesion | Unknown |
| Effects on a $\alpha$-catenin | No effect | Increase in neural isoform by transcriptional mechanism |
| Effects on *engrailed* expression | Induced in *Drosophila* | Altered distribution pattern by incubation with E-cadherin antibodies |

tional). Despite these differences, it is interesting to note that inhibition of E-cadherin-mediated adhesion in embryonic mouse brain in organ culture altered *engrailed* gene expression patterns (Shimamura and Takeichi, 1992). Additional investigation will be required to sort out these differences in *Wnt* signaling and how the different pathways converge on the *engrailed* target (Table II).

Homozygous null mutants for *Wnt-4* do not complete differentiation of the metanephric mesenchyme, and do not form mature kidneys (Stark *et al.,* 1994). Detailed analysis showed that the metanephric mesenchyme was induced by the ureteric bud to form condensates of future epithelial cells, the initial step in tubule formation. However, the developmental process did not proceed to the next stage, aggregation of the mesenchymal tissues that leads to epithelial cell differentiation. Because *Wnt-1* modulates cadherin-mediated cell adhesion (Hinck *et al.,* 1994c; Shimamura *et al.,* 1994) and E-cadherin expression normally follows the onset of *Wnt-4* expression in the kidney (Vestweber *et al.,* 1985), it is possible that mesenchymal aggregation and tubulogenesis may be controlled by *Wnt-4* through the regulation of cadherin/catenin complex activity in cell–cell adhesion.

Finally, *Wnt-7a* null mutant mice have been generated. The phenotype implicates *Wnt-7a* in controlling the polarity of both dorsal–ventral and anterior–posterior axes, and integrating signals between both dorsal–ventral and anterior–posterior systems during limb development (Parr and McMahon, 1995). No specific cadherin/catenin complex effects were reported, but it is interesting that embryonic axis polarity is also regulated by *Wnt*/β-catenin signaling in *Xenopus* (McMahon and Moon, 1989; McCrea *et al.,* 1993; Funayama *et al.,* 1995; He *et al.,* 1995).

## XI. Summary and Perspectives

In this article, we have outlined significant experimentation that illustrates the various roles of cadherin cell adhesion molecules during development. New roles for cadherin adhesion molecules during embryogenesis are likely to be discovered. For example, the demonstration that cadherin-mediated cell adhesion regulates cell division rates (Watabe *et al.,* 1994) suggests that cadherins may also control cell growth during embryogenesis. The mechanisms used by the cadherin/catenin complex for pattern formation and determination of cell fates are only beginning to be understood. Determining the molecular mechanisms for modulating cadherin function (e.g., by growth factors and other signaling molecules) during development are also important questions. Our understanding of these mechanisms is crucial for explaining how cadherin function is regulated during complex morphogenetic movements. Experiments have demonstrated that cadherins regulate differentation (Marrs *et al.,* 1995). Considering that the number of cadherin gene superfamily members is growing steadily, and that developmentally regulated expression patterns have been described for many cadherins, it is likely that examples of cadherin-mediated differentiation events will also be numerous. Specific roles for desmosomal cadherins in development remain an open question, but these molecules are positioned within the embryo to perform important functions, especially during differentiation events.

Several disease processes are influenced by the uncoupling of cadherin cell adhesion molecule attachment to membrane–cytoskeleton. Desmosomal cadherin function may be an important factor used by the keratin filament network for maintenance of tissue integrity in the epidermis. Desmosome/keratin filament association is disrupted in certain hereditary blistering diseases that disrupt keratin filament assembly (Fuchs, 1994). The potential role of desmosome/keratin filament association in hereditary blistering disease is supported by the observed pathophysiological consequences of blistering diseases that result from production of autoantibodies against desmosomal cadherins (Stanley, 1993). Regulation of classical cadherin function during tumorigenesis and metastatic progression has been well documented. Increased metastatic potential results from downregulation of cadherin cell adhesion molecule expression, or by mutations that disrupt the connection of cadherins with the membrane–cytoskeleton (Takeichi, 1991, 1993; Hirano *et al.,* 1992; Matsuyoshi *et al.,*1992; Shimoyama *et al.,* 1992; Birchmeier *et al.,* 1993; Breen *et al.,* 1993; Oyama *et al.,* 1994; Watabe *et al.,* 1994). The future holds significant promise for theraputic intervention to limit pathogenesis in certain diseases by modulating cell adhesion molecule function. However, more basic understanding of factors regulating cadherin function will be required for design, and testing the efficacy of potential theraputic agents.

Finally, genetic approaches will continue to reveal new information about cadherin function. With the discovery of *Drosophila* cadherins (Mahoney *et al.,* 1991; Oda *et al.,* 1994), powerful genetic techniques available in this organism (isolation of new mutant alleles, mosaic analysis, genetic transformation) will allow investigators to address important structure–function issues, and assess the role of the cadherin/catenin complex in the *wingless*-signaling pathway. The demonstration of the absolute requirement for E-cadherin expression during early embryonic development by producing E-cadherin-deficient mice using targeted gene disruption techniques (Larue *et al.,* 1994) will surely be followed by targeted disruption of genes encoding other cadherins. Combining these genetic approaches with detailed phenotypic analysis, using our growing insight into cadherin cell biology, will provide a complete understanding of the consequences of cadherin expression on the determination of cell types and morphogenetic movements in the developing embryo.

## Acknowledgments

We thank our colleagues, Drs. Birchmeir, Buxton, Collins, Garrod, Green, Grunwald, Knudson, Nagafuchi, Peifer, Ranscht, Kemler, Stanley, Suzuki, Takeichi, Tsukita, Watt, and Weis, for providing reprints and preprints of their articles. These were extremely helpful for preparing this article. We also thank Dr. Simon Atkinson for helpful and critical comments.

## References

Aberle, H., Butz, S., Stappert, J., Weissig, H., Kemler, R., and Hoschuetzky, H. (1994). Assembly of the cadherin–catenin complex in vitro with recombinant proteins. *J. Cell Sci.* **107,** 3655–3663.

Aoyama, H., Delouvee, A., and Thiery, J. P. (1985). Cell adhesion mechanisms in gangliogenesis studied in avian embryo and in a model system. *Cell Differ.* **17,** 247–260.

Behrens, J., Mareel, M. M., Van, R. F., and Birchmeier, W. (1989). Dissecting tumor cell invasion: Epithelial cells acquire invasive properties after the loss of uvomorulin-mediated cell–cell adhesion. *J. Cell Biol.* **108,** 2435–2447.

Behrens, J., Vakaet, L., Friis, R., Winterhager, E., Van Roy, F., Mareel, M. M., and Birchmeier, W. (1993). Loss of epithelial differentiation and gain of invasiveness correlates with tyrosine phosphorylation of the E-cadherin/$\beta$-catenin complex in cells transformed with temperature-sensitive v-*src* gene. *J. Cell Biol.* **120,** 757–766.

Birchmeier, W., Weidner, K. M., and Behrens, J. (1993). Molecular mechanisms leading to loss of differentiation and gain of invasiveness in epithelial cells. *J. Cell Sci. Suppl.* **17,** 159–164.

Bixby, J. L., Grunwald, G. B., and Bookman, R. J. (1994). $Ca^{2+}$ influx and neurite growth in response to purified N-cadherin and laminin. *J. Cell Biol.* **127,** 1461–1475.

Bixby, J. L., and Zhang, R. (1990). Purified N-cadherin is a potent substrate for the rapid induction of neurite outgrowth. *J. Cell Biol.* **110,** 1253–1260.

Bloom, T. L. (1989). The effects of phorbol ester on mouse blastomeres: A role for protein kinase C in compaction? *Development* **106,** 159–171.

Bok, D. (1982). Autoradiographic studies on the polarity of plasma membrane receptors in RPE cells. *In* "The Structure of the Eye" (J. Hollyfield, ed.) pp. 247–256. Elsevier North Holland, New York.

Boller, K., Vestweber, D., and Kemler, R. (1985). Cell-adhesion molecule uvomorulin is localized in the intermediate junctions of adult intestinal epithelial cells. *J Cell Biol.* **100,** 327–332.

Bradley, R. S., Cowin, P., and Brown, A. M. (1993). Expression of Wnt-1 in PC12 cells results in modulation of plakoglobin and E-cadherin and increased cellular adhesion. *J. Cell Biol.* **123,** 1857–1865.

Breen, E., Clark, A., Steele, G., and Mercurio, A. M. (1993). Poorly differentiated colon carcinoma cell lines deficient in $\alpha$-catenin expression express high levels of surface E-cadherin but lack $Ca^{+2}$-dependent cell–cell adhesion. *Cell Adhes. Commun.* **1,** 239–250.

Brieher, W. M., and Gumbiner, B. M. (1994). Regulation of C-cadherin function during activin induced morphogenesis of *Xenopus* animal caps. *J. Cell Biol.* **126,** 519–527.

Bryant, P. J., Huettner, B., Held, L. I., Ryerse, J., and Szidonya, J. (1988). Mutations at the *fat* locus interfere with cell proliferation control and epithelial morphogenesis in *Drosophila. Dev. Biol.* **129,** 541–554.

Bryant, P. J., and Schmidt, O. (1990). The genetic control of cell proliferation in *Drosophila* imaginal discs. *J. Cell Sci. Suppl.* **17,** 171–181.

Buck, C. A. (1992). Immunoglobulin superfamily: Structure, function and relationship to other receptor molecules. *Semin. Cell Biol.* **3,** 179–188.

Butz, S., Stappert, J., Weissig, H., and Kemler, R. (1992). Plakoglobin and beta-catenin: Distinct but closely related. *Science* **257,** 1142–1143.

Buxton, R. S., Cowin, P., Franke, W. W., Garrod, D. R., Green, K. J., King, I. A., Koch, P. J., Magee, A. I., Rees, D. R., Stanley, J. R., and Steinberg, M. S. (1993). Nomenclature of the desmosomal cadherins. *J. Cell Biol.* **121,** 481–483.

Buxton, R. S., and Magee, A. I. (1992). Structure and interactions of desmosomal and other cadherins. *Semin. Cell Biol.* **3,** 157–167.

Claverie, J. M., Hardelin, J. P., Legouis, R., Levilliers, J., Bougueleret, L., Marrei, M.-G., and Petit, C. (1993). Characterization and chromosomal assignment of a human cDNA encoding a protein related to the murine 102-kDa cadherin-associated protein ($\alpha$-catenin). *Genomics* **15,** 13–20.

Collins, J. E., Lorimer, J. E., Garrod, D. R., Pidsley, S. C., Buxton, R. S., and Fleming, T. P. (1995). Regulation of desmosomal transcription in mouse preimplantation embryos. *Development* **121,** 743–753.

Coulombe, P. A., Hutton, M. E., Vassar, R., and Fuchs, E. (1991). A function for keratins and a common thread among different types of epidermolysis bullosa simplex diseases. *J. Cell Biol.* **115,** 1661–1674.

Cowin, P., Kapprell, H.-P., and Franke, W. W. (1985). The complement of desmosomal plaque proteins in different cell types. *J. Cell Biol.* **101,** 1442–1454.

Cowin, P., Kapprell, H.-P., Franke, W. W., Tamkun, J., and Hynes, R. O. (1986). Plakolobin: A protein common to different kinds of intercellular adhering junctions. *Cell* **46,** 1063–1073.

Cowin, P., Mattey, D., and Garrod, D. (1984). Identification of desmosomal surface components (desmocollins) and inhibition of desmosome formation by specific Fab′. *J. Cell Sci.* **70,** 41–60.

Davis, J., and Bennett, V. (1990). The anion exchanger and $Na^+$,$K^+$-ATPase interact with distinct sites on ankyrin in in vitro assays. *J. Biol. Chem.* **265,** 17252–17256.

Detrick, R. J., Dickey, D., and Kintner, C. R. (1990). The effects of N-cadherin misexpression on morphogenesis in *Xenopus* embryos. *Neuron* **4,** 493–506.

Devarajan, P., Scaramuzzino, D. A., and Morrow, J. S. (1994). Ankyrin binds to two distinct cytoplasmic domains of Na,K-ATPasea subunit. *Proc. Natl. Acad. Sci. USA* **91,** 2965–2969.

Doherty, P., Ashton, S. V., Moore, S. E., and Walsh, F. S. (1991a). Morphoregulatory activities of NCAM and N-cadherin can be accounted for by G protein-dependent activation of L- and N-type neuronal $Ca^{2+}$ channels. *Cell* **67,** 21–33.

Doherty, P., Rowett, L. H., Moore, S. E., Mann, D. A., and Walsh, F. S. (1991b). Neurite outgrowth in response to transfected N-CAM and N-cadherin reveals fundamental differences in neuronal responsiveness to CAMs. *Neuron* **6,** 247–258.

Drazba, J., and Lemmon, V. (1990). The role of cell adhesion molecules in neurite outgrowth on Muller cells. *Dev. Biol.* **138,** 82–93.

Duband, J.-L., Volberg, T., Sabany, I., Thiery, J. P., and Geiger, B. (1988). Spatial and temporal distribution of the adherens-junction-associated adhesion molecule A-CAM during avian embryogenesis. *Development* **103,** 325–344.

Dufour, S., Saint-Jeannet, J.-P., Broders, F., Wedlich, D., and Thiery, J. P. (1994). Differential perturbations in the morphogenesis of anterior structures induced by overexpression of truncated XB- and N-cadherins in *Xenopus* embryos. *J. Cell Biol.* **127,** 521–535.

Edelman, G. M. (1985). Cell adhesion and the molecular process of morphogenesis. *Annu. Rev. Biochem.* **54,** 135–169.

Edelman, G. M. (1986). Cell-adhesion molecules in the regulation of animal form and tissue pattern. *Annu. Rev. Cell Biol.* **2,** 81–116.

Edelman, G. M., and Crossin, K. L. (1991). Cell adhesion molecules: Implications for a molecular histology. *Annu. Rev. Biochem.* **60,** 155–190.

Edelman, G. M., Murray, B. A., Mege, R., Cunningham, B. A., and Gallin, W. J. (1987). Cellular expression of liver and neural cell adhesion molecules after transfection with their cDNAs results in specific cell-cell binding. *Proc. Natl. Acad. Sci. USA* **84,** 8502–8506.

Farquhar, M. G., and Palade, G. E. (1963). Junctional complexes in various epithelia. *J. Cell Biol.* **17,** 375–412.

Fleming, T. P., Garrod, D. R., and Elsmore, A. J. (1991). Desmosome biogenesis in the mouse preimplantation embryo. *Development* **112,** 527–539.

Fleming, T. P., Javed, Q., Collins, J., and Hay, M. (1993).Biogenesis of structural intercellular junctions during cleavage in mouse embryo. *J. Cell Sci. Suppl.* **17,**119–125.

Fleming, T. P., and Johnson, M. H. (1988). From egg to epithelium. *Ann. Rev. Cell Biol.* **4,** 459–485.

Franke, W. W., Goldshmidt, M. D., Zimbelmann, R., Mueller, H. M., Schiller, D. L., and Cowin, P. (1989). Molecular cloning and amino acid sequence of human plakoglobin, the common junctional plaque protein. *Proc. Natl. Acad. Sci. U.S.A.* **86,** 4027–4031.

Franke, W. W., Kapprell, H., and Cowin, P. (1987a). Plakoglobinis a component of the filamentous subplasmalemmal coat of lens cells. *Eur. J. Cell Biology* **43,** 301–315.

Franke, W. W., Kapprell, H. P., and Cowin, P. (1987b). Immunolocalization of plakoglobin in endothelial junctions: Identification as a special type of *Zonulae adhaerentes. Biol. Cell.* **59,** 205–218.

Friedlander, D. R., Mege, R. M., Cunningham, B. A., and Edelman, G. M. (1989). Cell sorting-out is modulated by both the specificity and amount of different cell adhesion molecules (CAMs) expressed on cell surfaces. *Proc. Natl. Acad. Sci. U.S.A.* **86,** 7043–7047.

Frisch, S. M., and Francis, H. (1994). Disruption of epithelial cell–matrix interactions induces apoptosis. *J. Cell Biol.* **124,** 619–626.

Fuchs, E. (1994). Intermediate filaments and disease: Mutations that cripple cell strength. *J. Cell Biol.* **125,** 511–516.

Fujimori, T., Miyatani, S., and Takeichi, M. (1990). Ectopic expression of N-cadherin perturbs histogenesis in *Xenopus* embryos. *Development* **110,** 97–104.

Fujimori, T., and Takeichi, M. (1993). Disruption of epithelial cell-cell adhesion by exogenous expression of a mutated nonfunctional N-cadherin. *Mol. Biol. Cell.* **4,** 37–47.

Funayama, N., Fagotto, F., McCrea, P., and Gumbiner, B. M. (1995). Embryonic axis induction by the armadillo repeat domain of $\beta$-catenin: Evidence for intracellular signaling. *J. Cell Biol.* **128,** 959–968.

Furukawa, Y., Nakatsuru, S., Nagafuchi, A., Tsukita, S., Muto, T., Nakamura, Y., and Horii, A. (1994). Structure, expression and chromosomal assignment of the human $\alpha$-catenin gene. *Cytogenet. Cell Genet.* **65,** 74–78.

Geiger, B. (1989). Cytoskeleton-associated cell contacts. *Curr. Opin. Cell Biol.* **1,** 103–109.

Geiger, B., Avnur, Z., Volberg, T., and Volk, T. (1985). Molecular domains of adherens junctions. *In* "The Cell in Contact" (G. M. Edelman and J. P. Thiery, eds.), pp. 461–469. John Wiley & Sons, New York.

Geiger, B., and Ayalon, O. (1992). Cadherins. *Annu. Rev. Cell Biol.* **8,** 307–332.

Goodwin, L., Hill, J. E., Raynor, K., Raszi, L., Manabe, M., and Cowin, P. (1990). Desmoglein shows extensive homology to the cadherin family of cell adhesion molecules. *Biochem. Biophys. Res. Commun.* **173,** 1224–1230.

Goridis, C., and Brunet, J.-F. (1992). NCAM: Structural diversity, function and regulation of expression. *Semin. Cell Biol.* **3,** 189–197.

Grunwald, G. B. (1991). The conceptual and experimental foundations of vertebrate embyronic cell adhesion research. *In* "A Conceptual History of Modern Embryology" (S. F. Gilbert, ed.), pp. 129–158. Plenum Press, New York.

Gumbiner, B., Stevenson, B., and Grimaldi, A. (1988). The role of the cell adhesion molecule uvomorulin in the formation and maintenance of the epithelial junctional complex. *J. Cell Biol.* **107,** 1575–1587.

Gundersen, D., Orlowski, J., and Rodriguez-Boulan, E. (1991). Apical polarity of Na,K-ATPase in retinal pigment epithelium is linked to a reversal of the ankyrin-fodrin submembrane cytoskeleton. *J. Cell Biol.* **112,** 863–872.

Gundersen, D., Powell, S. K., and Rodriguez-Boulan, E. (1993). Apical polarization of N-CAM in retinal pigment epithelium is dependent on contact with the neural retina. *J. Cell Biol.* **121,** 335–343.

Hamaguchi, M., Matsuyoshi, N., Ohnishi, Y., Gotoh, B., Takeichi,M., and Nagai, Y. (1993). $p60^{v\text{-}src}$ causes tyrosine phosphorylation and inactivation of the N-cadherin-catenin cell adhesion system. *EMBO J.* **12,** 307–314.

Hammerton, R. W., Krzeminski, K. A., Mays, R. W., Ryan, T. A., Wollner, D. A., and Nelson, W. J. (1991). Mechanism forregulating cell surface distribution of Na,K-ATPase in polarized epithelial cells. *Science* **254,** 847–850.

Hashimoto, M., Niwa, O., Nitta, Y., Takeichi, M., and Yokoro, K. (1989). Unstable expression of E-cadherin adhesion molecules in metastatic ovarian tumor cells. *Jpn. J. Cancer Res.* **80,** 459–463.

Hatta, K., Okada, T. S., and Takeichi, M. (1985). A monoclonal antibody disrupting calcium-dependent cell-cell adhesion of brain tissues: Possible role of its target antigen in animal pattern formation. *Proc. Natl. Acad. Sci. U.S.A.* **82,** 2789–2793.

Hatta, K., Takagi, S., Fujisawa, H., and Takeichi, M. (1987). Spatial and temporal expression pattern of N-cadherin cell adhesion molecules correlated with morphogenetic processes of chicken embryos. *Dev. Biol.* **120,** 215–227.

Hatta, K., and Takeichi, M. (1986). Expression of N-cadherin adhesion molecules associated with early morphogenetic events in chick development. *Nature* **320,** 447–449.

Hatzfeld, M., Kristjansson, G. I., Plessmann, U., and Weber, K. (1994). Band 6 protein, a major constituent of desmosomes from stratified epithelia, is a novel member of the *armadillo* multigene family. *J. Cell Sci.* **107,** 2259–2270.

He, X., Saint-Jeannet, J.-P., Woodgett, J. R., Varmus, H., and Dawid, I. B. (1995). Glycogen synthase kinase-3 and dorsoventral patterning in *Xenopus* embryos. *Nature* (*London*) **374,** 617–622.

Hermiston, M. L., and Gordon, J. I. (1995). In vivo analysis of cadherin function in mouse intestinal epithelium: Essential rolesin adhesion, maintenance of differentiation, and regulation of programmed cell death. *J. Cell Biol.* **129,** 489–506.

Herrenknecht, K., Ozawa, M., Eckerskorn, C., Lottspeich, F., Lenter, M., and Kemler, R. (1991). The uvomorulin-anchorage protein alpha catenin is a vinculin homologue. *Proc. Natl. Acad. Sci. U.S.A.* **88,** 9156–9160.

Hinck, L., Nathke, I. S., Papkoff, J., and Nelson, W. J. (1994a). Beta-catenin: A common target for regulation of cell adhesion by Wnt-1 and src signalling pathways. *Trends Biochem. Sci.* **19,** 538–542.

Hinck, L., Näthke, I. S., Papkoff, J., and Nelson, W. J. (1994b). Dynamics of cadherin/catenin complex formation: Novel protein interactions and pathways of complex formation. *J. Cell Biol.* **125,** 1327–1340.

Hinck, L., Nelson, W. J., and Papkoff, J. (1994c). Wnt-1 modulates cell–cell adhesion in mammalian cells by stabilizing beta-catenin binding to the cell adhesion protein cadherin. *J. Cell Biol.* **124,** 729–741.

Hirai, Y., Nose, A., Kobayashi, S., and Takeichi, M. (1989).Expression and role of E- and P-cadherin adhesion molecules in embryonic histogenesis II. Skin morphogenesis. *Development* **105,** 271–277.

Hirano, S., Kimoto, N., Shimoyama, Y., Hirohashi, S., and Takeichi, M. (1992). Identification of a neural α-catenin as a key regulator of cadherin function and multicellular organization. *Cell* **70,** 293–301.

Hirano, S., Nose, A., Hatta, K., Kawakami, A., and Takeichi, M.(1987). Calcium-dependent cell–cell adhesion molecules (cadherins): Subclass specificities and possible involvement ofactin bundles. *J. Cell. Biol.* **105,** 2501–2510.

Hodivala, K. J., and Watt, F. M. (1994). Evidence that cadherins play a role in the downregulation of integrin expression that occurs during keratinocyte terminal differentiation. *J. Cell Biol.* **124,** 589–600.

Hoffmann, I., and Balling, R. (1995). Cloning and expression analysis of a novel mesodermally expressed cadherin. *Dev. Biol.* **169,** 337–346.

Holtfreter, J. (1944). Experimental studies on the development of the pronephros. *Rev. Can. Biol.* **3,** 220–250.

Hooper, J. E. (1994). Distinct pathways for autocrine and paracrine Wingless signalling in *Drosophila* embryos. *Nature* (*London*) **372,** 461–464.

Hoschuetzky, H., Aberle, H., and Kemler, R. (1994). β-Catenin mediates the interaction of the cadherin–catenin complex with epidermal growth factor receptor. *J. Cell Biol.* **127,** 1375–1380.

Hülsken, J., Birchmeier, W., and Behrens, J. (1994). E-cadherin and APC compete for the interaction with β-catenin and the cytoskeleton. *J. Cell Biol.* **127,** 2061–2069.

Hyafil, F., Babinet, C., and Jacob, F. (1981). Cell–cell interactions in early embryogenesis: A molecular approach to the role of calcium. *Cell* **21,** 927–934.

Inuzuka, H., Redies, C., and Takeichi, M. (1991). Differential expression of R- and N-cadherin in neural and mesodermal tissues during early chicken development. *Development* **113,** 959–967.

Johnson, M. H., Maro, B., and Takeichi, M. (1988). The role of cell adhesion in the synchronization and orientation of polarization in 8-cell blastomeres. *J. Embryol. Exp. Morphol.* **93,** 239–255.

Jones, P. H., Harper, S., and Watt, F. M. (1995). Stem cell patterning and fate in human epidermis. *Cell* **80,** 83–93.

Jongen, W. M. F., Fitzgerald, D. J., Asamoto, M., Piccoli, C., Slaga, T. J., Gros, D., Takeichi, M., and Yamasaki, H. (1991). Regulation of connexin 43-mediated gap junctional intercellular communication by $Ca^{2+}$ in mouse epidermal cells is controlled by E-cadherin. *J. Cell Biol.* **114,** 545–555.

Jou, T.-S., Stewart, D. B., Stappart, J., Nelson, W. J., and Marrs, J. A. (1995). Genetic and biochemical dissection of protein linkages in the cadherin–catenin complex. *Proc. Natl. Acad. Sci. U.S.A.* **92,** 5067–5071.

Kemler, R. (1992). Classical cadherins. *Semin. Cell Biol.* **3,** 149–155.

Kemler, R. (1993). From cadherins to catenins: Cytoplasmic protein interactions and regulation of cell adhesion. *Trends Genet.* **9,** 317–321.

Kemler, R., Babinet, C., Eisen, H., and Jacob, F. (1977). Surface antigen in early differentiation. *Proc. Natl. Acad. Sci. U.S.A.* **74,** 4449–4452.

Kintner, C. (1992). Regulation of embryonic cell adhesion by the cadherin cytoplasmic domain. *Cell* **69,** 225–236.

Kimura, Y., Matsunami, H., Inoue, T., Shimamura, K., Uchida, N., Ueno, T., Miyazaki, T., and Takeichi, M. (1995). Cadherin-11 expressed in association with mesenchymal morphogenesis in the head, somite, and limb bud of early mouse embryos. *Dev. Biol.* **169,** 347–358.

Knudsen, K. A., Peralta Soler, A., Johnson, K. R., and Wheelock, M. J. (1995). Interaction of α-actinin with the cadherin/catenin cell–cell adhesion complex via α-catenin. *J. Cell Biol.* **130,** 67–77.

Knudsen, K. A., and Wheelock, M. J. (1992). Plakoglobin, or an 83-kD homologue distinct from β-catenin, interacts with E-cadherin and N-cadherin. *J. Cell Biol.* **118,** 671–679.

Koch, P. J., Walsh, M. J., Schmelz, M., Goldschmidt, M. D., Zimbelmann, R., and Franke, W. W. (1990). Identification of desmoglein, a constitutive desmosomal glycoprotein, as a member of the cadherin family of cell adhesion molecules. *Eur. J. Cell Biol.* **53,** 1–12.

Koob, R., Zimmerman, M., Schoner, W., and Drenckhahn, D. (1988). Colocalization and coprecipitation or ankyrin and Na,K-ATPase in kidney epithelial cells. *Eur. J. Cell Biol.* **45,** 230–237.

Korman, N. J., Eyre, R. W., Klaus-Kovtun, V., and Stanley, J. R. (1989). Demonstration of an adhering junction molecule (plakoglobin) in the autoantigens of pemphigus foiaceus and pemphigus vulgaris. *N. Engl. J. Med.* **321,** 631–635.

Kouklis, P. D., Hutton, E., and Fuchs, E. (1994). Making the connection: Direct binding between keratin intermediate filaments and desmosomal proteins. *J. Cell Biol.* **127,** 1049–1060.

Kowalczyk, A. P., Palka, H. L., Luu, H. H., Nilles, L. A., Anderson, J. E., Wheelock, M. J., and Green, K. J. (1995). Posttranslational regulation of plakoglobin expression: Influence of the desmosomal cadherins on plakoglobin metabolic stability. *J. Biol. Chem.* **269,** 31214–31223.

Kowalczyk, A. P., Stappenbeck, T. S., Parry, D. A. D., Palka, H. L., Virata, M. L. A., Bornslaeger, E. A., Nilles, L. A., and Green, K. J. (1994). Structure and function of desmosomal transmembrane core and plaque molecules. *Biophys. Chem.* **50,** 97–112.

Lampugnani, M. G., Corada, M., Caveda, L., Breviario, F., Ayalon, O., Geiger, B., and Dejana, E. (1995). The molecular organization of endothelial cell to cell junctions: Differential association of plakoglobin, $\beta$-catenin, and $\alpha$-catenin with vascular endothelial cadherin (VE-cadherin). *J. Cell Biol.* **129,** 203–217.

Larue, L., Ohsugi, M., Hirchenhain, J., and Kemler, R. (1994). E-cadherin null mutant embryos fail to form a trophectoderm epithelium. *Proc. Natl. Acad. Sci. U.S.A.* **91,** 8263–8267.

Levine, E., Lee, C. H., Kintner, C., and Gumbiner, B. M. (1994). Selective disruption of E-cadherin function in early *Xenopus* embryos by a dominant negative phenotype. *Development* **120,** 901–909.

Lombardo, C. R., Rimm, D. L., Koslov, E., and Morrow, J. S. (1994). Human recombinant $\alpha$-catenin binds to spectrin. *Mol. Biol. Cell* **5,** 47a.

Mahoney, P. A., Weber, U., Ononfrechuk, P., Biessmann, H., Bryant, P. J., and Goodman, C. S. (1991). The fat tumor suppressor gene in *Drosophila* encodes a novel member of the cadherin gene superfamily. *Cell* **67,** 853–868.

Marrs, J. A., Andersson-Fisone, C., Jeong, M. C., Cohen-Gould, L., Zurzolo, C., Nabi, I. R., Rodriguez-Boulan, E., and Nelson, W. J. (1995). Plasticity in epithelial cell phenotype: Modulation by expression of different cadherin cell adhesion molecules. *J. Cell Biol.* **129,** 507–519.

Marrs, J. A., Napolitano, E. W., Murphy-Erdosh, C., Mays, R. W., Reichardt, L. F., and Nelson, W. J. (1993). Distinguishing roles of the membrane-cytoskeleton and cadherin mediated cell–cell adhesion in generating different $Na^+,K^+$-ATPase distributions in polarized epithelia. *J. Cell Biol.* **123,** 149–164.

Mathur, M., Goodwin, L., and Cowin, P. (1994). Interactions of the cytoplasmic domain of the desmosomal cadherin Dsg1 with plakoglobin. *J. Biol. Chem.* **269,** 14075–14080.

Matsunaga, M., Hatta, K., and Takeichi, M. (1988). Role of N-cadherin cell adhesion molecules in the histogenesis of neural retina. *Neuron* **1,** 289–295.

Matsuyoshi, N., Hamaguchi, M., Taniguchi, S., Nagafuchi, A., Tsukita, S., and Takeichi, M. (1992). Cadherin-mediated cell–cell adhesion is perturbed by v-*src* tyrosine phosphorylation in metastatic fibroblasts. *J. Cell Biol.* **118,** 703–714.

McCrea, P. D., Brieher, W. M., and Gumbiner, B. M. (1993). Induction of a secondary body axis in *Xenopus* by antibodies to $\beta$-catenin. *J. Cell Biol.* **123,** 477–484.

McCrea, P. D., Turck, C. W., and Gumbiner, B. (1991). A homolog of the *armadillo* protein in *Drosophila* (plakoglobin) associated with E-cadherin. *Science* **254,** 1359–1361.

McMahon, A. P., and Bradley, A. (1990). The *Wnt*-1 (*int*-1) proto-oncogene is required for development of a large region of the mouse brain. *Cell* **62,** 1073–1085.

McMahon, A. P., and Moon, R. T. (1989). Ectopic expression of the proto-oncogene *int-1* in *Xenopus* embryos leads to duplication of the embyronic axis. *Cell* **58,** 1075–1084.

McNeill, H., Ozawa, M., Kemler, R., and Nelson, W. J. (1990). Novel function of the cell adhesion molecule uvomorulin as an inducer of cell surface polarity. *Cell* **62,** 309–316.

McNeill, H., Ryan, T., Smith, S. J., and Nelson, W. J. (1993). Spatial and temporal dissection of immediate and early events following cadherin-mediated epithelial cell adhesion. *J. Cell Biol.* **120,** 1217–1226.

Miyatani, S., Shimamura, K., Hatta, M., Nagafuchi, A., Nose, A., Matsunaga, M., Hatta, K., and Takeichi, M. (1989). Neural cadherin: Role in selective cell–cell adhesion. *Science* **245,** 631–635.

Montgomery, A. M. P., Reisfeld, R. A., and Cheresh, D. A. (1994). Integrin $a_vb_3$ rescues melanoma cells from apoptosis in three-dimensional dermal collagen. *Proc. Natl. Acad. Sci. U.S.A.* **91,** 8856–8860.

Morrison, A. I., Keeble, S., and Watt, F. M. (1988). The peanut lectin-binding glycoproteins of human epidermal keratinocytes. *Exp. Cell Res.* **177,** 247–256.

Morrow, J. S., Cianci, C. D., Ardito, T., Mann, A. S., and Kashgarian, M. (1989). Ankyrin links fodrin to the alpha subunit of $Na^+$,$K^+$-ATPase in Madin-Darby canine kidney cells and in intact renal tubule cells. *J. Cell Biol.* **108,** 455–465.

Moscona, A., and Moscona, H. (1952). Dissociation and aggregation of cells from organ rudiments of the early chick embryos. *J. Anat.* **86,** 287–301.

Moscona, A. A. (1952). Cell suspensions from organ rudiments of chick embryos. *Exp. Cell. Res.* **3,** 536–539.

Murphy-Erdosh, C., Napolitano, E. W., and Reichardt, L. F. (1994). The expression of B-cadherin during embryonic chick development. *Dev. Biol.* **161,** 107–125.

Murphy-Erdosh, C., Yoshida, C. K., Paradies, N., and Reichardt, L. F. (1995). The cadherin-binding specificities of B-cadherin and LCAM. *J. Cell Biol.* **129,** 1379–1390.

Musil, L. S., Cunningham, B. A., Edelman, G. M., and Goodenough, D. A. (1990). Differential phosphorylation of the gap junction protein Connexin43 in junctional communication-competent and -deficient cell lines. *J. Cell Biol.* **111,** 2077–2088.

Nabi, I. R., Mathews, A. P., Cohen-Gould, L., Gundersen, D., and Rodriguez-Boulan, E. (1993). Immortalization of polarized ratretinal pigmented epithelium. *J. Cell Sci.* **104,** 37–49.

Nagafuchi, A., Ishihara, S., and Tsukita, S. (1994). The roles of catenins in the cadherin-mediated cell adhesion: Function alanalysis of E-cadherin-$\alpha$ catenin fusion molecules. *J. Cell Biol.* **127,** 235–246.

Nagafuchi, A., Shirayoshi, Y., Okazaki, K., Yasuda, K., and Takeichi, M. (1987). Transformation of cell adhesion properties by exogenously introduced E-cadherin cDNA. *Nature (London)* **329,** 340–343.

Nagafuchi, A., and Takeichi, M. (1988). Cell binding function of E-cadherin is regulated by the cytoplasmic domain. *EMBO J.* **7,** 3679–3684.

Nagafuchi, A., and Takeichi, M. (1989). Transmembrane control of cadherin-mediated cell adhesion: A 94 kd protein functionally associated with a specific region of the cytoplasmic domain of E-cadherin. *Cell Regul.* **1,** 37–44.

Nagafuchi, A., Takeichi, M., and Tsukita, S. (1991). The 102 kd cadherin-associated protein: Similarity to vinculin and posttranscriptional regulation of expression. *Cell* **65,** 849–857.

Nagafuchi, A., and Tsukita, S. (1994). Loss of the expression of $\alpha$ catenin, the 102 kD cadherin associated protein, in central nervous tissues during development. *Dev. Growth Differ.* **36,** 59–71.

Nakagawa, S., and Takeichi, M. (1995). Neural crest cell–cell adhesion controlled by sequential and subpopulation-specific expression of novel cadherins. *Development* **121,** 1321–1332.

Napolitano, E. W., Venstrom, K., Wheeler, E. F., and Reichardt, L. F. (1991). Molecular cloning and characterization of B-cadherin, a novel chicken cadherin. *J. Cell Biol.* **113,** 893–905.

Näthke, I. S., Hinck, L., Swedlow, J. R., Papkoff, J., and Nelson, W. J. (1994). Defining steady state interactions and distributions of E-cadherin and catenin complexes in polarized epithelial cells. *J. Cell Biol.* **125,** 1341–1352.

Nelson, W. J. (1992). Regulation of cell surface polarity from bacteria to mammals. *Science* **258,** 948–955.

Nelson, W. J., and Hammerton, R. W. (1989). A membrane–cytoskeletal complex containing $Na^+$,$K^+$-ATPase, ankyrin, and fodrin in Madin-Darby canine kidney (MDCK) cells: Implications for the biogenesis of epithelial cell polarity. *J. Cell Biol.* **108,** 893–902.

Nelson, W. J., Shore, E. M., Wang, A. Z., and Hammerton, R. W. (1990). Identification of a membrane–cytoskeletal complex containing the cell adhesion molecule uvomorulin (E-cadherin), ankyrin, and fodrin in Madin-Darby canine kidney epithelial cells. *J. Cell Biol.* **110,** 349–357.

Nelson, W. J., and Veshnock, P. J. (1986). Dynamics of membrane–skeleton (fodrin) organization during development of polarity in Madin-Darby canine kidney epithelial cells. *J. Cell Biol.* **103,** 1751–1765.

Nelson, W. J., and Veshnock, P. J. (1987). Ankyrin binding to $Na^+$,$K^+$-ATPase and implications for the organization of membrane domains in polarized cells. *Nature* (*London*) **328,** 533–536.

Neugebauer, K. M., Tomaselli, K. J., Lilien, J., and Reichardt, L. F. (1988). N-cadherin, NCAM, and integrins promote retinal neurite outgrowth on astrocytes *in vitro*. *J. Cell Biol.* **107,** 1177–1187.

Noordermeer, J., Kingensmith, J., Perrimon, N., and Nusse, R. (1994). *dishevelled* and *armadillo* act in the Wingless signalling pathway in *Drosophila. Nature* (*London*) **367,** 80–83.

Nose, A., Nagafuchi, A., and Takeichi, M. (1987). Isolation of placental cadherin cDNA: Identification of a novel gene family of cell–cell adhesion molecules. *EMBO J.* **6,** 3655–3661.

Nose, A., Nagafuchi, A., and Takeichi, M. (1988). Expressed recombinant cadherins mediate cell sorting in model systems. *Cell* **54,** 993–1001.

Nose, A., and Takeichi, M. (1986). A novel cadherin cell adhesion molecule: Its expression patterns associated with implantation and organogenesis of mouse embryos. *J. Cell Biol.* **103,** 2649–2658.

Nusse, R., and Varmus, H. (1992). *Wnt* genes. *Cell* **69,** 1073–1087.

Oda, H., Uemura, T., Harada, Y., Iwai, Y., and Takeichi, M. (1994). A *Drosophila* homolog of cadherin associated with Armadillo and essential for embryonic cell–cell adhesion. *Dev. Biol.* **165,** 716–726.

Oda, H., Uemura, T., Shiomi, K., Nagafuchi, A., Tsukita, S., and Takeichi, M. (1993). Identification of a *Drosophila* Homologue of $\alpha$-catenin and its association with the *armadillo* protein. *J. Cell Biol.* **121,** 1133–1140.

Olson, D. J., Christian, J. L., and Moon, R. T. (1991). Effect of *Wnt*-1 and related proteins on gap junctional communication in *Xenopus* embryos. *Science* **252,** 1173–1176.

Overduin, M., Harvey, T. S., Bagby, S., Tong, K. I., Yau, P., Takeichi, M., and Ikura, M. (1995). Solution structure of the epithelial cadherin domain responsible for selective cell adhesion. *Science* **267,** 386–389.

Owaribe, K., Kartenbeck, J., Rungger-Brandle, E., and Franke, W. W. (1988). Cytoskeletons of retinal pigment epithelial cells: Interspecies differences of expression patterns indicate independence of cell function from the specific complement of cytoskeletal proteins. *Cell Tissue Res.* **254,** 301–315.

Oyama, T., Kanai, Y., Ochiai, A., Akimoto, S., Oda, T., Yanagihara, K., Nagafuchi, A., Tsukita, S., Shibamoto, S., Ito, F., Takeichi, M., Matsuda, H., and Hirohashi, S. (1994). Atruncated $\beta$-catenin disrupts the interaction between E-cadherin and $\alpha$-catenin: A cause of loss of intercellular adhesiveness in human cancer cell lines. *Cancer Res.* **54,** 6282–6287.

Ozawa, M., Baribault, H., and Kemler, R. (1989). The cytoplasmic domain of the cell adhesion molecule uvomorulin associates with three independent proteins structurally related in different species. *EMBO J.* **8,** 1711–1717.

Ozawa, M., and Kemler, R. (1990). Correct proteolytic cleavage is required for the cell adhesive function of uvomorulin. *J. Cell Biol.* **111,** 1645–1650.

Parker, A. E., Wheeler, G. N., Arnemann, J., Pidsley, S. C., Ataliotis, P., Thomas, C. L., Rees, D. A., Magee, A. I., and Buxton, R. S. (1991). Desmosomal glycoproteins II and III. Cadherin-like junctional molecules generated by alternative splicing. *J. Biol. Chem.* **266,** 10438–10445.

Parr, B. A., and McMahon, A. P. (1995). Dorsalizing signal Wnt-7a required for normal polarity of D-V and A-P axes of mouse limb. *Nature* (*London*) **374,** 350–353.

Peifer, M., Berg, S., and Reynolds, A. B. (1994a). A repeating amino acid motif shared by proteins with diverse cellular roles. *Cell* **76,** 789–791.

Peifer, M., McCrea, P. D., Green, K. J., Wieschaus, E., and Gumbiner, B. M. (1992). The vertebrate adhesive junction proteins β-catenin and plakoglobin and the *Drosophila* segmentpolarity gene *armadillo* form a multigene family with similar properties. *J. Cell Biol.* **118,** 681–691.

Peifer, M., Orsulic, S., Pai, L.-M., and Loureiro, J. (1993a). A model system for cell adhesion and signal transduction in *Drosophila. Development* **1993** (Suppl.), 163–176.

Peifer, M., Orsulic, S., Sweeton, D., and Wieschaus, E. (1993b). A role for the *Drosophila* segment polarity gene *armadillo* in cell adhesion and cytoskeletal integrity during oogenesis. *Development* **118,** 1191–1207.

Peifer, M., Pai, L.-M., and Casey, M. (1995). Phosphorylation of the *Drosophila* adherens junction protein Armadillo: Roles for Wingless signal and Zeste white-3 kinase. *Dev. Biol.* **166,** 543–546.

Peifer, M., Sweeton, D., Casey, M., and Wieschaus, E. (1994b). *wingless* signal and Zeste-white 3 kinase trigger opposing changes in the intracellular distribution of Armadillo. *Development* **120,** 369–380.

Peifer, M., and Wieschaus, E. (1990). The segment polarity gene *armadillo* encodes a functionally modular protein that is the *Drosophila* homolog of human plakoglobin. *Cell* **63,** 1167–1178.

Peyrieras, N., Louvard, D., and Jacob, F. (1985). Characterization of antigens recognized by monoclonal and polyclonal antibodies directed against uvomorulin. *Proc. Natl. Acad. Sci. U.S.A.* **82,** 8067–8071.

Piepenhagen, P., and Nelson, W. J. (1993). Defining E-cadherin-associated protein complexes in epithelial cells: Plakoglobin, β- and γ-catenin are distinct components. *J. Cell Sci.* **104,** 751–762.

Pierce, S. B., and Kimelman, D. (1995). Regulation of Spemann organizer formation by the intracellular kinase Xgsk-3. *Development* **121,** 755–765.

Quinton, P. M., Wright, E. M., and Tormey, J. M. (1973). Localization of sodium pumps in the choroid plexus epithelium. *J. Cell Biol.* **58,** 724–730.

Ranscht, B., and Bronner-Fraser, M. (1991). T-cadherin expression alternates with migrating neural crest cells in the trunk of the avian embryo. *Development* **111,** 15–22.

Ranscht, B., and Dours-Zimmermann, M. T. (1991). T-cadherin, a novel cadherin cell adhesion molecule in the nervous system lacks the conserved cytoplasmic region. *Neuron* **7,** 391–402.

Redies, C., Inuzuka, H., and Takeichi, M. (1992). Restricted expression of N- and R-cadherin on neurites of the developing chicken CNS. *J. Neurosci.* **12,** 3525–3534.

Redies, C., and Takeichi, M. (1993a). Expression of N-cadherin mRNA during development of the mouse brain. *Dev. Dynam.* **197,** 26–39.

Redies, C., and Takeichi, M. (1993b). N- and R-cadherin expression in the optic nerve of the chicken embryo. *Glia* **8,** 161–171.

Reynolds, A. B., Herbert, L., Cleveland, J. L., Berg, S. T., and Gaut, J. R. (1992). p120, a novel substrate of protein tyrosinekinase receptors and $pp60^{v\text{-}src}$, is related to cadherin-binding factors β-catenin, plakoglobin and *armadillo*. *Oncogene* **7,** 2439–2445.

Riggleman, B., Schedl, P., and Wieschaus, E. (1990). Spatial expression of the *Drosophila* segment polarity gene *armadillo* is posttranscriptionally regulated by *wingless*. *Cell* **63,** 549–560.

Rimm, D. L., Koslov, E. R., Kebriaei, P., Cianci, C. D., and Morrow, J. S. (1994). (E) α-catenin is an actin binding and -bundling protein mediating the attachment of F-actin to the membrane adhesion complex. *Proc. Natl. Acad. Sci. U.S.A* **92,** 8813–8817.

Ringwald, M., Schuh, R., Vestweber, D., Eistetter, H., Lottspeich, F., Engel, J., Dolz, R., Jahnig, F., Epplen, J., Mayer, S., Muller, C., and Kemler, R. (1987). The structure of the cell adhesion molecule uvomorulin. Insights into the molecular mechanisms of $Ca^{2+}$-dependent cell adhesion. *EMBO J.* **6,** 3647–3653.

Rizzolo, L. J. (1990). The distribution of Na,K-ATPase in the retinal pigmented epithelium from chicken embryo is polarized *in vivo* but not in primary cell culture. *Exp. Eye Res.* **51,** 435–446.

Rizzolo, L. J. (1991). Basement membrane stimulates the polarized distribution of integrins but not the Na,K-ATPase in the retinal pigment epithelium. *Cell Regul.* **2,** 939–949.

Rodriguez-Boulan, E., and Nelson, W. J. (1989). Morphogenesis of the polarized epithelial cell phenotype. *Science* **245,** 718–725.

Rubinfeld, B., Souza, B., Albert, I., Muller, O., Chamberlain, S. H., Masiarz, F. R., Munemitsu, S., and Polakis, P. (1993). Association of the APC gene product with beta-catenin. *Science* **262,** 1731–1734.

Rubinfeld, B., Souza, B., Albert, I., Munemitsu, S., and Polakis, P. (1995). The APC protein and E-cadherin form similar but independent complexes with α-catenin, β-catenin and plakoglobin. *J. Biol. Chem.* **270,** 5549–5555.

Sacristán, M. P., Vestal, D. J., Dours-Zimmermann, M. T., and Ranscht, B. (1993). T-cadherin 2: Molecular characterization, function in cell adhesion, and coexpression with T-cadherin and N-cadherin. *J. Neurosci. Res.* **34,** 664–680.

Sano, K., Tanihara, H., Heimark, R. L., Obata, S., Davidson, M., St. John, T., Taketanin, S., and Suzuki, S. (1993). Protocadherins: A large family of cadherin-related molecules in central nervous system. *EMBO J.* **12,** 2249–2256.

Schuh, R., Vestweber, D., Riede, I., Ringwald, M., Eosenberg, U. B., Jackle, H., and Kemler, R. (1986). Molecular cloning of the mouse cell adhesion molecule uvomorulin: cDNA contains B1-related segment. *Proc. Natl. Acad. Sci. U.S.A.* **93,** 1364–1368.

Schwarz, M. A., Owaribe, K., Kartenbeck, J., and Franke, W. W. (1990). Desmosomes and hemidesmosomes: Constitutive molecular components. *Ann. Rev. Cell Biol.* **6,** 461–491.

Seimers, K., Wilson, R., Mays, R., Ryan, T. A., Wollner, D. A., and Nelson, W. J. (1993). Delivery of $Na^+$,$K^+$-ATPase in polarized epithelial cells. *Science* **260,** 554–556.

Shapiro, L., Fannon, A. M., Kwong, P. D., Thompson, A., Lehmann, M. S., Grubel, G., Legrand, J.-F., Als-Nielsen, J., Colman, D. R., and Hendrickson, W. A. (1995). Structural basis of cell–cell adhesion by cadherins. *Nature* (*London*) **374,** 327–337.

Shibamoto, S., Hayakawa, M., Takeuchi, K., Hori, T., Miyazawa, K., Kitamura, N., Johnson, K., Wheelock, M., Matsuyoshi, N., Takeichi, M., and Ito, F. (1995). Association of p120, a tyrosine kinase substrate, with E-cadherin/catenin complexes. *J. Cell Biol.* **128,** 949–957.

Shibamoto, S., Hayakawa, M., Takeuchi, K., Hori, T., Oku, N., Miyazawa, K., Kitamura, N., Takeichi, M., and Ito, F. (1994). Tyrosine phosphorylation of β-catenin and plakoglobin enhanced by hepatocyte growth factor and epidermal growth factor in human carcinoma cells. *Cell Adhesion Commun.* **1,** 295–305.

Shimamura, K., Hirano, S., McMahon, A. P., and Takeichi, M. (1994). *Wnt-1*-dependent regulation of local E-cadherin and αN-catenin expression in the embryonic mouse brain. *Development* **120,** 2225–2234.

Shimamura, K., and Takeichi, M. (1992). Local and transient expression of E-cadherin involved in mouse embryonic brain morphogenesis. *Development* **116,** 1011–1019.

Shimoyama, Y., Nagafuchi, A., Fujita, S., Gotoh, M., Takeichi, M., Tsukita, S., and Hirohashi, S. (1992). Cadherin dysfunction in a human cancer cell line: Possible involvement of loss of $\alpha$-catenin expression in reduced cell–cell adhesiveness. *Cancer Res.* **52,** 5770–5774.

Shirayoshi, Y., Hatta, K., Hosoda, M., Tsunasawa, S., Sakiyama, F., and Takeichi, M. (1986). Cadherin cell adhesion molecules with distinct binding specificities share a common structure. *EMBO J.* **5,** 2485–2488.

Shore, E. M., and Nelson, W. J. (1991). Biosynthesis of the cell adhesion molecule uvomorulin (E-cadherin) in Madin-Darby canine kidney (MDCK) epithelial cells. *J. Biol. Chem.* **266,** 19672–19680.

Siegfried, E., Chou, T., and Perrimon, N. (1992). *wingless* signaling acts through *zeste-white 3,* the *Drosophila* homolog of *glycogen synthase kinase-3,* to regulate *engrailed* and establish cell fate. *Cell* **71,** 1167–1179.

Siegfried, E., Wilder, E. L., and Perrimon, N. (1994). Components of *wingless* signalling in *Drosophila. Nature* (*London*) **367,** 76–80.

Sokol, S. Y., Klingensmith, J., Perrimon, N., and Itoh, K. (1995). Dorsalizing and neuralizing properties of Xdsh, a maternally expressed *Xenopus* homolog of *dishevelled. Development* **121,** 1637–1647.

Sorkin, B. C., Gallin, W. J., Edelman, G. M., and Cunningham, B. A. (1991). Genes for two calcium-dependent cell adhesion molecules have similar structures and are arranged in tandem in the chicken genome. *Proc. Natl. Acad. Sci. U.S.A.* **88,** 11545–11549.

Staddon, J. M., Smales, C., Schulze, C., Esch, F. S., and Rubin, L. L. (1995). p120, a p120-related protein (p100), and the cadherin/catenin complex. *J. Cell Biol.* **130,** 369–381.

Stanley, J. R. (1993). Cell adhesion molecules as targets of autoantibodies in pemphigus and pemphigoid, bullous diseases due to defective epidermal cell adhesion. *Adv. Immunol.* **53,** 291–325.

Stappenbeck, T. S., Bornslaeger, E. A., Corcoran, C. M., Luu, H. H., Virata, M. L. A., and Green, K. J. (1993). Functional analysis of desmoplakin domains: Specification of the interaction with keratin versus vimentin intermediate filament networks. *J. Cell Biol.* **123,** 691–705.

Stappenbeck, T. S., and Green, K. J. (1992). The desmoplakin carboxyl terminus coaligns with and specifically disrupts intermediate filament networks when expressed in cultured cells. *J. Cell Biol.* **116,** 1197–1209.

Stappenbeck, T. S., Lamb, J. A., Corcoran, C. M., and Green, K. J. (1994). Phosphorylation of the desmoplakin C-terminus negatively regulates its interaction with keratin intermediate filament networks. *J. Biol. Chem.* **269,** 29351–29354.

Stappert, J., and Kemler, R. (1994). A short region of E-cadherin is essential for catenin binding and is highly phosphorylated. *Cell Adhes. Commun.* **2,** 319–323.

Stark, K., Vainio, S., Vassileva, G., and McMahon, A. (1994). Epithelial transformation of metanephric mesenchyme in thedeveloping kidney regulated by *Wnt-4. Nature* (*London*) **372,** 679–683.

Steinberg, M. S. (1964). The problem of adhesive selectivity in cellular interactions. *In* "Cellular Membranes in Development" (M. Locke, ed.), pp. 321–366. Academic Press, New York.

Steinberg, M. S., and Takeichi, M. (1994). Experimental specification of cell sorting, tissue spreading, and specific spatial patterning by quantitative differences in cadherin expression. *Proc. Natl. Acad. Sci. U.S.A.* **91,** 206–209.

Steinberg, R. M., and Miller, S. S. (1979). Transport and membrane properties of the retinal pigment epithelium. *In* "The Retinal Pigment Epithelium" (K. M. Zinn and M. F. Marmor, eds.), pp. 205–225. Harvard University Press, Cambridge, Massachusetts.

Su, L. K., Vogelstein, B., and Kinzler, K. W. (1993). Association of the APC tumor suppressor protein with catenins. *Science* **262,** 1734–1737.

Suzuki, S., Sano, K., and Tanihara, H. (1991). Diversity of the cadherin family: Evidence for eight new cadherins in nervous tissue. *Cell Regul.* **2,** 261–270.

Takeichi, M. (1977). Functional correlation between cell adhesive properties and some cell surface molecules. *J. Cell Biol.* **75,** 464–474.

Takeichi, M. (1988). The cadherins: Cell–cell adhesion molecules controlling animal morphogenesis. *Development* **102,** 639–655.

Takeichi, M. (1990). Cadherins: A molecular family important in selective cell–cell adhesion. *Ann. Rev. Biochem.* **59,** 237–252.

Takeichi, M. (1991). Cadherin cell adhesion receptors as a morphogenetic regulator. *Science* **251,** 1451–1455.

Takeichi, M. (1993). Cadherins in cancer: Implications for invasion and metastasis. *Curr. Opin. Cell Biol.* **5,** 806–811.

Takeichi, M., Atsumi, T., Yoshida, C., Uno, K., and Okada, T. S. (1981). Selective adhesion of embryonal carcinoma cells and differentiated cells by $Ca^{2+}$-dependent sites. *Dev. Biol.* **87,** 340–350.

Thiery, J. P., Delouvee, A., Gallin, W. J., Cunningham, B. A., and Edelman, G. M. (1984). Ontogenetic expression of cell adhesion molecules: L-CAM is found in epithelia derived from the three primary germ layers. *Dev. Biol.* **102,** 61–78.

Thomas, K. R., and Capecchi, M. R. (1990). Targeted disruption of the murine *int-1* proto-oncogene resulting in severe abnormalities in midbrain and cerebellar development. *Nature* (*London*) **346,** 847–850.

Tomaselli, K. J., Neugebauer, K. M., Bixby, J. L., Lilien, J., and Reichardt, L. F. (1988). N-cadherin and integrins: Two receptor systems that mediate neuronal process outgrowth on astrocyte surfaces. *Neuron* **1,** 33–43.

Townes, P. L., and Holtfreter, J. (1955). Directed movements and selective adhesion of embryonic amphibian cells. *J. Exp. Zool.* **128,** 53–118.

Trinkaus, J. P. (1965). Mechanisms of morphogenetic movements. *In* "Organogenesis" (R. L. DeHaan and H. Ursprung, eds.), pp. 55–104. Holt, Rinehart and Winston, New York.

Troyanovsky, S. M., Eshkind, L. G., Troyanovsky, R. B., Leube, R. E., and Franke, W. W. (1993). Contributions of cytoplasmic domains of desmosomal cadherins to desmosome assembly and intermediate filament anchorage. *Cell* **72,** 561–574.

Troyanovsky, S. M., Troyanovsky, R. B., Eshkind, L. G., Krutovskikh, V. A., Leube, R. E., and Franke, W. W. (1994). Identification of the plakoglobin-binding domain in desmoglein and its role in plaque assembly and intermediate filament anchorage. *J. Cell Biol.* **127,** 151–160.

Tsukita, S., Oishi, K., Akiyama, T., Yamanashi, Y., Yamamoto, T., and Tsukita, S. (1991). Specific proto-oncogene tyrosine kinases of src family are enriched in cell-to-cell junctions where the level of tyrosine phosphorylation is elevated. *J. Cell Biol.* **113,** 867–879.

Tsukita, S., Tsukita, S., Nagafuchi, A., and Yonemura, S.(1992). Molecular linkage between cadherins and actin filaments in cell–cell adherens junction. *Curr. Opin. Cell Biol.* **4,** 834–839.

Uchida, N., Shimamura, K., Miyatani, S., Copeland, N. G., Gilbert, D. J., Jenkins, N. A., and Takeichi, M. (1994). Mouse $\alpha$N-catenin: Two isoforms, specific expression in the nervous system, and chromosomal localization of the gene. *Dev. Biol.* **163,** 75–85.

Vassar, R., Coulombe, P. A., Degenstein, L., Alpers, K., and Fuchs, E. (1991). Mutant keratin expression in transgenic mice causes marked abnormalities resembling a human genetic skin disease. *Cell* **64,** 365–380.

Vestal, D., and Ranscht, B. (1992). Glycosyl phosphatidylinositol-anchored T-cadherin mediates calcium-dependent, homophilic cell adhesion. *J. Cell Biol.* **119,** 451–461.

Vestweber, D., Gossler, A., Boller, K., and Kemler, R. (1987). Expression and distribution of the cell adhesion molecule uvomorulin in mouse preimplantation embryos. *Dev. Biol.* **124,** 451–456.

Vestweber, D., and Kemler, R. (1985). Identification of a putative cell adhesion domain of uvomorulin. *EMBO J.* **4,** 3393–3398.

Vestweber, D., and Kemler, R. (1984a). Rabbit antiserum against a purified surface glycoprotein decompacts mouse preimplantation embryos and reacts with specific adult tissues. *Exp. Cell Res.* **152,** 169–178.

Vestweber, D., and Kemler, R. (1984b). Some structural and functional aspects of the cell adhesion molecule uvomorulin. *Cell. Differ.* **15,** 269–273.

Vestweber, D., Kemler, R., and Ekblom, P. (1985). Cell-adhesion molecule uvomorulin during kidney development. *Dev. Biol.* **112,** 213–221.

Vleminckx, K., Vakaet, L., Mareel, M., Fiers, W., and Van Roy, F. (1991). Genetic manipulation of E-cadherin expression by epithelial tumor cells reveals an invasion suppressor role. *Cell* **66,** 107–120.

Volberg, T., Geiger, B., Dror, R., and Zick, Y. (1991). Modulation of intercellular adherens-type junctions and tyrosine phosphorylation of their components in RSV-transformed cultured chick lens cells. *Cell Regul.* **2,** 105–120.

Walsh, F. S., and Doherty, P. (1993). Factors regulating the expression and function of calcium-independent cell adhesion molecules. *Curr. Opin. Cell Biol.* **5,** 791–796.

Walsh, F. S., and Doherty, P. (1992). Second messengers underlying cell-contact-dependent axonal growth stimulated by transfected N-CAM, N-cadherin, or L1. *Cold Spring Harbor Symp. Quant. Biol.* **57,** 431–440.

Wang, A. Z., Ojakian, G. K., and Nelson, W. J. (1990). Steps in the morphogensis of a polarized epithelium. I. Uncoupling the roles of cell–cell and cell substratum contact in establishing plasma membrane polarity in multicellular epithelial (MDCK) cysts. *J. Cell Sci.* **95,** 137–151.

Watabe, M., Matsumoto, K., Nakamura, T., and Takeichi, M. (1993). Effect of hepatocyte growth factor on cadherin-mediated cell–cell adhesion. *Cell Struct. Funct.* **18,** 117–124.

Watabe, M., Nagafuchi, A., Tsukita, S., and Takeichi, M. (1994). Induction of polarized cell–cell association and retardation of growth by activation of the E-cadherin-catenin adhesion system in a dispersed carcinoma line. *J. Cell Biol.* **127,** 247–256.

Watt, F. M. (1983). Involucrin and other markers of keratinocyte terminal differentiation. *J. Invest. Dermatol.* **81,** 100S-103S.

Watt, F. M., and Green, H. (1982). Stratification and terminal differentiation of cultured epidermal cells. *Nature (London)* **295,** 434–436.

Weis, W. (1995). Cadherin structure: A revealing zipper. *J. Struct.* **3,** 425–427.

Wheelock, M. J., and Jensen, P. J. (1992). Regulation of keratinocyte intercellular junction organization and epidermal morphogenesis by E-cadherin. *J. Cell Biol.* **117,** 415–425.

Woods, D. F., and Bryant, P. J. (1993). Apical junctions and cell signalling in epithelia. *J. Cell Sci. Suppl.* **17,** 171–181.

Wright, E. M. (1972). Mechanisms of ion transport across the choroid plexus. *J. Physiol.* **226,** 545–571.

Yonemura, S., Itoh, M., Nagafuchi, A., and Tsukita, S. (1995). Cell-to-cell adherens junction formation and actin filament organization: Similarities and differences between non-polarized fibroblasts and polarized epithelial cells. *J. Cell Sci.* **108,** 127–142.

# Escape and Migration of Nucleic Acids between Chloroplasts, Mitochondria, and the Nucleus

Peter E. Thorsness and Eric R. Weber
Department of Molecular Biology, University of Wyoming, Laramie, Wyoming 82071-3944

The escape and migration of genetic information between mitochondria, chloroplasts, and nuclei have been an integral part of evolution and has a continuing impact on the biology of cells. The evolutionary transfer of functional genes and fragments of genes from chloroplasts to mitochondria, from chloroplasts to nuclei, and from mitochondria to nuclei has been documented for numerous organisms. Most documented instances of genetic material transfer have involved the transfer of information from mitochondria and chloroplasts to the nucleus. The pathways for the escape of DNA from organelles may include transient breaches in organellar membranes during fusion and/or budding processes, terminal degradation of organelles by autophagy coupled with the subsequent release of nucleic acids to the cytoplasm, illicit use of nucleic acid or protein import machinery, or fusion between heterotypic membranes. Some or all of these pathways may lead to the escape of DNA or RNA from organellar compartments with subsequent uptake of nucleic acids from the cytoplasm into the nucleus. Investigations into the escape of DNA from mitochondria in yeast have shown the rate of escape for gene-sized fragments of DNA from mitochondria and its subsequent migration to the nucleus to be roughly equivalent to the rate of spontaneous mutation of nuclear genes. Smaller fragments of mitochondrial DNA may appear in the nucleus even more frequently. Mutations of nuclear genes that define gene products important in controlling the rate of DNA escape from mitochondria in yeast also have been described. The escape of genetic material from mitochondria and chloroplasts has clearly had an impact on nuclear genetic organization throughout evolution and may also affect cellular metabolic processes.

**KEY WORDS:** Mitochondria, Chloroplast, Nucleus, Genome, DNA escape.

## I. Introduction

Recombination, rearrangement, and transposition of genetic information play a large role in the biology of organisms. Typically, these processes are considered to occur within a genome of the organism, whether it be a single chromosome within a bacterium, the nuclear genome of a complex eukaryote, or the smaller genomes found in mitochondria and chloroplasts. The introduction of foreign nucleic acids via viral infection, conjugation, or transfection can have a profound impact on nuclear genetic organization and cellular physiology. Potentially as important as these well-documented processes, the integration into the nuclear genome of organellar DNA, escaping from compartments such as the mitochondrion or chloroplast contained within the cell, also can alter the organization of the nuclear genome. Nucleic acids escaping from organelles have had an undeniable impact on the evolution of eukaryotic cells and it ultimately may be found that the biology of individual cells also is affected on a continual basis by nucleic acids escaping from organelles.

The goal of this article is to describe concepts, examples, and experiments that detail the escape of nucleic acid from mitochondria and chloroplasts and its subsequent transfer to the nucleus. We discuss broad concepts in evolution and cell biology to establish a framework for understanding the evolutionary basis of this transfer. We particularly emphasize the experimental detection and analysis of DNA escape in order to put potential mechanisms of DNA escape from mitochondria and chloroplasts into perspective.

## II. The Endosymbiotic Hypothesis and Transfer of Genetic Material between Compartments

### A. Implications of the Endosymbiotic Hypothesis

In its simplest form, the endosymbiotic hypothesis proposes that modern-day mitochondria and chloroplasts arose from the colonization of one cell by another and subsequent establishment of a symbiotic relationship between the organisms (Margulis, 1981; Gray, 1989, 1992). Continual evolutionary pressure, which selects those cells that best utilized carbon sources and optimized growth potential, presumably results in the complex biology found in eukaryotic cells. Compartmentalization of metabolic activities is conceivably a fundamental property that allows cell differentiation in complex, multicellular organisms.

A corollary of the endosymbiotic hypothesis is that both the colonizing and host cells initially had a full complement of genetic material as both were, at some point, viable cells in the absence of the symbiotic relationship. While this is no longer true, at this time there is evidence for the absolute requirement for the presence of one such compartment, the mitochondrion, in eukaryotic cells. Yeast cells that fail to inherit mitochondria are inviable (McConnell *et al.*, 1990) and mutations that interfere with biogenesis of mitochondria by preventing the import of proteins also lead to inviability (Baker and Schatz, 1991). Likewise, mitochondria are dependent on proteins encoded by the nucleus and manufactured in the cytoplasm for biogenesis (Pon and Schatz, 1991).

Assuming the validity of the endosymbiotic hypothesis, these observations of eukaryotic cells lead to the following question: where has all the DNA gone? If the mitochondrion represents the vestigial remains of a once viable and free-living organism, what has become of the vast bulk of the mitochondrial genome, which presumably encoded all of the proteins necessary for the biogenesis of the organism? There are eight proteins [excluding the open reading frames (ORFs) found in optional introns], two rRNAs, and a 9S RNA that participate in the maturation of the 24 or 25 tRNAs encoded in the 75 kb of mitochondrial DNA found in yeast (Pon and Schatz, 1991). The human mitochondrial genome encodes 13 proteins, 2 rRNAs, and 22 tRNAs in 16.5 kb of DNA (Anderson *et al.*, 1981). Other mitochondrial genomes are larger and encode various selections of proteins and RNAs, but in all cases the full complement of genetic material required for the biogenesis of the compartment is not contained in the mitochondrial genome.

## B. Molecular Evidence for Transfer of Genetic Material from Organelles to Nuclei

### 1. Loss and Transfer of Organellar Genetic Information

There are three general explanations, none of which are mutually exclusive, that may account for the DNA "missing" from organelles. First, once the colonizing organism became a full-time endosymbiont a number of biological functions may have become superfluous and consequently would have been lost over time owing to lack of selection. It is only possible to speculate what types of functions might have become expendable when the protomitochondrion became a full-time endosymbiont, but such functions may have included certain types of environmental sensors, metabolic pathways for the transformation of nutrients not encountered inside the host cell, and any other proteins or chemicals synthesized by the free-living organism that increased its fitness outside of the host cell.

Second, a number of duplicate functions presumably existed in the protoeukaryote: both the host and colonizing cell presumably had proteins dedicated to essential housekeeping functions such as intermediary metabolism and protein synthesis. Perhaps some of these functions were ultimately taken over by the host cell and the genes encoding proteins involved in these cellular functions were simply lost from the colonizing cell when they no longer provided a selective advantage. As an example, the functional copy of the chloroplast ribosomal protein L23 is encoded in the nucleus in spinach and related plants (Bubunenko *et al.*, 1994). Unlike most chloroplast ribosomal proteins, L23 is a descendant of eukaryotic cytoplasmic ribosomes rather then prokaryotic-type ribosomes, indicating that the disrupted *L23* gene in the chloroplast of spinach has been replaced by the "host cell" gene and consequently allowing the inactivation and eventual loss of the chloroplast *L23* gene (Bubunenko *et al.*, 1994).

A third possible explanation invokes the transfer of essential genetic material from the colonizing cell to the genome of the host cell. This transfer of genetic information presumably increased the fitness of the protoeukaryote. Possible benefits for moving genetic information to the nucleus may include more efficient and faithful replication, increased regulatory control, and coordination with other cellular metabolic activities. There are several examples of genes that appear to have been transferred to the nucleus from organelles during evolution. The structural gene for the chloroplast protein synthesis factor Tu, *tufA,* is encoded by the chloroplast genome in green alga *Chlamydomonas reinhardtii* (Watson and Surzycki, 1982) and in *Euglena gracilis* (Montandon and Stutz, 1983). However, further up the evolutionary tree in land plants such as liverwort, rice, and tobacco the *tufA* gene is not found in the chloroplast (Ohyama *et al.,* 1986; Shimada and Sugirua, 1991). The *Arabidopsis thaliana* chloroplast genome also does not encode elongation factor Tu, rather, the *tufA* gene is found in the nucleus. It encodes a protein more closely related to chloroplast elongation factor Tu from *Chlamydomonas* then to cytoplasmic elongation factor Tu from *Arabidopsis* (Baldauf and Palmer, 1990). This then suggests that the chloroplast *tufA* gene was transferred to the nucleus before the emergence of land plants (Baldauf and Palmer, 1990). Discussed previously was the case of the chloroplast ribosomal protein L23 in spinach that was derived from the corresponding cytoplasmic ribosomal protein (Bubunenko *et al.,* 1994). In contrast to that example is the nuclear-encoded chloroplast ribosomal protein L21 in spinach that is clearly related to the chloroplast-encoded L21 protein in liverwort (Smooker *et al.,* 1990). It has been proposed that the L21-encoding gene was transferred to the nucleus some time after the evolutionary divergence of nonvascular plants (liverwort) and angiosperms (flowering plants) (Smooker *et al.,* 1990).

The process by which genetic information originally housed in organelles was transferred to the nucleus has been discussed in some detail (Obar and Green, 1985; Gellissen and Michaelis, 1987; Brennicke *et al.,* 1993). For the successful transfer of genetic material between separate genomes inside a cell, a number of important criteria must be met. First, the genetic information in the organelle must be present in multiple copies, such that the loss of one or more copies to the nucleus does not harm cellular metabolism. The presence of multiple copies of mitochondrial and chloroplast genomes in eukaryotic cells provides the redundancy of genetic information necessary for this first requirement in the transfer process. The escape of genetic material from organelles and its subsequent transfer to the nucleus is the next step in the process. On the basis of observations made in yeast (Thorsness and Fox, 1990, 1993; Schiestl *et al.,* 1993) this is an event that occurs much more frequently then previously assumed (see Section III). Conceivably a rate-limiting step in the transfer of genetic information to the nucleus is the integration and stable maintenance of escaped genetic material. As in transformation of eukaryotic cells with recombinant plasmids, where the frequency of transformation is several orders of magnitude higher if the DNA is capable of extrachromosomal replication (Hinnen *et al.,* 1978), a large fraction of the organelle-derived DNA that arrives in the nucleus will not be faithfully replicated or integrated into the nuclear genome. Organelle-derived DNA that does integrate into the nuclear genome is then faced with the toughest task of all: expression of the encoded gene product by the nuclear transcription apparatus, translation of the mRNA by cytoplasmic ribosomes in the case of protein gene products, and finally the targeting of the gene product back to the correct cellular compartment. The expression and correct targeting of the products of the transferred genetic material must take place in a relatively short period of time in order to avoid "inactivation" by random mutation. If the gene product encoded by the transferred organellar genetic information is successfully introduced back into the organelle, it then can compete and eventually displace the organelle-encoded version of the gene product. Ultimately, if the nuclear copy of the gene provides an increase in fitness for the cell, the organellar copy of the gene may be inactivated by mutation and/or ultimately lost.

### 2. Distribution of Organellar Sequences in Heterologous Genomes

The nuclear genomes of any eukaryotic cell that has been examined in detail have been found to include sequences that clearly originated from the mitochondrial genome and, if chloroplasts are present, from the genome of that compartment as well. In some cases these sequences include the entire coding region of mitochondrial or chloroplast genes and may even be functional. Often these sequences in the nuclear genome correspond to fragments of an organellar genome and are unlikely to represent functional genetic information.

The human nuclear genome contains sequences homologous to mitochondrial sequences (Kamimura *et al.,* 1989; Shay and Werbin,1992; Hu and Thilly, 1994), and estimates place the number of sites where integration may have occurred at as many as several hundred (Fukuda *et al.,* 1985). The nuclear genomes of locusts (Gellissen *et al.,* 1983), sea urchins (Jacobs *et al.,* 1983), yeast (Farrelly and Butow, 1983; Louis and Haber, 1991), and various species of plants (Fukuchi *et al.,* 1991; Nugent and Palmer, 1991; Grohmann *et al.,* 1992; Sun and Callis, 1993) all contain sequences that are clearly derived from sequences found in the mitochondrial genome. It is interesting to note that the transfer of sequences from plant mitochondria to the nucleus in some cases appears to have utilized an RNA intermediate, because the sequence found in the nuclear genome more closely resembles the edited sequence than the mitochondrial genomic sequence (Fukuchi *et al.,* 1991; Nugent and Palmer, 1991; Covello and Gray, 1992; Grohmann *et al.,* 1992). However, in at least one case the transfer of mitochondrial sequences to the nucleus presumably bypassed RNA because the transferred sequence included intron sequences (Sun and Callis, 1993).

Sequences homologous to chloroplast DNA can be found integrated in the nuclear genomes of tobacco (Ayliffe and Timmis, 1992) and tomato (Pichersky *et al.,* 1991). In addition, a considerable amount of documentation has been accumulated that describes sequences apparently derived from the chloroplast genome in the mitochondrial genome of a number of plants (Stern and Lonsdale, 1982; Stern and Palmer, 1986; Moon *et al.,* 1988; Joyce and Gray, 1989; Jubier *et al.,* 1990) and even in the mitochondrial genomes of ameboid and ciliated protozoans (Pritchard *et al.,* 1989; Lonergan and Gray, 1994). While the chloroplast genome has clearly contributed to the evolution and structure of mitochondrial genomes, there is as yet no evidence that nucleic acids have been transferred from mitochondrial or nuclear genomes to chloroplasts and integrated into the chloroplast genome. This may be a reflection of a relative lack of sequence information about chloroplast genomes, such that instances of genetic information transfer will ultimately be found, or there may be additional physical constraints faced by nucleic acids attempting to enter the chloroplast compartment that effectively limit the transfer of genetic material.

While there is ample evidence for the transfer of genetic material from mitochondria and chloroplast to the nucleus, the transfer of genetic material derived from the nuclear genome to the chloroplast and mitochondrial genomes is, at best, infrequent. The mitochondrial genome of *Oenothera* has a 528-nucleotide sequence that shows 91% homology with the nuclear-encoded 18S rRNA from maize (Schuster and Brennicke, 1987). This sequence includes a segment 160 nucleotides in length that is unique to the nuclear rRNA. Thus, while there is only limited evidence for the transfer of nuclear-encoded genetic information to mitochondria or chloroplasts, it has apparently happened. As more organellar genomes are sequenced it likely will be found to have occurred elsewhere.

## III. Detection of DNA Escape from Organelles in Living Cells

The molecular archaeology that has identified genes and pseudogenes of organellar origin in the nucleus of many different organisms has not defined either a mechanism or a rate of DNA escape from organelles. The malleable genetic nature of the yeast *Saccharomyces cerevisiae* allows several different experimental approaches to be taken that address the frequency and mechanism of DNA escape from mitochondria and its subsequent migration to the nucleus. Whereas the pathway for DNA escape from mitochondria remains elusive, it is clear that the escape of DNA from mitochondria occurs at a frequency roughly equal to the rate of mutational inactivation of a nuclear gene.

### A. Direct Detection of DNA Escape from Mitochondria of Yeast

#### 1. Escape of DNA from rho$^-$ Mitochondria

Extensive work with deleted and recombinant mitochondrial genomes has revealed that virtually any DNA sequence can be replicated in yeast mitochondria. Large deletions of the mitochondrial genome arise spontaneously in yeast and the remaining mitochondrial DNA is concatemerized so that the total size of individual mitochondrial chromosomes is roughly equal to wild-type chromosomes, approximately 75 kilobase pairs (kbp) (Pon and Schatz, 1991). Yeast strains containing such altered mitochondrial chromosomes are referred to as being rho$^-$. About 50 copies of these altered genomes (like normal mitochondrial genomes) are found per cell. Thus, if a deletion of a mitochondrial genome left behind 7.5 kb of sequence information, that sequence would be represented 500 times within the mitochondrial compartment.

Taking advantage of new developments in the technology of DNA transformation, a strain of *S. cerevisiae* that contained no mitochondrial DNA (such strains are referred to as rho$^0$) was transformed with a recombinant DNA capable of replication in both mitochondria and nuclei (Thorsness and Fox, 1990). This plasmid, pMK2, contained a 2$\mu$ origin for replication in the nucleus. This plasmid also contained genetic information that allowed its detection in either compartment. Plasmid pMK2 bore the commonly used nuclear marker *URA3,* encoding a gene product necessary for uracil biosynthesis, and a genetic marker of the mitochondrial compartment, *COX2,* encoding an essential subunit of cytochrome oxidase. The presence of pMK2 in the nucleus could be determined by the ability of the *URA3* gene on the plasmid to complement a *ura3* mutation at the chromosomal location and thus allow growth in the absence of uracil. The presence of pMK2 in mitochondria could be determined by mating the yeast to a tester strain of the opposite mating type that contained a deletion of the *COX2* gene in the mitochondrial genome. Diploid yeast that contained the complete mitochondrial genome were able to utilize a nonfermentable carbon source, in contrast to the haploids or diploids that contained only the recombinant plasmid or the deleted mitochondrial genome.

Yeast lacking the recombinant plasmid in the nucleus but containing it in mitochondria were unable to grow in the absence of uracil. However, these yeast spontaneously gave rise to uracil prototrophs at a rate of $2 \times 10^{-5}$ events per cell per generation. The appearance of uracil prototrophs in this yeast strain was dependent on the presence of the recombinant plasmid bearing the *URA3* and *COX2* genes in mitochondria. Yeast lacking the plasmid in both the nuclei and mitochondria never gave rise to uracil prototrophs. These observations led to the conclusion that the appearance of uracil prototrophs in a population of *ura3* yeast was due to the escape of DNA from mitochondria and its subsequent migration to the nucleus (Thorsness and Fox, 1990).

## 2. Escape of DNA from rho$^+$ Mitochondria

A second generation of yeast strains was engineered that validated the initial observation and allowed extensive genetic analysis of the escape of DNA from mitochondria. The nuclear gene *TRP1*, necessary for tryptophan biosynthesis, was integrated into an otherwise normal mitochondrial genome (referred to as rho$^+$) (Thorsness and Fox, 1993). This yeast strain behaved normally in every respect other then the spontaneous appearance of tryptophan prototrophs at a rate of $5 \times 10^{-6}$ events per cell per generation in a strain bearing a nonreverting deletion mutation of *TRP1* at the nuclear chromosomal location. A diagram of this experimental system is shown in Fig. 1. The rate of DNA escape observed in rho$^+$ yeast was about 30 times slower than the rate of DNA escape from a rho$^-$ yeast of the same genetic background. This difference in rate may be due to several factors. There is a difference in the number of *TRP1* genes present in mitochondria: roughly 50 copies in the rho$^+$ strain and 500 copies in the rho$^-$ strain. There are also observable differences in the structure of rho$^+$ and rho$^-$ mitochondrial compartments and clear differences in the metabolic activi-

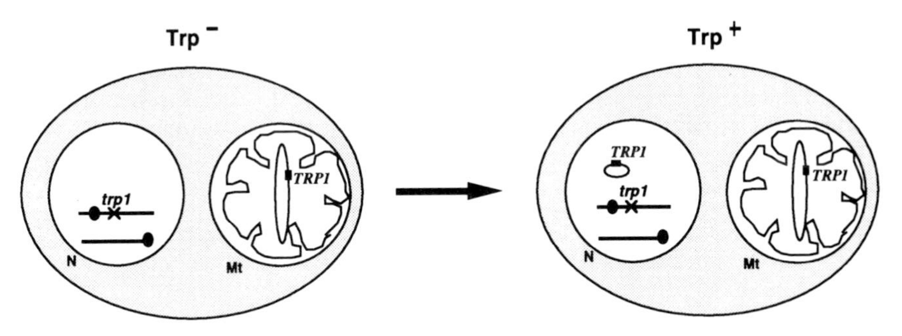


FIG. 1 Detecting escape of DNA from mitochondria in yeast. *Saccharomyces cerevisiae* containing a *trp1* mutation in the nucleus (N) and bearing an altered mitochondrial chromosome that contains a wild-type copy of the *TRP1* gene are phenotypically tryptophan auxotrophs ($Trp^-$). During mitotic growth, fragments of DNA escape mitochondria (Mt) and migrate to the nucleus. Escaped fragments of the mitochondrial genome that contain the wild-type *TRP1* gene and are also replicated in the nucleus allow the yeast to become a tryptophan prototroph ($Trp^+$).

ties carried out by $rho^+$ and $rho^-$ mitochondria. As seen for the escape of DNA from $rho^-$ mitochondria, the appearance of tryptophan prototrophs was dependent on the presence of mitochondrial DNA. The escaped mitochondrial DNA in the nucleus was analyzed by DNA blot hybridization and found to contain sequences corresponding to the *TRP1* gene and flanking mitochondrial sequences. The size of the escaped DNA varied from 7 kb to greater than 30 kb. Only 1 of the 18 independently isolated DNA escape events had a restriction map that was collinear with the restriction map of authentic mitochondrial DNA, indicating that significant rearrangement of sequences accompanied escape of the DNA from mitochondria and its subsequent migration to the nucleus.

There are several important facts to consider when comparing the frequent mitochondrial DNA escape observed in these experiments with the relatively infrequent appearance of mitochondrial DNA integrated into the nuclear genome. Fragments of escaped mitochondrial DNA that do not contain a selectable marker or that are not stably maintained in the nucleus are not detected in this assay. The recombinant plasmid bearing the *URA3* nuclear marker also contained an origin of replication for the nucleus, the $2\mu$ origin. Likewise, the *TRP1* gene has a closely associated autonomous replicating sequence, *ARS1,* that allows replication of DNA in the nucleus. Less than 1% of all DNA escape events scored in this assay resulted in integration of the escaped mitochondrial DNA into a nuclear chromosome (K. H. White and P. E. Thorsness, unpublished observations). DNA escaping from mitochondria is presumably subject to circularization via recombination, a highly active process in yeast nuclei, and the measured rate of

DNA escape is a reflection of that process. Indeed, the measurement of the rate at which DNA escapes from mitochondria is likely to be dependent on a number of complex processes: the escape event itself, localization of escaped DNA to the nucleus, recombination events within the nucleus, and replication and segregation of the escaped DNA.

## B. Indirect Detection of DNA Escape from Mitochondria of Yeast

In studies directed at investigating the nature of illegitimate integration of nonhomologous transforming DNA in yeast, Schiestl *et al.* (1993) detected the presence of mitochondrial DNA in the nucleus. Using a yeast strain that contained a complete deletion of the *URA3* locus and a linear transforming DNA containing the *URA3* gene, 4% of mitotically unstable transformants (uracil prototrophs) were plasmids formed from the ligation of transforming DNA to fragments of mitochondrial DNA. The mitochondrial portion of these plasmids localized in the nucleus ranged in size from 200 to 700 bp in length. Two of the four plasmids analyzed by sequence analysis contained a contiguous sequence from the mitochondrial genome and the other two contained two separate mitochondrial DNA fragments from different locations on the mitochondrial genome that had been joined with the transforming DNA to form the complementing plasmid. The ability of the complementing plasmids to replicate in the nucleus was a direct result of the mitochondrial DNA sequences contained in the plasmids. A nuclear ARS (autonomous replicating sequence) in yeast is largely composed of long tracts of adenine and thymidine with a degenerate core sequence (Marahrens and Sillman, 1992). Being composed of 80% adenine and thymidine (Pon and Schatz, 1991), the yeast mitochondrial genome contains many such A/T-rich stretches and various fragments of the mitochondrial genome have been recovered in assays for DNA sequences that have ARS activity (Hyman *et al.,* 1982; Delouya and Norbrega, 1991).

It has been noted that in yeast the efficiency of transformation with nonhomologous linear DNA fragments is 1% of that observed for transformation with homologous linear DNA fragments (Schiestl and Petes, 1991). Assuming that the nonhomologous sequences ligated to mitochondrial DNA with 100% efficiency and that treating yeast to make them competent for transformation does not change the fraction of cells containing mitochondrial DNA sequences in the nucleus, Schiestl *et. al.* estimate that mitochondrial DNA may be found in the nucleus in as many as 1 in 1000 cells (Schiestl *et al.,* 1993). As noted previously, detecting the escape and subsequent migration of a complete gene from mitochondria to the nucleus requires at least 1.5 kb of intact, collinear mitochondrial DNA, whereas detection of mitochondrial DNA in the nucleus via ligation to nonhomologous linear DNA lowers the size requirement for detection of escaped DNA by as much as 10-fold. In addition, within the mitochondrial genome the introduced *TRP1* gene comprises approximately 2% of the total mitochondrial genome, while 80% of the mitochondrial genome may potentially provide ARS function to nonhomologous linear DNA, so that virtually any escaped DNA fragment can be detected. Thus, the detection of escaped mitochondrial DNA in the nucleus by transformation with nonhomologous linear DNA is significantly less demanding in both size and sequence requirements than is the detection of escaped mitochondrial DNA by monitoring the transfer of a complete gene and thus may more accurately reflect the frequency at which mitochondrial DNA sequences migrate to the nucleus.

## C. Detection of Nucleic Acids That Have Escaped from Mitochondria in Higher Eukaryotes

The analysis of DNA that escapes from organelles of higher eukaryotes and migrates to the nucleus has largely been restricted to cloning and hybridization studies of events that happened perhaps more than 1 million years in the past. While no direct investigation into the escape of DNA from organelles of living cells has been reported, several experimental efforts not directly aimed at analyzing this phenomenon suggest that nucleic acids can and do escape from organelles on a routine basis in eukaryotic cells other than *S. cerevisiae.*

### 1. Detection of Intact Mitochondrial DNA in Isolated HeLa Cell Nuclei

Recently, Kristensen and Prydz (1986) reported the detection of intact mitochondrial DNA in preparations of nuclei from human cells, HeLa S3 cells. The mitochondrial DNA was detected in an *in vitro* DNA replication system by labeling newly synthesized DNA with [$^{32}$P]dCTP, and demonstrating that a discrete band detected after agarose gel electrophoresis and autoradiography corresponded to intact mitochondrial DNA. As noted by the authors, the presence of mitochondrial DNA in these nuclear preparations may be due to contamination by mitochondrial compartments or mitochondrial DNA copurifying with the nuclear preparations. However, as the addition of purified mitochondria or mitochondrial DNA to the nuclear extracts did not increase the labeling of mitochondrial DNA, the authors conclude that mitochondrial DNA is present in nuclei *in vivo.* If this observation withstands further investigation, it will be interesting to ascertain the number of mitochondrial DNA molecules per nuclei and the fraction of nuclei in a population of cells that contain mitochondrial DNA.

#### 2. Mitochondrially Encoded Large rRNA Outside of Mitochondria in Germ Plasm of *Drosophila*

While there is some disagreement on its exact role in the formation of pole cells in developing *Drosophila* embryos, it is clear that a nucleic acid encoded by mitochondrial DNA, the large 16S ribosomal RNA, is located outside of mitochondria in the pole plasm of embryos (Kobayashi and Okada, 1989; Kobayashi *et al.,* 1993; Ding *et al.,* 1994). Treatment of *Drosophila* embryos with ultraviolet light prevents the formation of pole cells. However, this inability to form pole cells can be compensated by the addition of the mitochondrially encoded large ribosomal RNA (Kobayashi and Okada, 1989). Furthermore, Northern blot analysis of fractionated embryos and *in situ* hybridization places a significant amount of the mitochondrially encoded large ribosomal RNA outside of mitochondria in the polar plasm (Kobayashi and Okada, 1989; Kobayashi *et al.,* 1993). The presence of the mitochondrially encoded large ribosomal RNA in the polar plasm is apparently developmentally regulated, as there is a rapid decrease in the amount of ribosomal RNA localized in the polar regions of blastodermal embryos (Kobayashi *et al.,* 1993; Ding *et al.,* 1994). It is interesting to note that the localization of the mitochondrially encoded large ribosomal RNA outside of mitochondria is fairly specific, as neither the mitochondrially encoded RNAs for the small ribosomal RNA or NADH dehydrogenase subunit 1 are found concentrated in the polar plasm (Kobayashi *et al.,* 1993). On the basis of these observations, it has been suggested that the mitochondrially encoded large ribosomal RNA is transported out of mitochondria into the cytosol only in the polar plasm (Kobayashi *et al.,* 1993). Also required is a mechanism to remove or degrade the ribosomal RNA as development of the embryo progresses. There is no evidence that mitochondrial DNA or mitochondrially encoded RNAs escape wholesale from mitochondria, as probes for other mitochondrial RNAs and antisense probes did not elicit signals *in situ* (Kobayashi *et al.,* 1993).

## IV. Mechanisms of Nucleic Acid Escape from Organelles

### A. Possible Mechanisms of Nucleic Acid Escape from Organelles

Much speculation on the mechanism by which DNA or RNA escapes organelles and migrates to the nucleus has been recorded (Gellissen and

Michaelis, 1987; Richter, 1988; Shay and Werbin, 1992; Brennicke *et al.*, 1993), but no direct experimental investigations were undertaken. Potential pathways for the escape of nucleic acids include transient breaches in organellar membranes during fusion, budding or morphological reorganization of the compartment, release of genome fragments during terminal digestion of damaged compartments by lysozomes or vacuoles, the illicit use of protein or nucleic acid transport systems in organellar membranes, or even the fusion of heterotypic membranes.

Mitochondria are dynamic organelles and morphological alterations during development and differentiation processes as well as changes in subcellular localization have been previously documented (Thorsness, 1992). Analysis of genetically marked mitochondrial DNAs (Strausberg and Perlman, 1978; Lockshon *et al.*, 1995) and physically marked mitochondrial compartments (Azpiroz and Butow, 1993) in yeast and physically marked mitochondrial compartments in animal cells (Hayashi *et al.*, 1994) have clearly demonstrated that mitochondrial compartments bud and fuse rapidly. Some of these processes have been well defined utilizing electron and light microscopy techniques (Bereiter-Hahn and Voth, 1994; Kuroiwa *et al.*, 1994). While the actual mechanisms of and reasons for organelle movement remain largely unknown, the dynamic nature of mitochondria provides ample opportunity for transient breaches in the fusing and budding membranes to form and allow the escape of mitochondrial DNA. It is also possible that environmental conditions could cause such transient breaches in the membranes of compartments. For example, it was noted that the escape of mitochondrial DNA in yeast and its subsequent migration to the nucleus could be induced in nongrowing cells by a freeze/thaw cycle or prolonged incubation in 15% glycerol (Thorsness and Fox, 1990). Such treatments could conceivably create such breaches and allow the escape of mitochondrial DNA.

An alternative explanation for the effect that a freeze/thaw cycle or incubation in glycerol has on the escape of mitochondrial DNA in yeast might be that such treatments damage mitochondrial compartments in a way that subsequently targets them for degradation, and DNA is released during that process. That turnover of mitochondrial compartments and cytoplasmic constituents by autophagic vesicles does occur has been documented for both yeast and animal cells (Knecht *et al.*, 1988; Chiang and Schekman, 1991; Takeshige *et al.*, 1992; Baba *et al.*, 1994). It is not known what environmental or genetic factors might increase or stimulate the turnover of mitochondrial or chloroplast compartments, but an increase in the rate of dissolution of these compartments may increase the chances for the escape of nucleic acids to the cytoplasm. This putative escape pathway is complicated by the additional need of the nucleic acid to escape the lysozome or vacuole after autophagy. There is some genetic data (described in Section IV,C) from yeast that imply a role for vacuoles in mediating DNA escape in certain mutant backgrounds.

As noted in Section II,B,2, some sequences of mitochondrial origin found in the nucleus of plants appear to have escaped and migrated to the nucleus via an RNA intermediate, as the nuclear sequences more closely resemble the edited RNA then the mitochondrial genomic sequence (Fukuchi *et al.*, 1991; Nugent and Palmer, 1991; Covello and Gray, 1992; Grohmann *et al.*, 1992). These observations have led to the speculation that perhaps some or all DNA sequences of organellar origin found in the nucleus proceeded to the nucleus by way of an RNA intermediate. This process would require that a reverse transcriptase activity be present and active somewhere along the escape route to generate the DNA for integration into the nuclear genome. The wide distribution of reverse transcriptase activities indigenous to eukaryotic retrotransposons (Flavell, 1995) and mitochondrial introns (Kennell *et al.*, 1993) suggests that such a conversion of RNA to DNA during the escape process is possible. Furthermore, the apparent presence of naturally occurring RNA transport systems in organelles allows the invocation of illicit use of such transport systems by escaping RNA. The import of tRNAs encoded by nuclear genes into mitochondria occurs in plants (Chiu *et al.*, 1975; Marechal-Drouard *et al.*, 1988, 1990; Small *et al.*, 1992) and trypanosomes (Schneider *et al.*, 1994). The nucleus of animal and yeast cells encodes an essential RNA component of a site-specific ribonucleoprotein endoribonuclease that cleaves an RNA sequence complementary to mammalian or yeast mitochondrial origins of replication in a manner consistent with a role in primer RNA metabolism (Chang and Clayton, 1989; Schmitt and Clayton, 1992; Li *et al.*, 1994). It has also been found that the protein import machinery of isolated yeast mitochondria can import a short, double-stranded DNA molecule if the DNA has been covalently linked to a peptide that can interact with the protein import apparatus (Vestweber and Schatz, 1989). Thus, with mechanisms occurring naturally that can support the import of nucleic acids across two membranes and into mitochondria, it is a small step to suggest that infrequent illicit use by RNA or DNA may allow their export to the cytoplasm.

Fusion of homotypic membranes is a means by which the cell can regulate the growth, localization, and inheritance of organelles (Rothman and Warren, 1994). This fusion between like membranes has been defined to some extent at the molecular level for the endoplasmic reticulum–nuclear envelope–Golgi membrane network of yeast (Latterich and Schekman, 1994). As noted previously, homotypic fusion of mitochondrial and chloroplast compartments has been described only at a cytological level. There are also some cytological studies that suggest that heterotypic membrane fusions may also occur, and this suggests the possibility of direct transfer of organellar DNA to the nucleus, or between organelles. In particular, the direct physical association of mitochondrial and nuclear membranes (Mota, 1963), the actual encapsulation of mitochondrial compartments inside nuclei (Jensen *et al.,* 1976; Shay and Werbin, 1992), and continuities of chloroplast, mitochondrial, endoplasmic reticulum, and plasma membranes (Crotty and Ledbetter, 1973) have been described. While noting that only cytological evidence for fusion between DNA-containing organelles and nuclei exists, it would require only rare fusion events between these heterotypic membranes to generate ample opportunity for the transfer of genetic material between organelles and nuclei.

## B. Transfer of Escaped Organellar DNA to Nucleus

It is interesting to note that all proposed mechanisms concerning the escape and migration of nucleic acids from organelles to the nucleus have focused only on the aspect of escape of nucleic acids from organelles. There is some basis for this focus, as it is apparent that once DNA enters the cytoplasm, it rapidly and efficiently enters the nucleus. Transformation of bacteria (Smith *et al.,* 1992), fungi (Armaleo *et al.,* 1990), plants (Takeuchi *et al.,* 1992), and animals (Johnston and Tang, 1993) by microprojectile bombardment and other transformation protocols that rely largely on the physical disruption of the cell wall and/or plasma membrane in the presence of DNA in an osmotically supportive environment (Costanzo and Fox, 1988) has led to the conclusion that delivery of DNA to the cytoplasm is the rate-limiting step in nuclear transformation. Entry of DNA into animal cells via receptor-mediated endocytosis has been demonstrated (Zenke *et al.,* 1990), but the level of transformation is poor, presumably because of trapping of the DNA–ligand–receptor complex in the endosome. In fact, it is possible to enhance significantly the transformation efficiency of cells via receptor-mediated endocytosis by inclusion of viral particles that can disrupt endosomes and thus mediate release of the transforming DNA into the cytoplasm (Cotten *et al.,* 1992, 1993; Wagner *et al.,* 1992). Furthermore, the ability to transform yeast with large (up to several hundred kilobase pairs in length) artificial chromosomes known as YACs (Burke and Olson, 1991) is a compelling reason to believe in the existence of an active mechanism for the uptake of DNA into the nucleus of cells. Despite this circumstantial evidence for the effective targeting of DNA introduced into the cytoplasm to the nucleus, it is possible that once the cell wall and/or plasma membrane is breached, there may be other rate-limiting steps in the transfer of DNA to nucleus. For instance, microinjection of DNA directly into the nucleus was significantly more efficient at achieving expression of recombinant DNAs than was the microinjection of DNA into the cytoplasm (Mirzayans *et al.,* 1992). Once the DNA has entered the nucleus, the ability of that DNA to be expressed or integrated is also clearly sequence dependent. Even recombinant DNAs that are designed to be expressed in the nucleus have different transformation efficiencies. Transformation by microinjection of mouse cells with a recombinant DNA containing the thymidine kinase gene was 100 times less efficient then transformation by microinjection with a recombinant DNA bearing the neomycin gene (Strauss *et al.,* 1983).

## C. Investigations into Escape of DNA from Yeast Mitochondria

With the advent of the modified mitochondrial genomes in *S. cerevisiae* (Thorsness and Fox, 1990, 1993), it has been possible to begin an investigation into the pathways of DNA escape from an organelle. While no definitive pathways for DNA escape from mitochondria have yet been found, a number of interesting observations have been made. Initial characterization of yeast containing the altered rho$^-$ genome revealed that the rate at which this engineered mitochondrial genome escaped and migrated to the nucleus varied fivefold in response to changes in the nuclear genetic background and growth conditions. The escape of DNA from mitochondria could be induced in nongrowing cells by treatments such as freezing at −70°C in 15% glycerol or prolonged incubation in 15% glycerol at room temperature. These observations suggested that the process of DNA escape from mitochondria was under both genetic and environmental control and that further investigation might reveal genetic elements that, when mutated, could alter the rate at which DNA escaped from mitochondria, migrated to the nucleus, and was replicated and expressed.

It has been determined that the escape of DNA from mitochondria and the subsequent transfer to the nucleus are intracellular events. It was formally possible that mitochondrial DNA ending up in the nucleus could have been endocytosed from the extracellular medium and found its way to the nucleus. The lysis of a small fraction of yeast cells in a population could provide such a reservoir of mitochondrial DNA in the growth medium. However, when two strains of yeast with different nutritional markers were cocultured, only the strain that contained an altered mitochondrial genome bearing the nuclear gene *TRP1* as a marker gave rise to tryptophan prototrophs (K. White and P. Thorsness, unpublished data). Thus, DNA escapes from mitochondria inside a cell and migrates to the nucleus of that cell.

It has been speculated that the transfer of genetic information between organelles proceeds through an RNA intermediate (Gellissen and Michaelis, 1987; Nugent and Palmer, 1991; Shayand Werbin, 1992). A pathway involving an RNA intermediate seems most plausible for the case of the transfer of the *coxII* structural gene from mitochondria to the nucleus during flowering plant evolution, as the nuclear gene more closely resembles the edited RNA form of the mitochondrial *coxII* structural gene then the gene encoded on the mitochondrial chromosome (Nugent and Palmer, 1991). In yeast, an RNA intermediate does not seem to play a significant role in the escape of DNA from mitochondria. The mitochondrial RNA polymerase of yeast (Greenleaf *et al.,* 1986) was inactivated and it was found that DNA escaped from mitochondria at roughly the same rate as that found for strains containing the wild-type mitochondrial RNA polymerase activity (K. White and P. Thorsness, unpublished data). This experiment does not mean that RNA is not part of some pathway for the escape and migration of genetic material to the nucleus, just that it is not necessary for the transfer of genetic material by the most active pathways in yeast.

The observed differences in the rate of DNA escape from yeast mitochondria of different nuclear genetic backgrounds (Thorsness and Fox, 1990) suggested that the powerful tools of classic and modern genetic analysis could be used to investigate the pathway of DNA escape from mitochondria. In fact, a number of nuclear mutations have been isolated that alter the rate at which DNA escapes mitochondria and migrates to the nucleus in yeast (Thorsness and Fox, 1993). At least 10 different nuclear genes when mutated can increase the rate of mitochondrial DNA escape. Nine of these mutations are recessive, one is dominant, and a number of them exhibit collateral growth phenotypes in addition to the high rate of DNA escape from mitochondria (Thorsness and Fox, 1993; Thorsness *et al.,* 1993). These mutations have been termed *yme,* for yeast mitochondrial escape.

The *yme1* mutation has a number of interesting collateral phenotypes that may or may not have anything to do with the escape of DNA from mitochondria (Thorsness *et al.,* 1993). These phenotypes include an inability to utilize nonfermentable carbon sources at higher growth temperatures (37°C versus the normal 30°C) and very slow growth in the complete absence of mitochondria DNA (Weber *et al.,* 1996). The wild-type *YME1* gene was cloned by complementation of the temperature-sensitive respiratory-deficient growth phenotype. The protein encoded by this gene, Yme1p, has sequences characteristic of proteins that bind and hydrolyze ATP and also sequences characteristic of zinc-dependent proteases (Thorsness *et al.,* 1993; Campbell *et al.,* 1994). This protein is localized to the matrix side of the inner mitochondrial membrane and is likely to be directly responsible for the degradation of unassembled cytochrome oxidase subunit II (Weber *et al.,* 1996). This protease is likely to have additional substrates, as *yme1* mutant yeast strains totally lacking mitochondrial DNA, and hence cytochrome oxidase subunit II, have a slow-growth phenotype (Thorsness *et al.,* 1993; Weber *et al.,* 1996).

It is likely that the increased rate of DNA escape from mitochondria due to the *yme1* mutation stems from pleiotropic effects of the mutation. It is possible to isolate extragenic suppressors of the *yme1* phenotypes, some of which suppress all of the phenotypes (Campbell *et al.,* 1994) and some of which suppress only one or two of the phenotypes (D. Kominsky and P. Thorsness, unpublished observations). Suppression of the high rate of DNA escape in *yme1* strains can be accomplished by inactivating important vacuolar proteases, which consequently inhibit vacuolar activities (Campbell and Thorsness, unpublished observations). During electron and fluorescent microscopy studies of *yme1* mutant yeast (Campbell *et al.,* 1994), it was observed that the morphology of mitochondria in the mutant strains was severely altered and mitochondrial compartments were often seen directly associated with the vacuole, the organelle responsible for autophagy and degradation of cellular components (Chiang and Schekman, 1991; Baba *et al.,* 1994). These data have led to the hypothesis that in *yme1* mutant cells mitochondria are defective in a way that leads to an increased turnover by vacuoles. This proposed increase in turnover of the mitochondrial compartment perhaps is reflected in the increased rate of DNA escape seen in *yme1* mutant yeast. Thus, in this model, if the vacuole is inactivated by mutation, damaged mitochondrial compartments will not be turned over and hence the rate of DNA escape will be lowered. However, not all DNA escape from mitochondria is mediated by vacuolar activity, as vacuolar mutations do not significantly affect the rate of DNA escape from wild-type yeast, nor do they affect the elevated rates of DNA escape found in other *yme* mutant strains (Campbell and Thorsness, unpublished observations). A pathway for DNA escape from mitochondria that invokes the terminal degradation or dissolution of the compartment is also appealing because it also accounts for what appears to be a unidirectional transfer of DNA, from mitochondria to nucleus (Thorsness and Fox, 1990).

Two other genes identified by *yme* mutations have been cloned. *YME2* was found to be identical to *RNA12,* a protein presumed to be involved in a dispensable methylation event of ribosomal RNA (Liang *et al.,* 1992). Mutation of both *yme2* and *yme1* lead to an inability to grow on nonfermentable carbon sources as well as to an increase in the rate at which DNA escapes mitochondria and migrates to the nucleus (Thorsness and Fox, 1993). Yme2p, the gene product of *YME2/RNA12,* is an integral membrane protein located in the inner mitochondrial membrane (Hanekamp and Thorsness, 1996). It seems likely that Yme2p is involved in functions other then methylation of ribosomal RNA. Also cloned was the *YME6* gene. It was found to be identical to *MMM1,* a gene necessary for maintaining normal mitochondrial morphology (Burgess *et al.,* 1994). The encoded protein is found in the outer membrane of mitochondria and yeast that lack it rapidly lose all of their mitochondrial DNA, have grossly swollen mitochondria, and are very slow growing. The *yme6-1* allele of *MMM1* is an attenuated version of this protein, as *yme6-1* strains are capable of respiratory growth and *yme6* deletion strains do not show an increased rate of DNA escape from mitochondria, principally because their mitochondria lack any DNA (T. Hanekamp and P. Thorsness, unpublished data).

While no detailed pathway of DNA escape from mitochondria has yet emerged from the genetic studies done in yeast, discussion of the mechanism of the transfer of genetic information between compartments can, for the first time, leave the realm of speculation. It is possible to draw a few simple conclusions concerning the mechanism of DNA escape from mitochondria of yeast: an RNA intermediate is not an obligate step in the transfer of DNA from mitochondria to the nucleus, the process is intracellular, and there are multiple pathways for the escape of DNA from mitochondria. With the identification of gene products that when absent or defective increase the rate at which DNA escapes mitochondria and migrates to the nucleus, further progress in defining the pathways of DNA migration should be forthcoming.

### D. Limits to Transfer of Genetic Information between Compartments

With the overwhelming evidence for the transfer of genetic material from organelles to the nucleus, why is there any genetic material left in organelles? This question is even more striking when we realize that by using recombinant DNA technology it is possible to express mitochondrial proteins *in vitro* and successfully import them into mitochondria (Gearing and Nagley, 1986; Farrell *et al.,* 1988). There may be, however, limits to the effective expression of genes in the nucleus that are normally found in organellar genomes. Variations of the genetic code between genomes (Fox, 1987) and the difficult task of expressing and targeting a transferred gene back to the correct compartment *in vivo* using natural processes are formidable barriers in and of themselves. It has also been suggested that gene products encoded by the mitochondrial genome would enter the secretory pathway via the endoplasmic reticulum if they were synthesized in the cytoplasm (von Heijne, 1986). There is also a clear barrier to the import of highly hydrophobic proteins into yeast mitochondria *in vivo* (Claros *et al.,* 1995). Apparently, large stretches of hydrophobic residues, corresponding to three or four transmembrane domains, can effectively inhibit import into mitochondria. A strong correlation between the types of proteins encoded in mitochondria and the projected inhibition of import into mitochondria was noted. The ultimate barrier to transfer of genetic material from one compartment to another may reside within the environmental niche filled by the organism in question. As noted previously, the selection of genes maintained in an organelle genome and those that are maintained in the nuclear genome varies between organisms. For reasons that may not be apparent without considerable investigation, certain genes may be best suited to encryption in one compartment instead of another for a particular organism. At the same time, a similar genetic organization may not be in the best interest of a related organism found in a slightly different environmental niche.

As noted in Section II,B,2, it appears that the preponderance of sequences transferred during evolution has been from mitochondria and chloroplasts to the nucleus rather than from the nucleus to mitochondria and chloroplasts. There thus appears to be a barrier to the transfer of genetic material, whether RNA orDNA, from the nucleus to organelles. In yeast, where the transfer of an intact gene from mitochondria to the nucleus occurs at a frequency as high as $2 \times 10^{-5}$ events per cell per generation, the transfer of DNA to the mitochondria has never been observed, and the rate at which DNA might be transferred to the nucleus is unlikely to be greater than $10^{-10}$ events per cell per generation (Thorsness and Fox, 1990). This relative lack of nucleic acid transfer from the nucleus to mitochondria or chloroplast is likely to be a reflection of the mechanisms employed in the observed transfer of genetic material from mitochondria and chloroplasts to nuclei. A mechanism invoking release of genetic material on terminal dissolution of a compartment and subsequent uptake of that material by the nucleus would make the transfer pathway effectively unidirectional. Perhaps the absence of an active process for localization of DNA to mitochondria or chloroplasts coupled with a corresponding rapid localization of DNA into the nucleus conspire to keep the concentration of migratory DNA low in the cytoplasm, so that even if mitochondrial or chloroplast compartments were capable of taking in exogenous DNA, there would be an extremely low probability of the event occurring. Finally, mitochondria and chloroplasts have more then one membrane that must be breached by the migrating nucleic acid, which may also limit the transfer of material from nuclei to mitochondria and chloroplasts.

## V. Consequences of Escaped and Migrated Genetic Material in Living Cells

Much of the discussion in previous sections has detailed the transfer of genetic information from one compartment to another, with the primary recipient being the nucleus. The effect that DNA migration has had on the genomes found in mitochondria, chloroplasts and nuclei has been extensively documented (Margulis, 1981; Weeden, 1981; Gray *et al.,* 1984; Palmer, 1990; Brennicke *et al.,* 1993; Gray, 1993; Hu and Thilly, 1994). Both in terms of the ongoing process of evolution and the vast number of organisms yet to be investigated, the last word on the effects of intracellular transfer of genetic information between organelles is yet to be written.

It has been proposed that DNA of organellar origin entering the nucleus may act as a mutagenic agent (Richter, 1988; Hadler, 1989; Shay and Werbin, 1992). Integration of escaped mitochondrial or chloroplast DNA in nuclear structural genes or regulatory regions has the potential to disrupt or alter vital cell processes. Despite the potential, there is no evidence that escaped nucleic acids from organelles has led to mutations or alterations of gene expression. In yeast, a mutation that increases 25-fold the rate at which DNA escapes mitochondria and migrates to the nucleus compared to the rate found in wild-type yeast (Thorsness and Fox, 1993) does not increase the rate of mutation of either of two nuclear genes (K. H. White and P. Thorsness, unpublished data). However, as yeast greatly prefer homologous to nonhomologous recombination events (Schiestl and Petes, 1991), the situation may be different in mammalian cells, where very few recombination events involving introduced DNAs occur at sites of homology (Capecchi, 1989). In yeast, escaping fragments of mitochondrial DNA are likely to be as biased toward A + T-rich sequences, as is the intact mitochondrial genome (Pon and Schatz, 1991). Thus, sites for integration of such escaped mitochondrial DNAs into the nuclear genome might be biased toward intragenic or promoter regions that are more likely to contain stretches of A + T-rich sequences than are structural genes. While escaped mitochondrial DNA does not appear to be a mutagenic agent of consequence in yeast, organisms less discriminating in their recombination sites may find that DNA escaping from organelles may pose a mutagenic risk.

Making a conservative estimate, Schiestl and co-workers have estimated that as many as 1 in 1000 yeast cells have small fragments of mitochondrial DNA in the nucleus (Schiestl *et al.,* 1993), so it seems possible that the appearance of mitochondrial- or chloroplast-derived DNAs in the nucleus or cytoplasm is an event that must be dealt with on a routine basis by eukaryotic cells. It is conceivable that the cell may monitor extracellular and intracellular events that may trigger an escape of DNA from mitochondria and/or chloroplasts by virtue of the appearance in the cytoplasm and nucleus of DNA that is not genomic in nature. The well-characterized recombination repair system in yeast (Petes *et al.,* 1991) clearly responds to the appearance of free double-stranded DNA ends in the nucleus and so must on occasion contend with the appearance of DNA originating from organelles. Other metabolic systems in the cell may also have to respond to a sudden influx of DNA from organelles on a regular basis.

## VI. Concluding Remarks

As noted throughout this article, DNA escape and migration between chloroplasts, mitochondria, and the nucleus has been a major player in the course of evolution of the eukaryotic cell. The process in some cases is ongoing, while in others a limit to the wholesale transfer of genetic information from organelles to the nucleus may have been reached. Until recently the investigation into the process by which genetic material escapes one compartment and migrates to another has been limited to molecular archaeology. The ability to transform organelles with recombinant DNAs has led to some advances in the field, namely determination of rates of escape and transfer and the identification of gene products and environmental conditions that can influence the process. While no pathway for escape has been clearly defined, there is likely to be more then one process by which nucleic acids escape one organelle and migrate to another. There is compelling evidence for both direct transfer of DNA from mitochondria to the nucleus and transfer of genetic information by a pathway that requires an RNA intermediate. The preponderance of pseudogenes of mitochondrial and/or chloroplast origin in the nucleus and the rates of DNA escape detected from mitochondria in yeast suggest that the appearance of DNA in the nucleus of cells is a frequent occurrence and one that may influence the nuclear genome directly via integration or indirectly via metabolic responses to nongenomic DNA.

## Acknowledgments

Research from this laboratory has been supported by a U. S. Public Health Service grant (GM47390) and a Junior Faculty Research Award from the American Cancer Society (JFRA-404). We thank Mary Thorsness, Theodor Hanekamp, Tom Fox, and Tom McMullin for many helpful discussions.

## References

Anderson, S., Bankier, A. T., Barrell, B. G., de Bruijn, M. H. L., Coulson, A. R., Drouin, J., Eperon, I. C., Nierlich, D. P., Roe, B. A., Sanger, F., Schreier, P. H., Smith, A. J. H., Staden, R., and Young, I. G. (1981). Sequence and organization of the human mitochondrial genome. *Nature* (*London*) **290,** 457–465.

Armaleo, D., Ye, G. N., Klein, T. M., Shark, K. B., Sanford, J. C., and Johnston, S. A. (1990). Biolistic nuclear transformation of *Saccharomyces cerevisiae* and other fungi. *Curr. Genet.* **17,** 97–103.

Ayliffe, M. A., and Timmis, J. N. (1992). Plastid DNA sequence homologies in the tobacco nuclear genome. *Mol. Gen. Genet.* **236,** 105–112.

Azpiroz, R., and Butow, R. A. (1993). Patterns of mitochondrial sorting in yeast zygotes. *Mol. Biol. Cell.* **4,** 21–36.

Baba, M., Takeshige, K., Baba, N., and Ohsumi, Y. (1994). Ultrastructural analysis of the autophagic process in yeast: Detection of autophagosomes and their characterization. *J. Cell Biol.* **124,** 903–913.

Baker, K. P., and Schatz, G. (1991). Mitochondrial proteins essential for viability mediate protein import into yeast mitochondria. *Nature (London)* **349,** 205–208.

Baldauf, S. L., and Palmer, J. D. (1990). Evolutionary transfer of the chloroplast *tufA* gene to the nucleus. *Nature (London)* **344,** 262–265.

Bereiter-Hahn, J., and Voth, M. (1994). Dynamics of mitochondriain living cells: Shape changes, dislocations, fusion, and fission of mitochondria. *Microsc. Res. Tech.* **27,** 198–219.

Brennicke, A., Grohmann, L., Hiesel, R., Knoop, V., and Schuster, W. (1993). The mitochondrial genome on its way to the nucleus: Different stages of gene transfer in higher plants. *FEBS Lett.* **325,** 140–145.

Bubunenko, M. G., Schmidt, J., and Subramanian, A. R. (1994). Protein substitution in chloroplast ribosome evolution. A eukaryotic cytosolic protein has replaced its organelle homologue (L23) in spinach. *J. Mol. Biol.* **240,** 28–41.

Burgess, S. M., Delannoy, M., and Jensen, R. E. (1994). MMM1 encodes a mitochondrial outer membrane protein essential for establishing and maintaining the structure of yeast mitochondria. *J. Cell Biol.* **126,** 1375–1391.

Burke, D. T., and Olson, M. V. (1991). Preparation of clone libraries in yeast artificial-chromosome vectors. *Methods Enzymol.* **194,** 251–270.

Campbell, C. L., Tanaka, N., White, K. H., and Thorsness, P. E.(1994). Mitochondrial morphological and functional defects in yeast caused by *yme1* are suppressed by mutation of a 26S protease subunit homologue. *Mol. Biol. Cell* **5,** 899–905.

Capecchi, M. R. (1989). Altering the genome by homologous recombination. *Science* **244,** 1288–1292.

Chang, D. D., and Clayton, D. A. (1989). Mouse RNase MRP RNA is encoded by a nuclear gene and contains a decamer sequence complementary to a conserved region of mitochondrial RNA substrate. *Cell* **56,** 131–139.

Chiang, H.-L., and Schekman, R. (1991). Regulated import and degradation of a cytosolic protein in the yeast vacuole. *Nature (London)* **350,** 313–318.

Chiu, N., Chiu, A., and Suyama, Y. (1975). Native and imported transfer RNA in mitochondria. *J. Mol. Biol.* **99,** 37–50.

Claros, M. G., Perea, J., Shu, Y., Samatey, F. A., Popot, J. L., and Jacq, C. (1995). Limitations to *in vivo* import of hydrophobic proteins into yeast mitochondria. The case of a cytoplasmically synthesized apocytochrome *b. Eur. J. Biochem.* **228,** 762–771.

Costanzo, M. C., and Fox, T. D. (1988). Transformation of yeast by agitation with glass beads. *Genetics* **120,** 667–670.

Cotten, M., Wagner, E., Zatloukal, K., and Birnstiel, M. L. (1993). Chicken adenovirus (CELO virus) particles augment receptor-mediated DNA delivery to mammalian cells and yield exceptional levels of stable transformants. *J. Virol.* **67,** 3777–3785.

Cotten, M., Wagner, E., Zatloukal, K., Phillips, S., Curiel, D. T., and Birnstiel, M. L. (1992). High-efficiency receptor-mediated delivery of small and large (48 kilobase) gene constructs using the endosome-disruption activity of defective or chemically inactivated adenovirus particles. *Proc. Natl. Acad. Sci. U.S.A.* **89,** 6094–6098.

Covello, P. S., and Gray, M. W. (1992). Silent mitochondrial and active nuclear genes for subunit 2 of cytochrome *c* oxidase (cox2) in soybean: Evidence for RNA-mediated gene transfer. *EMBO J.* **11,** 3815–3820.

Crotty, W. J., and Ledbetter, M. C. (1973). Membrane continuities involving chloroplasts and other organelles in plant cells. *Science* **182,** 839–841.

Delouya, D., and Norbrega, F. G. (1991). Mapping of the ARS-like activity and transcription initiation sites in the non-canonical yeast mitochondrial *ori6* region. *Yeast* **7,** 51–60.

Ding, D., Whittaker, K. L., and Lipshitz, H. D. (1994). Mitochondrially encoded 16S large ribosomal RNA is concentrated in the posterior polar plasm of early *Drosophila* embryos but is not required for pole cell formation. *Dev. Biol.* **163,** 503–515.

Farrell, L. B., Gearing, D. P., and Nagley, P. (1988). Reprogrammed expression of subunit 9 of the mitochondrial ATPase complex of *Saccharomyces cerevisiae.* Expression *in vitro* from a chemically synthesized gene and import into isolated mitochondria. *Eur. J. Biochem.* **173,** 131–137.

Farrelly, F., and Butow, R. (1983). Rearranged mitochondrial genes in the yeast nuclear genome. *Nature* (*London*) **301,** 296–301.

Flavell, A. J. (1995). Retroelements, reverse transcriptase and evolution. *Comp. Biochem. Physiol. Biochem. Mol. Biol.* **110,** 3–15.

Fox, T. D. (1987). Natural variation in the genetic code. *Annu. Rev. Genet.* **21,** 67–91.

Fukuchi, M., Shikanai, T., Kossykh, V. G., and Yamada, Y. (1991). Analysis of nuclear sequences homologous to the B4 plasmid-like DNA of rice mitochondria; evidence for sequence transfer from mitochondria to nuclei. *Curr. Genet.* **20,** 487–494.

Fukuda, M., Wakasugi, S., Tsuzuki, T., Nomiyama, H., Shimada, K., and Miyata, T. (1985). Mitochondrial DNA-like sequences in the human nuclear genome. Characterization and implications in the evolution of mitochondrial DNA. *J. Mol. Biol.* **186,** 257–266.

Gearing, D. P., and Nagley, P. (1986). Yeast mitochondrial ATPase subunit 8, normally a mitochondrial gene product, expressed *in vitro* and imported back into the organelle. *EMBO J.* **5,** 3651–3655.

Gellissen, G., Bradfield, J. Y., White, B. N., and Wyatt, G. R. (1983). Mitochondrial DNA sequences in the nuclear genome of a locust. *Nature* (*London*) **301,** 631–634.

Gellissen, G., and Michaelis, G. (1987). Gene transfer. Mitochondria to nucleus. *Ann. N.Y. Acad. Sci.* **503,** 391–401.

Gray, M. W. (1989). The evolutionary origins of organelles. *Trends Genet.* **5,** 294–299.

Gray, M. W. (1992). The endosymbiont hypothesis revisited. *Int. Rev. Cytol.* **141,** 233–357.

Gray, M. W. (1993). Origin and evolution of organelle genomes. *Curr. Opin. Genet. Dev.* **3,** 884–890.

Gray, M. W., Sankoff, D., and Cedergren, R. J. (1984). On the evolutionary descent of organisms and organelles: A global phylogeny based on a highly conserved structural core in small subunit ribosomal RNA. *Nucleic Acids Res.* **12,** 5837–5852.

Greenleaf, A. L., Kelly, J. L., and Lehman, I. R. (1986). Yeast *RPO41* gene product is required for transcription and maintenance of the mitochondrial genome. *Proc. Natl. Acad. Sci. U.S.A.* **83,** 3391–3394.

Grohmann, L., Brennicke, A., and Schuster, W. (1992). The mitochondrial gene encoding ribosomal protein S12 has been translocated to the nuclear genome in *Oenothera. Nucleic Acids Res.* **20,** 5641–5646.

Hadler, H. I. (1989). Comment: Mitochondrial genes and cancer. *FEBS Lett.* **256,** 230–232.

Hanekamp, T., and Thorsness, P. E. (1996). Inactivation of *YME2,* an integral inner mitochondrial membrane protein, causes increased escape of DNA from mitochondria to the nucleus in *Saccharomyces cerevisiae.* (Submitted.)

Hayashi, J.-I., Takemitsu, M., Goto, Y.-i., and Nonaka, I. (1994). Human mitochondria and mitochondrial genome function as a single dynamic cellular unit. *J. Cell Biol.* **125,** 43–50.

Hinnen, A., Hicks, J. B., and Fink, G. R. (1978). Transformation of yeast. *Proc. Natl. Acad. Sci. U.S.A.* **75,** 1929–1933.

Hu, G., and Thilly, W. G. (1994). Evolutionary trail of the mitochondrial genome as based on human 16S rDNA pseudogenes. *Gene* **147,** 197–204.

Hyman, B. C., Cramer, J. H., and Rownd, R. H. (1982). Properties of a *Saccharomyces cerevisiae* mtDNA segment conferring high-frequency yeast transformation. *Proc. Natl. Acad. Sci. U.S.A.* **79,** 1578–1582.

Jacobs, H. T., Posakony, J. W., Grula, J. W., Roberts, J. W., Xin, J. H., Britten, R. J., and Davidson, E. H. (1983). Mitochondrial DNA sequences in the nuclear genome of *Strongylocentrotus purpuratus. J. Mol. Biol.* **165,** 609–632.

Jensen, H., Engedal, H., and Saetersdal, S. (1976). Ultrastructure of mitochondria-containing nuclei in human myocardial cells. *Virchows Arch. B Cell. Pathol.* **21,** 1–12.

Johnston, S. A., and Tang, D. C. (1993). The use of microparticle injection to introduce genes into animal cells *in vitro* and *in vivo. Genet. Eng.* (*N.Y.*) **15,** 225–236.

Joyce, P. B., and Gray, M. W. (1989). Chloroplast-like transfer RNA genes expressed in wheat mitochondria. *Nucleic Acids Res.* **17,** 5461–5476.

Jubier, M. F., Lucas, H., Delcher, E., Hartmann, C., Quetier, F., and Lejeune, B. (1990). An internal part of the chloroplast *atpA* gene sequence is present in the mitochondrial genome of *Triticum aestivum:* Molecular organisation and evolutionary aspects. *Curr. Genet.* **17,** 523–528.

Kamimura, N., Ishii, S., Ma, L. D., and Shay, J. W. (1989). Three separate mitochondrial DNA sequences are contiguous in human genomic DNA. *J. Mol. Biol.* **210,** 703–707.

Kennell, J. C., Moran, J. V., Perlman, P. S., Butow, R. A., and Lambowitz, A. M. (1993). Reverse transcriptase activity associated with maturase-encoding group II introns in yeast mitochondria. *Cell* **73,** 133–146.

Knecht, E., Martinez-Ramon, A., and Grisolia, S. (1988). Autophagy of mitochondria in rat liver assessed by immunogold procedures. *J. Histochem. Cytochem.* **27,** 1433–1140.

Kobayashi, S., Amikura, R., and Okada, M. (1993). Presence of mitochondrial large ribosomal RNA outside mitochondria in germ plasm of *Drosophila melanogaster. Science* **260,** 1521–1524.

Kobayashi, S., and Okada, M. (1989). Restoration of pole-cell-forming ability to u.v.-irradiated *Drosophila* embryos by injection of mitochondrial lrRNA. *Development* **107,** 733–742.

Kristensen, T., and Prydz, H. (1986). The presence of intact mitochondrial DNA in HeLa cell nuclei. *Nucleic Acids Res.* **14,** 2597–2609.

Kuroiwa, T., Ohta, T., Kuroiwa, H., and Shigeyuki, K. (1994). Molecular and cellular mechanisms of mitochondrial nuclear division and mitochondriokinesis. *Microsc. Res. Tech.* **27,** 220–232.

Latterich, M., and Schekman, R. (1994). The karyogamy gene *KAR2* and novel proteins are required for ER-membrane fusion. *Cell* **78,** 87–98.

Li, K., Smagula, C. S., Parsons, W. J., Richardson, J. A., Gonzalez, M., Hagler, H. K., and Williams, R. S. (1994). Subcellular partitioning of MRP RNA assessed by ultrastructural and biochemical analysis. *J. Cell. Biol.* **124,** 871–882.

Liang, S., Alksne, L., Warner, J. R., and Lacroute, F. (1992). rna12+, a gene of *Saccharomyces cerevisiae* involved in pre-rRNA maturation. Characterization of a temperature-sensitive mutant, cloning and sequencing of the gene. *Mol. Gen. Genet.* **232,** 304–312.

Lockshon, D., Zweifel, S. G., Freeman-Cook, L. L., Lorimer, H.E., Brewer, B. J., and Fangman, W. L. (1995). A role for recombination junctions in the segregation of mitochondrial DNA in yeast. *Cell* **81,** 947–955.

Lonergan, K. M., and Gray, M. W. (1994). The ribosomal RNA gene region in *Acanthamoeba castellanii* mitochondrial DNA. A case of evolutionary transfer of introns between mitochondria and plastids? *J. Mol. Biol.* **239,** 476–499.

Louis, E. J., and Haber, J. E. (1991). Evolutionarily recent transfer of a group I mitochondrial intron to telomere regions in *Saccharomyces cerevisiae. Curr. Genet.* **20,** 411–415.

Marahrens, Y., and Sillman, B. (1992). A yeast chromosomal origin of DNA replication defined by multiple functional elements. *Science* **255,** 817–823.

Marechal-Drouard, L., Neuburger, M., Guillemaut, P., Douce, R., Weil, J. H., and Dietrich, A. (1990). A nuclear-encoded potato (*Solanum tuberosum*) mitochondrial tRNA(Leu) and its cytosolic counterpart have identical nucleotide sequences. *FEBS Lett.* **262,** 170–172.

Marechal-Drouard, L., Weil, J. H., and Guillemaut, P. (1988). Import of several tRNAs from the cytoplasm into the mitochondria in bean Phaseolus vulgaris. *Nucleic Acids Res.* **16,** 4777–4788.

Margulis, L. (1981). "Symbiosis in Cell Evolution: Life and Its Environment on the Early Earth," W. H. Freeman, San Francisco.

McConnell, S. J., Stewart, L. C., Talin, A., and Yaffe, M. P.(1990). Temperature-sensitive yeast mutants defective in mitochondrial inheritance. *J. Cell. Biol.* **111,** 967–976.

Mirzayans, R., Aubin, R. A., and Paterson, M. C. (1992). Differential expression and stability of foreign genes introduced into human fibroblasts by nuclear versus cytoplasmic microinjection. *Mutat. Res.* **281,** 115–122.

Montandon, P. E., and Stutz, E. (1983). Nucleotide sequence of a *Euglena gracilis* chloroplast genome region coding for the elongation factor Tu; evidence for a spliced mRNA. *Nucleic Acids Res.* **11,** 5877–5892.

Moon, E., Kao, T. H., and Wu, R. (1988). Rice mitochondrial genome contains a rearranged chloroplast gene cluster. *Mol. Gen. Genet.* **213,** 247–253.

Mota, M. (1963). Electron microscope study of the relationship between the nucleus and mitochondria in *Chlorophytum capense* (L.) Kuntze. *Cytologia* **28,** 409–416.

Nugent, J. M., and Palmer, J. D. (1991). RNA-mediated transfer of the gene coxII from the mitochondrion to the nucleus during flowering plant evolution. *Cell* **66,** 473–481.

Obar, R., and Green, J. (1985). Molecular archaeology of the mitochondrial genome. *J. Mol. Evol.* **22,** 243–251.

Ohyama, K., Fukuzawa, H., Kohchi, T., Shirai, H., Sano, T., Sano, S., Umesono, K., Shiki, Y., Takeuchi, M., Chang, Z., Aota, S.-i., Inokuchi, H., and Ozeki, H. (1986). Chloroplast gene organization deduced from complete sequence of liverwort *Marchantia polymorpha* chloroplast DNA. *Nature* (*London*) **322,** 572–574.

Palmer, J. D. (1990). Contrasting modes and tempos of genome evolution in land plant organelles. *Trends Genet.* **6,** 115–120.

Petes, T. D., Malone, R. E., and Symington, L. S. (1991). Recombination in yeast. *In* "The Molecular and Cellular Biology of the Yeast *Saccharomyces:* Genome Dynamics, Protein Synthesis, and Energetics" (J. R. Broach, J. R. Pringle, and E. W. Jones, eds.), pp. 407–521. Cold Spring Harbor, New York.

Pichersky, E., Logsdon, J. M., Jr., McGrath, J. M., and Stasys, R. A. (1991). Fragments of plastid DNA in the nuclear genome of tomato: Prevalence, chromosomal location, and possible mechanism of integration. *Mol. Gen. Genet.* **225,** 453–458.

Pon, L., and Schatz, G. (1991). Biogenesis of yeast mitochondria. *In* "The Molecular and Cellular Biology of the Yeast *Saccharomyces:* Genome Dynamics, Protein Synthesis and Energetics" (J. R. Broach, J. R. Pringle, and E. W. Jones, eds.), pp. 333–406. Cold Spring Harbor Laboratory Press, Cold Spring Harbor, New York.

Pritchard, A. E., Venuti, S. E., Ghalambor, M. A., Sable, C. L., and Cummings, D. J. (1989). An unusual region of *Paramecium* mitochondrial DNA containing chloroplast-like genes. *Gene* **78,** 121–134.

Richter, C. (1988). Do mitochondrial DNA fragments promote cancer and aging? *FEBS Lett.* **241,** 1–5.

Rothman, J. E., and Warren, G. (1994). Implications of the SNARE hypothesis for intracellular membrane topology and dynamics. *Curr. Biol.* **4,** 220–233.

Schiestl, R. H., Dominska, M., and Petes, T. D. (1993). Transformation of *Saccharomyces cerevisiae* with nonhomologous DNA: Illegitimate integration of transforming DNA into yeast chromosomes and *in vivo* ligation of transforming DNA to mitochondrial DNA sequences. *Mol. Cell. Biol.* **13,** 2697–2705.

Schiestl, R. H., and Petes, T. D. (1991). Integration of DNA fragments by illegitimate recombination in *Saccharomyces cerevisiae. Proc. Natl. Acad. Sci. U.S.A.* **88,** 7585–7589.

Schmitt, M. E., and Clayton, D. A. (1992). Yeast site-specific ribonucleoprotein endoribonuclease MRP contains an RNA component homologous to mammalian RNase MRP RNA and essential for cell viability. *Genes Dev.* **6,** 1975–1985.

Schneider, A., McNally, K. P., and Agabian, N. (1994). Nuclear-encoded mitochondrial tRNAs of *Trypanosoma brucei* have a modified cytidine in the anticodon loop. *Nucleic Acids Res.* **22,** 3699–3705.

Schuster, W., and Brennicke, A. (1987). Plastid, nuclear and reverse transcriptase sequences in the mitochondrial genome of *Oenother:* Is genetic information transferred between organelles via RNA? *EMBO J.* **6,** 2857–2863.

Shay, J. W., and Werbin, H. (1992). New evidence for the insertion of mitochondrial DNA into the human genome: Significance for cancer and aging. *Mutat. Res.* **275,** 227–235.

Shimada, H., and Sugirua, M. (1991). Fine structural features of the chloroplast genome: Comparison of the sequenced chloroplast genomes. *Nucleic Acids Res.* **19,** 983–995.

Small, I., Marechal-Drouard, L., Masson, J., Pelletier, G., Cosset, A., Weil, J. H., and Dietrich, A. (1992). *In vivo* import of a normal or mutagenized heterologous transfer RNA into the mitochondria of transgenic plants: Towards novel ways of influencing mitochondrial gene expression? *EMBO J.* **11,** 1291–1296.

Smith, F. D., Harpending, P. R., and Sanford, J. C. (1992). Biolistic transformation of prokaryotes: Factors that affect biolistic transformation of very small cells. *J. Gen. Microbiol.* **138,** 239–248.

Smooker, P. M., Kruft, V., and Subramanian, A. R. (1990). A ribosomal protein is encoded in the chloroplast DNA in a lower plant but in the nucleus in angiosperms. Isolation of the spinach L21 protein and cDNA clone with transit and an unusual repeat sequence. *J. Biol. Chem.* **265,** 16,699–16,703.

Stern, D. B., and Lonsdale, D. M. (1982). Mitochondrial and chloroplast genomes of maize have a 12-kilobase DNA sequence incommon. *Nature* (*London*) **299,** 698–702.

Stern, D. B., and Palmer, J. D. (1986). Tripartite mitochondrial genome of spinach: Physical structure, mitochondrial gene mapping, and locations of transposed chloroplast DNA sequences. *Nucleic Acids Res.* **14,** 5651–5666.

Strausberg, R. L., and Perlman, P. S. (1978). The effect of zygote bud positions on the transmission of mitochondrial genes in *Saccharomyces cerevisiae. Mol. Gen. Genet.* **163,** 131–144.

Strauss, M., Kiessling, U., Zavision, B. A., Povitza, O. N., Tikhonenko, T. I., and Geissler, E. (1983). The efficiency of genetic transformation of mammalian cells by transfection and microinjection depends on the transferred gene. *Biomed. Biochim. Acta* **42,** K27–34.

Sun, C. W., and Callis, J. (1993). Recent stable insertion of mitochondrial DNA into an *Arabidopsis* polyubiquitin gene by nonhomologous recombination. *Plant Cell* **5,** 97–107.

Takeshige, K., Baba, M., Tsuboi, S., Noda, T., and Ohsumi, Y. (1992). Autophagy in yeast demonstrated with proteinase-deficient mutants and conditions for its induction. *J. Cell Biol.* **119,** 301–311.

Takeuchi, Y., Dotson, M., and Keen, N. T. (1992). Plant transformation: A simple particle bombardment device based on flowing helium. *Plant Mol. Biol.* **18,** 835–839.

Thorsness, P. E. (1992). Structural dynamics of the mitochondrial compartment. *Mutat. Res.* **275,** 237–241.

Thorsness, P. E., and Fox, T. D. (1990). Escape of DNA from mitochondria to the nucleus in *Saccharomyces cerevisiae. Nature* (*London*) **346,** 376–379.

Thorsness, P. E., and Fox, T. D. (1993). Nuclear mutations in *Saccharomyces cerevisiae* that affect the escape of DNA from mitochondria to the nucleus. *Genetics* **134,** 21–28.

Thorsness, P. E., White, K. H., and Fox, T. D. (1993). Inactivation of *YME1,* a gene coding a member of the *SEC18, PAS1, CDC48* family of putative ATPases, causes increased escape of DNA from mitochondria in *Saccharomyces cerevisiae. Mol. Cell. Biol.* **13,** 5418–5426.

Vestweber, D., and Schatz, G. (1989). DNA–protein conjugates can enter mitochondria via the protein import pathway. *Nature* (*London*) **338,** 170–172.

von Heijne, G. (1986). Why mitochondria need a genome. *FEBS Lett.* **198,** 1–4.

Wagner, E., Zatloukal, K., Cotten, M., Kirlappos, H., Mechtler, K., Curiel, D. T., and Birnstiel, M. L. (1992). Coupling of adenovirus to transferrin-polylysine/DNA complexes greatly enhances receptor-mediated gene delivery and expression of transfected genes. *Proc. Natl. Acad. Sci. U.S.A.* **89,** 6099–6103.

Watson, J. C., and Surzycki, S. J. (1982). Extensive sequence homology in the DNA coding for elongation factor Tu from *Escherichia coli* and the *Chlamydomonas reinhardtii* chloroplast. *Proc. Natl. Acad. Sci. U.S.A.* **79,** 2264–2267.

Weber, E. R., Hanekamp, T., and Thorsness, P. E. (1996). Biochemical and functional analysis of Yme1p; an ATP and zinc-dependent mitochondrial protease from *Saccharomyces cerevisiae. Mol. Biol. Cell* in press.

Weber, E. R., Rooks, R. S., Shafer, K. S., Chase, J. W., and Thorsness, P. E. (1995). Mutations in the mitochondrial ATP synthase gamma subunit suppress a slow-growth phenotype of *yme1* yeast lacking mitochondrial DNA. *Genetics* **140,** 435–442.

Weeden, N. F. (1981). Genetic and biochemical implications of the endosymbiotic origin of the chloroplast. *J. Mol. Evol.* **17,** 133–139.

Zenke, M., Steinlein, P., Wagner, E., Cotten, M., Beug, H., and Birnstiel, M. L. (1990). Receptor-mediated endocytosis of transferrin-polycation conjugates: An efficient way to introduce DNA into hematopoietic cells. *Proc. Natl. Acad. Sci. U.S.A.* **87,** 3655–3659.

# Cytoplasmic Mechanisms of Axonal and Dendritic Growth in Neurons

Steven R. Heidemann
Department of Physiology, Michigan State University, East Lansing, Michigan 48824-1101

The structural mechanisms responsible for the gradual elaboration of the cytoplasmic elongation of neurons are reviewed. In addition to discussing recent work, important older work is included to inform newcomers to the field how the current perspective arose. The highly specialized axon and the less exaggerated dendrite both result from the advance of the motile growth cone. In the area of physiology, studies in the last decade have directly confirmed the classic model of the growth cone pulling forward and the axon elongating from this tension. Particularly in the case of the axon, cytoplasmic elongation is closely linked to the formation of an axial microtubule bundle from behind the advancing growth cone. Substantial progress has been made in understanding the expression of microtubule-associated proteins during neuronal differentiation to stiffen and stabilize axonal microtubules, providing specialized structural support. Studies of membrane organelle transport along the axonal microtubules produced an explosion of knowledge about ATPase molecules serving as motors driving material along microtubule rails. However, most aspects of the cytoplasmic mechanisms responsible for neurogenesis remain poorly understood. There is little agreement on mechanisms for the addition of new plasma membrane or the addition of new cytoskeletal filaments in the growing axon. Also poorly understood are the mechanisms that couple the promiscuous motility of the growth cone to the addition of cytoplasmic elements.

**KEY WORDS:** Neurogenesis, Neuron, Axon, Dendrite, Cytoskeleton, Membrane addition, Cell shape, Cell motility.

## I. Introduction

In a professional environment awash in review journals, full of currents, trends, and opinions, one might well ask what role is served by the lengthier style of review in a venerable serial such as *International Review of Cytology*. One answer is to serve research workers who are new to the field and need a synopsis of an established bioresearch niche, in particular postdoctorates entering a new laboratory, graduate students, and senior undergraduates beginning a research project in the area, who are not always well served by short reviews highlighting the most recent advances. To address this audience, an effort is made to integrate the phenomenology of neural outgrowth into a coherent story. In addition to bringing a current perspective to the topic, this article also strives to achieve a balance by citing much important older work. In doing so, original experimental reports are cited to permit the reader better access to data, not just interpretations. Reviews are referred to for two purposes: to provide a source for more information concerning topics that are not central to this subfield of neural and cellular biology, and to allow the reader to obtain more detailed information about topics that are central, but whose coverage in these pages is particularly simplified owing to the complexity and volume of the material [e.g., the role of microtubule-associated proteins (MAPs) in neurite outgrowth]. Regrettably, stylistic conventions do not guarantee coherence, and this effort is a highly uncertain undertaking for scientific reasons. Few aspects of the mechanisms underlying neuronal outgrowth are satisfyingly understood at present, despite the fact that the basic cytological phenomena associated with neuronal outgrowth are well described and seem to be widely shared among neuronal types. Some topics (e.g., transport of the cytoskeleton within the axon) have been highly controversial for many years, with no end in sight. Thus, the putative coherence of the discussion will often depend on explicit recognition of questions that remain unanswered.

The vast majority of studies on neurogenesis at the cellular level have been pursued from one of two different points of view. From the extrinsic frame of reference, process outgrowth is seen as a problem of navigation involving cues, such as substrate preferences, and trophic factors that allow the outgrowths to steer toward the appropriate targets. Investigations of neurogenesis undertaken by the extrinsic school are primarily interested in the axon developing along the appropriate path. Experimental interventions are aimed at determining what aspects of the environment reflect navigational signals, and what is required to stay on the correct path, or to deviated from it. The intrinsic point of view wishes to understand the structural cytoplasmic mechanisms underlying the exaggerated growth of the cell processes. Here the focus is on what is required for the axon to develop at all, in any direction. Experiments are usually aimed at coaxing out new elongations, accelerating their growth, and determining what subcellular interventions cause elongation to stop. The artificial but useful dichotomy between extrinsic and intrinsic references to neurogenesis is adopted here, and the discussion focuses primarily on the structural mechanisms that enable neurons to produce their extraordinary shapes. It should be noted at the outset that although some progress has been made on both the intrinsic and extrinsic problems, the necessary connections between the two (i.e., how external signals actually alter the structural mechanisms to steer growth) are very poorly understood indeed.

The neuron is, arguably, the animal cell whose function is most intimately dependent on its unusual shape. Neurons extend from their cell body two functionally and morphologically distinct types of processes: axons and dendrites. The better studied of these is the axon. In vertebrates, the large majority of axons are less than 5 $\mu$m in diameter but can be many meters long in large animals such as whales. These highly exaggerated cell processes are the result of a close integration of mechanisms for cellular motility with growth (in the sense of mass addition) mechanisms. Indeed, a picturesque description of axonal elongation is that of a "leucocyte on a leash" (Pfenninger, 1986). As this phrase suggests, a highly motile front compartment, the growth cone, locomotes forward in its environment and the axon progressively elaborates from behind the advancing growth cone.

Given the importance of the cytoskeleton in motility and cell shape, it is not surprising that molecular and subcellular studies of neural outgrowth have tended to focus on the structure and dynamics of the neuronal cytoskeleton. Indeed, axonal growth is frequently described in the literature by explicit models of cytoskeletal assembly and reorganization leading to process elongation (Goldberg and Burmeister, 1986; Buxbaum and Heidemann, 1988; Mitchison and Kirschner, 1988; Smith, 1988). Because of the difficulty of continuous, *in situ* observation to assess cell behavior accurately and the difficulty of experimental manipulations of subcellular structures within the animal, much of the work on the mechanism of axonal elongation has been done in culture. In principle, the use of tissue culture could be the source of serious artifacts; tissue culture plastic is a very different environment from that found in animal embryos. However, studies over many decades suggest that events occurring in cultured neurons reasonably reflect a basic intrinsic mechanism of axonal elongation, which is also observed *in situ* (Speidel, 1933; Tennyson, 1965; Harris *et al.,* 1987). Thus far, the major limitation of *in vitro* work appears to be that normal extrinsic inputs to growth are not observed (e.g., changes in growth cone morphology and advance rate accompanying changes in the local environment) (Tosney and Landmesser, 1985; Bovolenta and Mason, 1987; Harris *et al.,* 1987). Of course, the limited number of studies of intrinsic mechanisms *in situ* is of particular relevance and importance. The axons and dendrites of cultured neurons are called "neurites," in part because it was originally difficult to determine whether a cultured outgrowth was axonal or dendritic. That is no longer problematic and the word "neurite" here, and in most literature, refers to axon-like growths from cultured neurons, and we specifically identify dendrites of cultured neurons as such.

## II. Overview

Most vertebrate neurons have one relatively long axon and, typically, a number of considerably shorter dendrites. The available evidence suggests that the dendrites grow out by a similar but less exaggerated mechanism than that of axons. For example, both types of processes arise from the same "minor processes" seen at very early times of development, and dendrites can develop into axons if the initial axon is lost (Dotti *et al.,* 1988). The cytoplasm of dendrites is similar to that of the cell body in noncytoskeletal content; it contains ribosomes, Golgi elements, smooth and rough endoplasmic reticulum, and large numbers of vesicular elements (Peters *et al.,* 1976). In some neurons, it is very difficult to determine where the cell body stops and the dendrites begin; hence the frequent inclusion of dendrites and cell body into a single "somatodendritic compartment." In contrast, the axonal cytoplasm is highly specialized in structure. In addition to their exceptional density of cytoskeletal elements, axons generally lack the structures required for macromolecular synthesis and assembly, including ribosomes, rough endoplasmic reticulum, and Golgi compartments (Peters *et al.,* 1976).

As a result of this unusual lack of organelles involved in synthesis, the axon is supplied with proteins and membranous elements by transport processes, which are discussed in Section V. The mechanisms and roles of these transport processes in axonal elongation are an area of considerable progress and also of considerable controversy. Microtubules serve as "tracks" for the saltatory movement of membranous organelles traveling at 1–4 μm/sec in both directions within the axon, called *fast axonal transport.* This process provides the growing axon with membrane-bound organelles and new membrane for elongation. Cellular and biochemical analysis of axonal microtubules engaged in fast axonal transport led to an explosion of studies on the motor proteins mediating this traffic and much other microtubule-associated motility, as briefly summarized in Section V,A (McIntosh and Porter, 1989; Vale, 1990; Goldstein, 1993; Skoufias and Scholey, 1993; Vallee, 1993; Walker and Sheetz, 1993). A second type of axonal transport, *slow axonal transport,* occurs at rates about 100 to 1000 times slower than fast axonal transport (0.01–0.04 μm/sec or 1–4 mm/day) and moves cytoskeletal material and much soluble protein from the cell body toward the growth cone or synapse. Slow axonal transport is less well understood than fast transport and for more than a decade the nature of slow axonal transport has been the focus of intense controversy, which is the major topic of Section V,B.

Both axons and dendrites have an extensive cytoskeleton throughout their length. Directly beneath the axonal plasma membrane (the axolemma) is an actin-rich cortical network (Kuczmarski and Rosenbaum, 1979; Hirokawa, 1982; Schnapp and Reese, 1982), similar to that surrounding all animal cells (Bray *et al.,* 1986). This cortical actin network probably plays a role in yet another, but less well-studied, transport process observed along axons, the movement of "packets" in both directions along the outer margin of the axon (Koenig *et al.,* 1985; Hollenbeck and Bray, 1987). Short actin filaments throughout the axoplasm have also been described in squid giant axon (Fath and Lasek, 1988).

Microtubules are axially oriented as bundles throughout the central region of the axons and dendrites, and microtubules are especially dense in small-caliber axons (Hirokawa, 1982; Schnapp and Reese, 1982). The microtubules of axons, in particular, are highly specialized. The microtubules of the axon, and probably the dendrite, are less dynamic in their assembly and disassembly characteristics than other cytoplasmic microtubule arrays. The basic mechanism of the assembly of pure tubulin into microtubules is that of dynamic instability, in which the free ends of each microtubule cycle between rapid elongation and catastrophic disassembly phases on a stochastic basis (Mitchison and Kirschner, 1984; Horio and Hotani, 1986). The fluctuations of this "boom-or-bust" mechanism are moderated to varying extents in different cells and different microtubule structures. Axonal microtubules are specialized by being among the most stable of cytoplasmic microtubule populations. For example, the half-time for turnover of tubulin subunits within an axonal microtubule bundle is roughly 1 hr (Lim *et al.,* 1989; Okabe and Hirokawa, 1990). In contrast, a population of microtubules in epithelial cells has a turnover half-time of about 10 min (Pepperkod *et al.,* 1990), and the half-times of mitotic microtubules and of less stable microtubules in many interphase cells are just 2–3min (Saxton *et al.,*, 1984; Sammak *et al.,* 1987). In axons, the intrinsic molecular polarity of microtubules is uniform: the more assembly-active "+" ends are oriented away from the cell body and toward the growth cone or synapse (Heidemann *et al.,* 1981; Baas *et al.,* 1988). In contrast, microtubules in dendrites have a nonuniform polarity orientation, i.e., these microtubules point both ways (Bass *et al.,* 1988; Burton, 1988). In contrast to most microtubule arrays, the axonal and dendritic bundles of microtubules do not arise from a well-defined microtubule-organizing center (e.g., a centriole) at the point where the process arises from the cell body (Lyser, 1968; Sharp *et al.,* 1982; see discussion in Joshi and Baas, 1993). Thus, axonal microtubules appear to have a specialized mechanism for organization and spatial patterning.

As we shall see throughout this article, the axonal bundle of microtubules plays a central role in current thinking about many aspects of axonal and dendritic elongation. One area of major advance in this general area has been the understanding of the role of microtubule-associated proteins in axonal microtubule stability, bundle formation, and structural function, as discussed in Section III,C. Another area of biochemical advance has been the identification of microtubule-based motor proteins, as previously noted. Despite some progress, how the microtubule bundle actually forms and how it is coupled to growth cone motility remain unclear. That is, the physiology underlying the organization, maintenance, and spatial patterning of the microtubule bundle is poorly understood. For example, there are many different views on the extent to which this very orderly array of microtubules results from the reorganization and transport of extant microtubules and/or of their *de novo* assembly in the axon shaft or growth cone. This open question is closely related to the slow axonal transport controversy, discussed in Section V,B.

The axonal cytoskeleton also contains axially oriented neurofilaments, a type of intermediate filament specialized to neurons and composed of three polypeptides (Marotta, 1983; Shaw, 1991). Neurofilaments occur in large numbers in the axons of large-diameter, myelinated axons; in smaller numbers in small-caliber axons; and in smaller numbers still in dendrites (Peters *et al.,* 1976; Hirokawa, 1993). The sole well-established function of neurofilaments is to regulate the diameter of large-caliber axons *in situ* (Hoffman *et al.,* 1984, 1987; Sakaguchi *et al.,* 1993). The phosphorylation of neurofilament polypeptides alters the packing space of neurofilaments in mouse retinal ganglion cells and thus axon caliber (Nixon *et al.,* 1994). As axonal caliber is the major determinant of conduction velocity in myelinated axons, neurofilaments play an important structural role *vis-à-vis* the electrical function of neurons. However, it has proved difficult to assign a structural role to neurofilaments in the early phases of neuronal outgrowth. A favored speculation is that they are involved in stabilizing the form of mature axons (Glicksman and Willard, 1985; Dahl and Bignami, 1986; Donahue *et al.,* 1988), although this has proved difficult to confirm (Donahue *et al.,* 1988). Neurofilaments appear to be dispensible to the general function of neurons; crayfish neurons lack them (Phillips *et al.,* 1983) as do granule cells of the cerebellum (Palay and Chan-Palay, 1974). More recently, mutations that essentially eliminate neurofilaments from axons of quail and mouse neurons were shown to have an effect on axon caliber, but caused only subtle neurological deficits (Ohara *et al.,* 1993; Eyer and Peterson, 1994). However, overexpression of neurofilament proteins in transgenic mice leads to phenotypic alterations that closely resemble motor neuron disease, in particular amyotrophic lateral sclerosis ("Lou Gehrig's disease"). Overexpression of neurofilament polypeptides leads to massive accumulation of neurofilaments in motor neuron cell bodies, axonal degeneration, and subsequent muscle fiber degeneration (Cote *et al.,* 1993; Xu *et al.,* 1993), all of which are characteristic of motor neuropathies.

In addition to axons, dendrites, and the cell body, the growth cone is the other specialized compartment of neurons that shall concern us. As mentioned earlier, the growth cone is a highly motile compartment that locomotes by "ameboid movement." The growth cone pulls the neurite of cultured neurons forward (Lamoureux *et al.,* 1989), causing the neurite to elongate (Bray, 1984; Zheng *et al.,* 1991). The relationship between tension and axonal development seems unusually intimate in both time scale and simplicity of relationship (Heidemann and Buxbaum, 1994). For example, the rate of elongation of cultured chick neurons is a simple linear function of the applied force under a variety of culture conditions and at both physiological and greater-than-physiological elongation rates (Zheng *et al.,* 1991; Lamoureux *et al.,* 1992). Two decades of work with anti-actin drugs indicate that the motility of the growth cone is the result of the activity of actin (Yamada *et al.,* 1970; Forscher and Smith, 1988) and, presumably, myosin (Landis, 1983; Bridgman and Daily, 1989). The growth cone continuously extends and retracts cylindrical "microspikes" or "filopodia" $<0.5\ \mu m$ in diameter that contain a tight bundle of 15–20 axially oriented actin filaments (Tosney and Wessells, 1983; Bridgman and Daily, 1989; Lewis and Bridgman, 1992). The "contraction" (in the biophysical sense) of these filopodia is at least one mechanism that underlies the pulling and advance of the growth cone (Heidemann *et al.,* 1990, 1991; O'Connor *et al.,* 1990). In addition, these filopodia appear to palpate the surface in front of the growth cone, which is very likely to be involved with the "sensory" function of filopodia, i.e., seeking navigational signals in their environment (Bentley and Toroian-Raymond, 1986; Bentley and O'Connor, 1992; Chien *et al.,* 1993; Davenport *et al.,* 1993). Lamellipodia are another motile structure of growth cones, which as their name implies are "veils," very thin, sheetlike regions of cytoplasm. These structures are particularly characteristic of rapidly moving growth cones (Argiro *et al.,* 1984; Kleitman and Johnson, 1989). It is not clear how filopodial and lamellipodial movements are related and whether movements by the two structures have the same, different, or partly overlapping biochemical mechanisms. One superficial relationship is that both filopodia and lamellipodia engage in a great deal of seemingly futile motion, i.e., similar motions are observed whether the growth cone is advancing forward or not (Argiro *et al.,* 1984; Goldberg and Burmeister, 1986; Heidemann *et al.,* 1990; Tanaka *et al.,* 1995). In addition to the movements of filopodia and lamellipodia, the growth cone supports several additional types of motility, including (1) a continuous retrograde (toward the cell body) flow of cortical actin across the top of the growth cone in culture, which is related to a continuous process of actin assembly and disassembly within the growth cone (Forscher and Smith, 1988; Smith, 1988; Okabe and Hirokawa, 1991), (2) an anterograde movement of membrane proteins toward the forward edge of the growth cone (Sheetz *et al.*, 1990), and (3) the formation of filopodial-like structures induced by the assembly of actin (Forscher *et al.*, 1992). Virtually all aspects of the underlying mechanisms regulating growth cone function and behavior remain open questions but three major questions are as follows:

1. How is the advance of the growth cone tightly coupled to the addition of membrane, cytoskeleton, and other cytoplasmic processes underlying the growth of the axonal shaft?
2. Are there other mechanisms in addition to filopodial contraction for the forward advance of growth cones?
3. What are the molecular mechanisms underlying the complexity of growth cone motions? Are the movements related in mechanism and, if so, how?

## III. Neurite Initiation

Neurons arise from epithelial-like precursor cells in the embryo; in vertebrates most neurons arise from the neuroepithelial cells initially lining the lumen (the ventricle) of the neural tube (Purves and Lichtman, 1985; Jacobson, 1991). In a few cases, neuroepithelial cells have been shown to exhibit characteristic epithelial marker proteins such as keratin, which are lost during their commitment to differentiate into a neuron (Calof and Chikaraishi, 1989). Only following a final mitotic division is the precursor cell regarded as being a differentiated neuron, although it has not yet produced axons and dendrites.

Both the precursor and postmitotic cell are characterized by substantial motility. In culture, where the neurons can be observed continuously, it is generally agreed that motile, ameboid activity is initially characteristic of most, if not all, of the cell margin. The growth of neurites from the cell body appears to require locomotory activity to become localized to particular sites, which then elongate into neurites (Collins, 1978; Wessels, 1982; Dotti *et al.*, 1988). In fixed specimens of early stages of axonal outgrowth both in culture and *in situ,* cytoskeletal elements are seen to concentrate at these sites of incipient axon/dendrites (Lyser, 1964, 1968; Tennyson, 1965; Stevens *et al.*, 1988). The number density of microtubules and/or neurofilaments at these "nubs" is substantially greater than at other places in the cell body.

## A. Earliest Events of Process Outgrowth

Smith (1994a,b) has succeeded in routinely observing the earliest phases of neurite initiation in cultured embryonic chick sympathetic neurons. When grown on culture substrata coated with polyornithine, or polyornithine and laminin, it is possible to predict the occurrence of neurite initiation, which had not been possible in other culture systems. As have others, Smith observed general motility all along the cell margin. The formation of neurites in this system invariably occurs when the tips of actin-rich filopodia adhere strongly to another object, e.g., the dish, another cell, or a plastic bead. Following tip attachment, the filopodia apparently exert tension, causing an invasion of microtubule-containing cytoplasm. That is, a filopodium becomes dilated with cytoplasm, thus widening to become a short neurite. A similar conversion of an actin-rich filopodium to a neurite shaft has been observed during elongation of grasshopper neurons (O'Conner *et al.,* 1990), and Smith's observations suggest that neurite initiation occurs by cytoplasmic mechanisms similar to those functioning during growth cone-mediated elongation of the axon, as is described more fully in Section IV,B. Importantly, the microtubules that invade the newly formed neurite are translocated from the cell body, and no new microtubule assembly is required for neurite initiation (Smith, 1994b). (Whether microtubule assembly is required for continued elongation is discussed in Section V,B.) Smith's finding provides support for an attractive model to explain the high degree of axonal microtubule organization in the absence of a microtubule-organizing center at the base of the incipient axon.

Joshi and Baas (1993) have proposed an attractive model for the initiation and organization of axonal microtubules that depends on both microtubule assembly within the axon and transport of microtubules from the cell body to the axon. The authors hypothesize that axonal microtubules are ultimately dependent on initiation of microtubule assembly by the centrosome (the centriole with its surrounding "cloud" that nucleates *de novo* microtubule assembly) within the cell body. Further, no *de novo* initiation of microtubules occurs in the axon. Newly assembled microtubules are transported out of the cell body into the incipient axon as small pieces. This transport of microtubules is postulated to be uniformly polar, with their plus end leading. The transported microtubules are able to fragment and, more importantly, serve as the nuclei for all subsequent microtubule assembly occurring within the axon. The uniform polarity orientation of axonal microtubules, then, arises from the polarity of the microtubule transport and the postulated absence of *de novo* microtubule initiation within the axon. Additional microtubule transport also continues to supply microtubules to the axon.

One noteworthy feature of this model is its combination of microtubule polymer transport and assembly to explain microtubule organization. As is discussed in Section V,B, the controversy surrounding the issue of microtubule transport vs axonal assembly has been phrased primarily in terms of "either/or, but not both." Another attractive feature is that the postulated mechanism can function with a wide variety of differing contributions from assembly and transport, i.e., 99% assembly and 1% transport seems as workable in principle as 99% transport and 1% assembly. Several lines of evidence support the idea that the axon is deficient in the ability to nucleate microtubule assembly *de novo* but capable of supporting extensive elongation of extant polymer. For example, experiments in which the axon and cell body are severed by micromanipulation suggest a deficit in the nucleation capability within axons (Baas and Heidemann, 1986). More recently, it has been observed that γ-tubulin, a tubulin polypeptide specialized for, and probably required for, nucleation of microtubule assembly *de novo* (Joshi, 1994), is found only in the centrosomal regions of neurons and not in the axon (Baas and Joshi, 1992). Further, microinjection of antibodies against γ-tubulin into cultured rat sympathetic neurons inhibited reestablishment of centrosomal microtubules following drug-induced deploymerization, while also substantially inhibiting subsequent axonal growth and microtubule number in those axons that did elongate (Ahmad *et al.*, 1994). Some ultrastructural evidence also supports the idea that microtubules are released after being nucleated at the centrosome (Yu *et al.*, 1993). The stipulation that microtubule transport is uniformly polar was based in part on the finding that the orientation of microtubules is uniform in all the early "minor processes" that later differentiate into axons or dendrites (i.e., dendrites later develop microtubules of opposite polarity to produce a bipolar population) (Baas *et al.*, 1989).

## B. Development of Neuronal Polarity

In the intact animal, virtually all neurons have both axonal and dendritic processes. Understanding this polarity is a major goal of cell biological studies of early neural development (Craig and Banker, 1994). Several types of central nervous system cells in culture produce both axonal and dendritic neurites, including those from the hippocampus, cerebellum, and cerebrocortex (Bartlett and Banker, 1984; Kosik and Finch, 1987; Caceres *et al.*, 1991). The sequence of development to produce axons and dendrites seems similar for these cell types but is best studied in cultured hippocampal neurons, each of which develops a single axon and a number of dendrites. In this cell type, the development of axonal/dendritic polarity occurs by a reproducible sequence of changes over about a week following plating.

After the initial period of promiscuous motile activity around the entire cell margin, a number of short (10 μm), stable, microtubule-containing processes forms within 24 hr. These are the "minor processes" with uniformly polar microtubules alluded to earlier. At about 24–48 hr, one of these short processes begins relatively rapid growth (5–10 μm/hr) to become the sole axon of the cell (Dotti *et al.,* 1988). Nothing distinguishes the particular short neurite that becomes the axon; it is not the first or last neurite to develop and is not visually different from any of the others. Some molecular specializations of the axonal compartment begin shortly after this period of rapid growth. For example, proteins of the growth cone and synaptic vesicles become preferentially segregated into the axon at this time (Goslin and Banker, 1990; Jareb *et al.,* 1993). In addition, ribosomes are preferentially excluded very soon after processes can be identified as being axons (Deitch and Banker, 1993). At this early time of axonal specification, it appears that the fate of this process is relatively easy to change. For example, the initial "axon" occasionally stops elongating, another short neurite begins elongation to become the axon, and the older, incipient axon eventually becomes a dendrite (Dotti *et al.,* 1988).

Dendritic growth does not begin from minor processes of hippocampal neurons until about day 4 in culture. Dendrites grow more slowly than the axonal outgrowth, and several dendrites grow at the same time. It appears that the axonal outgrowth inhibits the other short processes from becoming axonal and channels them toward the dendritic fate: if the axon is cut after axonal outgrowth but before dendritic growth, a different short process begins axonal growth, and the stump of the old axon frequently becomes dendritic (Dotti and Banker, 1987). This again suggests that the fate of the early processes is somewhat plastic. As with axonal proteins, molecular specializations can be seen at about the same time minor processes first grow into dendrites. The dendritic array of bipolar microtubules develops after about 4 days (Baas *et al.,* 1989), as does the segregation of dendrite-specific proteins, such as neurotransmitter-receptor polypeptides (Killisch *et al.,* 1991; Craig *et al.,* 1993).

In general, the mechanisms underlying this developmental course of events are poorly understood. Why do some motile regions develop into a well-defined outgrowth while others simply fade away? What signals are responsible for specifying whether a minor process becomes an axon or a dendrite? What underlies the finding that axons develop before dendrites, and what signal(s) do axons send that apparently inhibit additional axonal development? What mechanisms actually sort the various molecular manifestations of polarity? Although no definitive answers can be given to any of these questions, rapidly accumulating evidence suggests that important aspects of the polarity of neuronal cells are underlain by the same mechanisms responsible for epithelial polarity in general. Work indicates that neurons share the membrane polarity first discovered in columnar epithelial cells and that epithelial cells share the polar microtubule-based transport system first characterized in neurons.

Epithelial cells have apical and basolateral polar domains, which differ with respect to membrane proteins and secretion, that are based on sorting mechanisms associated with the endomembrane system, including rough endoplasmic reticulum and the Golgi apparatus (Rodriquez-Boulan and Nelson, 1989; Rothman, 1994). Neurons share this epithelial membrane polarity as shown by a classic method for analyzing epithelial cell polarity. Different viruses are known to bud differentially from the apical and basolateral surfaces of epithelial cells (Rodriguez-Boulan and Sabatini, 1978), and viral envelope proteins can be used as markers for the two domains (Rodriguez-Boulan and Pendergast, 1980). Dotti and Simons (1990) showed that the G protein of vesicular stomatitis virus, which is delivered basolaterally in epithelial cells, is targeted primarily to the somatodendritic domain of cultured hippocampal neurons. Similarly, they found that a viral protein that is apically delivered in epithelial cells is targeted to the axonal domains of hippocampal cells. Additionally, a normal membrane protein that localizes to the apical membrane of several types of columnar epithelial cells, Thy-1, is also sorted exclusively to the axons of hippocampal neurons (Dotti *et al.*, 1991). This epithelial mechanism of membrane sorting is likely to explain the targeting of some, but certainly not all, synaptic vesicle proteins (Jung and Scheller, 1991).

As we noted earlier, axons are provided with membrane-bound organelles and new membrane by a fast axonal transport system based on microtubule tracks. Given the similarity of the axon to apical domains of epithelial cells as indicated by the membrane-sorting experiments, it is of considerable interest that membranous transport to the apical domain in epithelial cells shares fundamental similarities with fast axonal transport. Like fast axonal transport, which is discussed in more detail in Section V, initial studies showed that disruption of epithelial microtubules strongly inhibited membrane-sorting phenomena (Hugon *et al.*, 1987; Rindler *et al.*, 1987). Specifically, disruption of microtubules strongly inhibited transport of proteins to the apical membrane, but only weakly to the basolateral membrane (Eilers *et al.*, 1989; Matter *et al.*, 1990; Gilbert *et al.*, 1991). Using *in vitro* systems that reconstitute parts of the epithelial transport system, the motor proteins initially discovered as underlying fast axonal transport were shown to be involved in apical transport, both anterograde (from endoplasmic reticulum to plasma membrane) and retrograde (Bomsel *et al.*, 1990; Lafont *et al.*, 1994). More recently, Topp *et al.*, (1995) have shown that the bundle of microtubules that extends along the basal/apical axis of epithelial cells (Gilbert *et al.*, 1991) is of uniform polarity orientation, as is the bundle of axonal microtubules. However, the orientation of the epithelial polarity is opposite to that one would expect on the basis of 1:1 mapping of epithelial and axonal membrane transport. That is, the plus ends of microtubules were oriented toward the nucleus in epithelial cells, rather than to the apical anterograde surface, as would be expected on the basis of axonal microtubule polarity. This is not a major difficulty, as the microtubule motor proteins are themselves polar in their function, one type specializing in plus end-directed motility and the other toward the minus end of microtubules (see Section V). Thus, differences in use of motor proteins could reconcile the differences in microtubule polarity in axons and epithelial cells.

This latter discrepancy, however, illustrates that although epithelial and neuronal polarity share a fundamental similarity, this is not likely to be the whole story of neuronal polarity. The difference in synthetic function between axons and dendrites seems likely to play an important role in neuronal polarity (Steward and Banker, 1992). As noted, dendrites have long been known to contain ribosomes, and ultrastructural observations of brain neurons *in situ* showed that polyribosomes are preferentially found beneath the postsynpatic regions on dendritic spines (Steward and Fass, 1983). Consistent with this finding, subsequent autoradiographic localization of labeled RNA indicated a transport process that selectively delivers RNA to dendrites, but not axons (Davis *et al.,* 1990). *In situ* hybridization studies indicate that the transport process is selective for particular mRNAs (Bruckenstein *et al.,* 1990; Kleiman *et al.,* 1990; Tiedge *et al.,* 1991). For example, the mRNAs for tubulin subunits are found only in the cell body, while mRNA encoding a dendrite-specific protein (MAP-2; see the following section) was present in both cell body and dendrites. The mechanisms causing localizations of polyribosomes and particular mRNAs are essentially unknown. However, Bassel *et al.* (1994) have shown that mRNA of cultured cerebrocortical neurons is associated with microtubules, which is intriguing in view of a well-studied mRNA of *Xenopus* oocytes that localizes to the vegetal pole via microtubule-based transport systems (Yisreali *et al.,* 1990). So, microtubule-based transport is yet again a suspect in underlying neuronal polarity. Indeed, ribosomal exclusion has been postulated to arise from the inability of ribosomes to be transported from the cell body into the axon along the uniformly polar microtubule array; i.e., ribosomes are preferentially transported to the minus ends of microtubules (Black and Baas, 1989). In this model, the presence of dendritic microtubules with the minus ends pointing away from the cell body supports the transport of ribosomes into the dendrite. Deitch and Banker (1993), however, find that ribosomes are preferentially excluded from the axon and accumulate in dendrites of hippocampal cells before minus end-distal microtubules are found in the dendrites (Baas *et al.,* 1989).

In view of the importance of the microtubule cytoskeleton in the development of cell processes and in neuronal polarity, it is frustrating that, in general, no predictive change in microtubule organization has been noted prior to the actual appearance of outgrowths or the development of polarity (Lyser, 1964, 1968; Tennyson, 1965; Stevens *et al.,* 1988; Deitch and Banker, 1993). For example, Dietch and Banker (1993) found no ultrastructural feature that predicted whether a minor process would become an axon or remain a minor process during the first 2 days in hippocampal cultures. To date, it seems that an accumulation of cytoskeletal and other cytoplasmic components that predicts the site of axonal outgrowth has been detected only in a pioneer neuron of the grasshopper limb (Lefcort and Bentley, 1989). In this system, a punctate region that stains with antibodies against tubulin develops on one side of the cell, from which the axon later extends. The Golgi apparatus and an accumulation of actin also localize to this pole of the cell. Although cytoskeletal reorganizations seem at this point better regarded as an effect, rather than a cause, of process outgrowth and development, microtubule-associated proteins have been implicated in axonal development, as is discussed in the next section, III,C.

Finally, this article would be incomplete without the brief introduction of another intriguing aspect of process initiation and early outgrowth that remains frustratingly mysterious. The fate of outgrowths, whether to become axons or dendrites and at least some aspects of their early pattern of branching, is determined in part by intrinsic factors. *In situ,* each individual neuron has a unique branching pattern, but the overall geometry of branching is similar within a class of neuron. Of course, one would anticipate that environmental signals within the neuropil would play a major role in this patterning and that such patterning would be lost in the uniform environment of the culture dish. Although this is clearly true to some extent (e.g., see discussion in Craig and Banker, 1994), tantalizing reports of intrinsic patterning in culture appear from time to time. Early work showed that the characteristic branching pattern of a neuron *in situ* is, in some cases, maintained when placed into culture (Banker and Cowan, 1979). In the hippocampal culture system, a neuronal precursor cell that divided in culture prior to differentiation was followed. The pattern of outgrowth of the two sister cells arising from this division were found to be very similar (Dotti *et al.,* 1988). Similarly, Mattson *et al.* (1989) found that approximately one-third of daughter pairs of hippocampal neurons had strikingly similar morphologies. The best studied example of intrinsic specification of the pattern of process is presented by Solomon (1979, 1980) on neuroblastoma, a transformed cell line, generally neuron-like but that continues to divide after differentiation to neuronal morphology. Solomon showed that 60% of daughter cell pairs recapitulated the detailed, rather irregular, neurite geometry and pattern of the mother cell. The fidelity of recapitulation is quite striking, although it should be noted that the neurites involved are rather short, five or six cell body lengths at maximum. Solomon showed that the intrinsic factor responsible for this pattern determination is not encoded in a microtubule-based "memory." A similar recapitulation of morphology occurred after recovery of neuroblastoma cells from neurite-collapsing treatment with microtubule-depolymerizing drugs.

### C. Role of Microtubule-Associated Proteins

A major aspect in the differentiation of the neuroepithelial cell precursor into a cell capable of extending stable axons and dendrites is the specialization of the axonal microtubules to serve as robust structural elements. Long before the discovery of microtubules or neurofilaments, the "neurofibrils" that could be seen in histological sections of nervous tissue were thought to play a skeletal role as internal support elements (Young, 1944). Microtubules play the predominant cytoskeletal role in initiation and initial elongation of axons and dendrites, and it appears that this capacity depends on several specializations of microtubule dynamics in neurons. Microtubules are mechanically quite stiff (Mizushima-Sugano *et al.*, 1983; Gittes *et al.*, 1993) and in axons *in situ* they are very long, with average lengths of 100–500 $\mu$m (Bray and Bunge, 1981; Tsukita and Ishikawa, 1981). Biochemically, axonal microtubules are relatively stable compared to other cell types, as we noted earlier, and stability to microtubule depolymerization increases as the axon matures (Black and Greene, 1982; Morris and Lasek, 1982; Brady *et al.*, 1984; Black *et al.*, 1986). Axonal microtubules are also unusual in their formation of a tight axial bundle that fills the axoplasm of cultured axons and small-diameter axons *in situ.* For example, by assuming that the axon is a cylinder, measuring axonal area, and counting axonal microtubules in electron microscopic cross-sections of unmyelinated cat neurons, one obtains an estimated concentration of assembled tubulin of 40 mg/ml, essentially that of a packed pellet of microtubules *in vitro* (S. R. Heidemann, unpublished observations). As is discussed in Section IV,B, the formation of this bundle of microtubules at the distal end of the axon appears to be a key event in actually adding additional lengths of axonal cylinder from behind the advancing growth cone. At the molecular level, major contributors to the axonal microtubule specializations of length, stability, high concentration, and bundling are the microtubule-associated proteins, or MAPs.

These proteins were originally identified as proteins that associated with microtubules at constant stoichiometry through several cold–warm cycles during the purification procedure of tubulin from brain tissue. Several of these proteins were shown to stimulate markedly the assembly of tubulin in brain extracts (Murphy and Borisy, 1975; Weingarten *et al.,* 1975; Sloboda *et al.,* 1976). Some of these MAPs were subsequently shown to be motor proteins, which are discussed in Section V,A in conjunction with axonal transport. It is the "structural MAPs" that are responsible for the structural specializations of axonal microtubules just described. A truly immense literature has arisen on the biochemistry of structural MAPs, which proves to be complex both in reality and, unfortunately, in the jargon that has evolved to refer to this chemistry. A group of high molecular weight MAPs ( >200 kDa) were given systematic names (MAP1, MAP2, MAP3, etc.) only to require further subdivision into MAP1A, MAP1B, and so forth. Only a superficial summary of the role and mechanism of neuronal MAPs is offered here. Readers interested in more detailed coverage will be well served by the many reviews of this topic (Tucker, 1990; Wiche *et al.,* 1992; Matus, 1994; Schoenfeld and Obar, 1994; Mandelkow and Mandelkow, 1995; see also Burgoyne, 1991; Hirokawa, 1993; Hyams and Lloyd, 1994).

Here the focus is on two structural MAPs: tau and MAP2. These are the most abundant MAPs of neurons and are found primarily, if not exclusively, in neurons. Most important, the structure–function relationship of these two proteins with neuronal microtubules clearly illustrates the roles of MAPs in producing the microtubule array of the axon and the contribution of MAPs to such neuronal specializations as process elongation and stabilization, axonal/dendritic polarity, and in neuronal pathology. The most dramatic evidence for the important role of tau and MAP2 in axonal development is the formation of axon-like processes in nonneural cells experimentally induced to express tau or MAP2. For example, ovarian cells of the moth *Spodoptera frugiperda* (Sf9 cells) are normally highly rounded in culture. However, when cultured Sf9 cells are transfected with the gene for tau and express large amounts of the protein, most cells produce a single, long, uniform-caliber process that is remarkably axon-like (Knops *et al.,* 1991) (Fig. 1). The dense bundle of microtubules within these tau-induced processes has the same uniform polarity orientation of axonal microtubules, the plus end of the microtubule is at the distal end of the process (Baas *et al.,* 1991), and the microtubules are exceptionally stable to drug-induced depolymerization (Baas *et al.,* 1994). Similarly, expression of a MAP2-like protein in Sf9 cells also causes the formation of uniform-caliber processes containing dense bundles of microtubules; in this case the cells frequently elaborated multiple processes (LeClerc *et al.,* 1993).

Tau and MAP2 are related proteins; both bind to microtubules via a conserved 18-amino acid motif that is repeated 3–4 times near the carboxyl end of the protein (Lewis *et al.,* 1989). These 18-amino acid binding domains are separated by conserved 13-amino acid motifs that serve as flexible "hinges" allowing the multiple binding domains to attach to the tubulin lattice of the microtubule (Butner and Kirschner, 1991). Figure 2 shows a

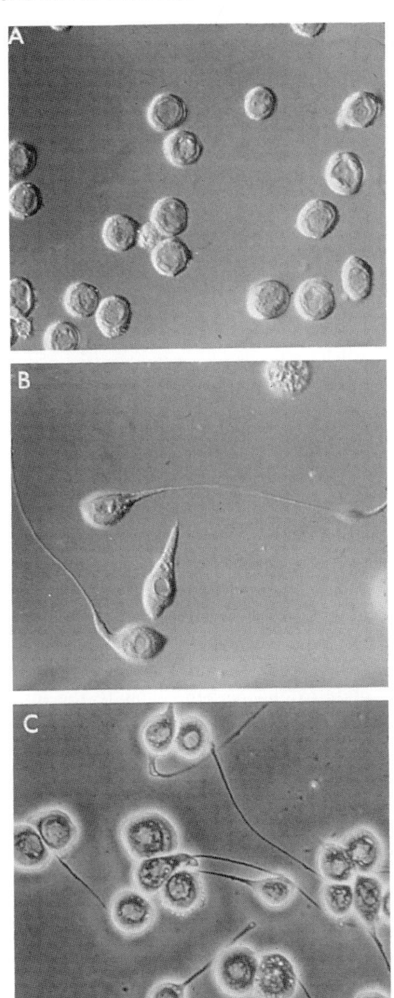


FIG. 1 Effect of high levels of expression of an axonal MAP on cell shape in cultured moth ovarian cells. Cultured Sf9 cells were induced to express large amounts of tau, a microtubule-associated protein typically restricted to the axonal compartments of neurons (see text). (A) Nomarski image of control cells infected with the baculovirus vector expressing $\beta$-galactosidase, and showing the normal shape of these cells. (B) Normarski image of Sf9 cells expressing high levels of tau with three repeated microtubule-binding domains, as shown in Fig. 2. (C) Phase image of Sf9 cells expressing high levels of tau with three repeated microtubule-binding domains, as in (B). Note in both (B) and (C) the unusually long, axon-like projections. [Reproduced from Knops *et al., The Journal of Cell Biology,* 1991, vol. 114, p. 727, by copyright permission of The Rockefeller University Press.]

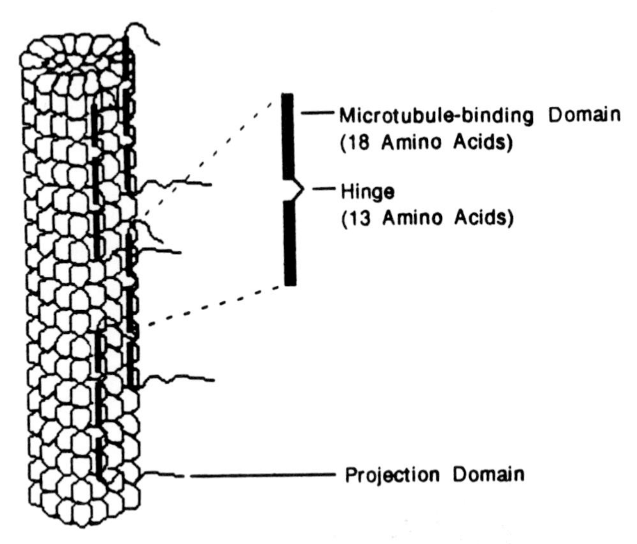


FIG. 2 Generalized structure of tau and MAP2 and their interaction with microtubules. MAP2 and tau share a conserved, repeated microtubule-binding domain consisting of 18 amino acids that is repeated 3 (shown here) or 4 times toward the carboxyl end of the peptide. The tandem repeats are separated by a 13-amino acid region that is postulated to serve as a hinge. As shown here, this family of MAP polypeptides binds axially along the microtubule, and projection domains of the polypeptide extend perpendicular to the long axis from the microtubule surface.

model similar to those found in the literature (Butner and Kirschner, 1991; Dye *et al.,* 1993; Matus, 1994) that provides an attractive basis for thinking about the multiple structural roles of neuronal MAPs. As illustrated in Fig. 2, tau and MAP2 bind together large numbers of tubulin subunits along the longitudinal axis of the microtubule. The stabilization of tubulin subunits within the lattice by MAP binding undoubtedly underlies the promotion and stabilization of microtubule assembly that have been repeatedly described both *in vitro* (Sloboda *et al.,* 1976; Cleveland *et al.,* 1977; Horio and Hotani, 1986) and *in vivo* (Drubin and Kirschner, 1986; Kanai *et al.,* 1989; Lewis *et al.,* 1989; Knops *et al.,* 1991; Weisshaar *et al.,* 1992; LeClerc *et al.,* 1993; Umeyama *et al.,* 1993). The longitudinal binding of MAP2 to the microtubule lattice has also been shown to stiffen microtubules (Dye *et al.,* 1993).

The bundling of microtubules is another well-described function of MAPs, including tau and MAP2. In all the experiments in which a transfected tau or MAP2 gene is expressed in nonneuronal cells, the cells show quite dramatic and abnormal microtubule bundles (Kanai *et al.,* 1989; Lewis *et al.,* 1989; Knops *et al.,* 1991; Weisshaar *et al.,* 1992; LeClerc *et al.,* 1993).

The microtubule bundles within axons appear to be cross-linked by wispy projections seen between the microtubules in electron micrographs (Hirokawa, 1982; Schnapp and Reese, 1982), and these may mediate the bundling of microtubules. On the one hand, structural MAPs have extensive domains that project from the surface of microtubules (Fig. 2; Valle, 1980; Hirokawa *et al.,* 1988a), and the cross-links observed in axonal microtubule bundles have been shown to contain MAP2 and tau by immunoelectron microscopy (Shiomura and Hirokawa, 1987; Hirokawa *et al.,* 1988a,b). In the Sf9 MAP expression system, expression of MAP2, composed of about 1700 amino acids in addition to the 100–120 amino acids of the microtubule-binding domain, produces bundles of microtubules with wide spacing of 50–100 nm. In contrast, tau (with a similar microtubule-binding domain but composed of a total of about 400 amino acids) produces much closer spacing within the bundle, 15–30 nm (Chen *et al.,* 1992). These data would seem to be a direct demonstration that MAPs determine bundle spacing, possibly by steric hindrance. But there is disagreement about whether the projection domain also provides a unique bundling domain. Certainly the data described above lend themselves to the interpretation that the projection domains of tau and of MAP2 bind and overlap with one another to cross-link and thus bundle axonal microtubules (Kanai *et al.,* 1989; Lewis *et al.,* 1989). Against this idea, however, are studies from the Matus laboratory with deletion mutants of a particular form of MAP2 (MAP2C; discussed below), which suggest that no single domain of this molecule can be shown to be involved in bundling (Burgin *et al.,* 1994). Rather, bundling appeared to be inseparable from microtubule binding per se: both bundling and binding increased in concert as the number of microtubule-binding domains was increased from zero to four (Ferralli *et al.,* 1994). These data argue that bundling is a direct result of microtubule stabilization, although spacing within the bundle may be due to the projection domains.

The developmental timing of the appearance and abundance of MAPs also supports their central role in specializing the microtubule cytoskeleton to form a stout cable. Tau protein, for example, is not found in neurons prior to axon formation, and its abundance rises sharply after the onset of axon formation in cultured cells (Drubin and Kirschner, 1986; Couchie *et al.,* 1986) and in embryonic and neonatal central nervous tissue (Mareck *et al.,* 1980; Francon *et al.,* 1982; Tucker *et al.,* 1988a,b). In the brain, tau also changes qualitatively from a juvenile isoform (molecular mass 50–60 kDa) to an adult form (60–70 kDa) that stimulates greater microtubule assembly than the juvenile form (Francon *et al.,* 1978, 1982). Similarly, MAP2 changes its expression from a relatively impotent juvenile form (MAP2C, 50–60 kDa) at the earliest times of process outgrowth to more potent, adult forms (MAP2B, 200–250 kDa) both in cultured cells (Binder *et al.,* 1984; Bernhardt *et al.,* 1985; Couchie *et al.,* 1986) and in embryonic and neonatal nerve tissue (Burgoyne and Cumming, 1984; Reiderer and Matus, 1985; Tucker *et al.,* 1988a,b). Both forms of MAP2 are transcribed from the same gene and arise through mRNA processing; the juvenile form retains the conserved microtubule-binding domain but is missing a large internal segment that forms the projection domain (Papandrikopoulo *et al.,* 1989). The increase in overall expression of MAPs and the change in expression from weak to strong isoforms is entirely consistent with the biochemical change in neurons from small amounts of easily reversed microtubule assembly in early stages of culured axonal growth to larger numbers of stable microtubules at later times (Black and Greene, 1982; Morris and Lasek, 1982; Brady *et al.,* 1984). The change in expression of MAPs is also entirely consistent with the developmental change from unstable "minor processes" into established axons (with their immense capacity for elongation) and dendrites (with their shorter, less robust shape). Indeed, the role of MAPs in establishing the early events of axonal/dendritic polarization is of long-standing interest.

Soon after the discovery of MAPs, it became clear that these proteins are differentially arrayed in axonal and dendritic compartments: MAP2B is found in the somatodendritic compartment of mature neurons and tau is observed in the axonal compartment in both cultured cells and *in situ,* as before (Caceres *et al.,* 1984; DeCamilli *et al.,* 1984; Bernhardt *et al.,* 1985; Dotti and Banker, 1987). MAP2C is observed only during the early stages of neurite outgrowth and is found in all processes. In the hippocampal culture system, juvenile MAP2C in dendrites is replaced by adult MAP2B at 5 days in culture, at the earliest stage of axon dendrite specification (Dotti and Banker, 1987). *In situ,* MAP2C is replaced by MAP2B at postnatal day 10–20 in rats or at metamorphosis in *Xenopus* tadpoles (Reiderer and Matus, 1985; Viereck *et al.,* 1988). In both cases, this corresponds to the end of the period of growth cone-mediated neurite outgrowth and synaptogenesis and the beginning of process stabilization and further elongation by expansion of the embryo. Similarly, tau is localized to the axons of hippocampal neurons only after initial specification of one process as the axon (Kosik and Finch, 1987).

This developmental timing seems to suggest that compartmentation of the two proteins is an effect, not a cause, of axonal/dendritic polarity orientation. But experiments using antisense RNA to block MAP2 and tau expression indicate that the expression of the various MAPs appears to be necessary for the normal outgrowth of all three types of processes: early neurites, axons, and dendrites. Absence of MAP2 expression (probably both forms) disables the formation of even minor processes in cultured cells (Dinsmore and Solomon, 1991; Caceres *et al.,* 1992). Supression of tau prior to neurite outgrowth permits minor process formation, but neither dendritic nor axonal growth was observed in cultured brain neurons (Caceres and Kosik, 1990). Suppression of tau after neurite had begun to grow caused axon-like projections to be lost, while dendrite-like neurites continued to elaborate (Caceres *et al.*, 1991). Most recently, however, mice lacking tau protein through gene knockout were shown to have rather normal nervous systems, and hippocampal cells cultured from these tau$^-$ mice were indistinguishable from cultured neurons of normal littermates (Harada *et al.*, 1994). However, lack of tau protein did cause a decrease in the density of microtubules in small-caliber axons and also of microtubule spacing and stability. These mice also showed an increase in another structural MAP not highlighted here, MAP1A, which is itself a potent stabilizer and cross-linker/spacer of axonal microtubules (Muller *et al.*, 1994). Thus, one presumes that proteins with the convergent functional activities of stabilization, packing, and stiffening are required for process outgrowth, but that one can substitute for another. Indeed, given the large number of neuronal MAP proteins with similar functional activities, it may be that these proteins form a network. The loss of any one member on the net does not much affect the overall function (Bray and Vasiliev, 1989; Buxbaum, 1995).

Finally, the importance of MAPs to neuronal function is illustrated by their involvement in neuropathologies. By far the majority of pathophysiological attention has been devoted to tau, which is implicated in Alzheimer's disease. Brain neurons in Alzheimer's disease are characterized by two cytological abnormalities: neuritic plaques, which are extracellular precipitates of a small fragment of a normal integral membrane protein (Selkoe, 1994), and neurofibrillary tangles composed principally of an altered form of tau (Goedert *et al.*, 1989; Lee *et al.*, 1991; Wille *et al.*, 1992). The altered tau binds to itself via the repeated tubulin-binding domains to form an abnormal type of cytoskeletal filament, paired helical filaments, that forms the neurofibrillary tangle. The formation of paired helical filaments from tau is the result of the hyperphosphorylation of tau (Grundke-Iqbal *et al.*, 1986; Goedert *et al.*, 1992; Bramblett *et al.*, 1993). Given the complexity of cellular phosphorylation reactions and their control, it is not surprising that the role of neurofibrillary tangles and tau in Alzheimer's disease is controversial and poorly understood. It is not clear whether neurofibrillary tangles are themselves a cause of neuronal death or simply an effect of Alzheimer's disease. A source of interest concerning the role of tau in the disease is the finding that a particular isoform for apolipoprotein E, a protein that plays a major role in the trafficking of lipids in blood serum, is correlated with the risk and mean age of onset of Alzheimer's disease (Strittmatter and Roses, 1995). Apolipoprotein E is found intracellularly in some neurons, and some isoforms bind to tau while other isoforms do not (Strittmatter *et al.*, 1994). One speculative hypothesis is that isoform-specific interactions of apolipoprotein E with tau protein may regulate tau activity in neurons and thus play a fairly direct role in the pathogenesis of Alzheimer's disease.

## IV. Growth Cone-Mediated Axonal Elongation

Following the initiation of axons and dendrites, the processes elongate as a result of growth cone motility until the tip reaches its target. This is by far the most intensively studied phase of neuronal outgrowth because of its crucial relevance to both extrinsic and instrinsic schools of neurogenesis. It is during this period that the growth cone must sense environmental signals, possibly biochemical signals exclusively, to navigate properly within the embryo. The most plausible physical mechanism for axonal guidance, differential adhesion of growth cones to the environment, has been damaged by experimental evidence, as is discussed in Section IV,C. For the intrinsic school, the changes in the neuron during the period of growth cone-mediated outgrowth provide an unequaled opportunity to investigate simultaneously the role of membrane and cytoplasm in cell shape, cell polarity, and cell motility functions of many kinds. Insofar as the growth cone leads this parade, it seems appropriate to begin the discussion with growth cone motility.

### A. Growth Cone Motility

The notion that growth cone locomotion is the result of muscle-like contractions by the filopodia that garnish its front end is at least as old as Harrison's first studies of cultured neurons (Harrison, 1910). Yet it is only in the last few years that direct experimental evidence has accumulated to confirm this as a mechanism underlying growth locomotion. These studies have also suggested a "duty cycle" of growth cone advance and have begun to clarify the functions of the various cytological components of the growth cone. However, these same studies provided evidence for very active motility mechanisms in addition to filopodial contraction that may also underlie growth cone advance.

Lamoureux *et al.* (1989) provided the first direct evidence that the growth cone is pulling forward. They showed that advance of the growth cone is accompanied by increasing tension in the axon shaft, as measured by calibrated glass needles attached to the cell bodies of elongating cultured neurons. Subsequently, Heidemann *et al.* (1990) provided the first extensive evidence that growth cone filopodia are contractile and can exert substantial pulling forces. The forces exerted by the filopodia were highly intermittent and seemed closely associated with the characteristically halting, jerky advance of the growth cones being observed. Observations of growth cones of individual, identified pioneer neurons in the "filleted" grasshopper limb bud also support the importance of filopodia in growth cone advance. O'Connor *et al.* (1990) observed a dramatic steering event mediated by contact of a single filopodial tip of the Ti1 pioneer neuron to a guidepost (navigational signal) cell. After contact the filopodium becomes engorged with cytoplasm moving up from the base of the filopodium to the tip in a single saltation. This filopodial dilation now becomes an additional neurite shaft and leads to a new direction of growth cone advance. Observations on the navigation of another pioneer neuron of grasshopper limb confirm dilation as a mechanism of neurite formation following filopodial contact with a navigational signal (Myers and Bastiani, 1993). Further, as discussed earlier, Smith (1994a) also observed dilation of cytoplasm into filopodia that had attached at their tips during neurite initiation. During this dilation, several of the observations suggested that the filopodium was exerting tension. Dilation did not occur immediately after the filopodial tip attached to an object, but only after the filopodium began to straighten out. Most telling, the movement of cytoplasm reversed if the object to which the filopodium was attached subsequently detached from the substrate and moved. And neurites were not initiated in this system by attachment of filopodia to unattached beads. This is entirely consistent with the observations of growth cone traction force by Heidemann *et al.* (1990), in which movable obstacles were placed in front of the growth cone to reveal the contractile activity of the filopodia. That is, the contracting filopodia will pull movable objects into the growth cone, but if attached to an immovable object at the distal end will pull cytoplasm into the filopodia.

A corollary to the scenario that pulling filopodia are responsible for growth cone advance is the importance of growth cone adhesion at the tips of filopodia, but not elsewhere. Most studies of growth cone adhesion have focused on the adhesion of the broad, palmate region of the growth cone to culture substrata (Letourneau, 1979; Gundersen, 1987, 1988; Lemmon *et al.*, 1992; Zheng *et al.*, 1994). However, a flurry of reports on the structure and activity of tips of growth cone filopodia suggests that this tiny region of environmental interaction may be of unusual significance to growth cone advance and guidance. As previously mentioned, several groups have shown strong functional correlations between navigational and initiation events with the attachment of single filopodial tips to their environment (O'Connor *et al.*, 1990; Myers and Bastiani, 1993; Smith, 1994a). In terms of the actual adhesions at such filopodial tips, they are exceptionally resistant to mechanical detachment (Zheng *et al.*, 1994). In addition, several studies have now focused on structural specializations at the tips of growth cone filopodia. Tsui *et al.* (1985) have observed structural specializations for adhesion at the tips of filopodia. Two other studies have focused on the formation of actin-rich regions at the tips of filopodia. O'Connor and Bentley (1993) showed for the grasshopper limb system that their reproducible navigational event by single filopodial contacts is accompanied by an accumulation of actin at or near the tip of the filopodia after initial contact. Similarly, Lin and Forscher (1995b) observed an accumulation of actin at the site of contact between two *Aplysia* growth cones, at which the growth cones were pulling on one another. Wu and Goldberg (1993) have shown an abundance of phosphotyrosine residues at the tips of filopodia in *Aplysia* neurons, using anti-phosphotyrosine antibodies. Slowly growing growth cones had more phosphotyrosine labeling than the growth cones stimulated to grow by a substrate ("adhesion") treatment. At this point, connections between or among these various filopodial tip phenomena are speculative, but there does seem to be an unusual congression of interest in this region of the growth cone. Regrettably, the small size and dynamic behavior of filopodial tips will make them rather difficult to study in detail.

Pulling filopodia are probably not the only mechanism by which growth cones advance or steer, however. Growth cones lacking filopodia are observed to advance in culture (Marsh and Letourneau, 1984) and *in situ* (Bentley and Toroian-Raymond, 1986; Chien *et al.*, 1993), albeit slowly and abnormally. Another possible motile mechanism in addition to filopodial pulling reflects a dramatic motility phenomenon on the dorsal surface (at least) of cultured growth cones: a retrograde flow of cortical actin in waves. In high-resolution, time-lapse images of the broad, flat, lamellipoidal growth cone of *Aplysia* neurons, the membrance surface looks like a set of escalator steps arising from the distal edge of the lammellipodia and then gracefully moving backward at constant speed to disappear near the "neck" of the growth cone (Forscher and Smith, 1988). This "escalator" is maintained by a continuous assembly of actin filaments at the distal, leading edge of the growth cone, and marked actin filaments at the distal edge are observed to translocate rearward toward the base of the growth cone (Forscher and Smith, 1988; Okabe and Hirokawa, 1991). In this regard, a pertinent question that has not been addressed concerns the stable location of actin filament assembly at the leading edge of the growth cone despite the forward movement of the leading edge relative to the substrate and the translocation of edge material toward the rear. In chick sensory neurons, the "escalator" of cortical actin ridges can attach to obstacles on the dorsal surface and move them from the periphery to the center of the growth cone, and exert substantial forces on the obstacles (Heidemann *et al.*, 1990).

Although this escalator could serve directly as a means of motility in a manner similar to the continuously moving belt of a Caterpillar tractor (Smith, 1988), in chick neurons this escalator seemed not to have a close relationship to growth cone advance. For example, chick sensory growth cones advance in a jerky, "two steps forward and one step back" manner, while the retrograde actin escalator produces very smooth movements (Heidemann *et al.,* 1990). However, the interesting observations of Lin and Forscher (1995a) suggest that this retrograde escalator and filopodial pulling may be linked. These workers observed the actin cytoskeleton during the interaction of pairs of *Aplysia* growth cones as they collided and pulled on one another. In their isolated state, these growth cones are unusually symmetric with broad lamellipodia. When growth cone lamellae meet, they form a thin filopodium that apparently pulls and recruits cytoplasm from the thicker, central region of the growth cone to widen the filopodium. Similar to the filopodial dilation discussed earlier, this newly fortified filopodium becomes the central direction of growth cone orientation thereafter. Observations of the retrograde actin escalator during these interactions indicate that growth cone advance by this mechanism is inversely proportional to the actin escalator. This leads to the intriguing possibility that the retrograde actin flow reflects a free-running motor with no load (like an automobile in neutral gear). A clutch mechanism engages this motor with filopodial actin [or filopodium-like actin bundles within the lamella (Lewis and Bridgman, 1992)] "putting it in gear" to exert a pulling force on the environment. Lin and Forscher (1995a) postulate a clutch at the distal end to mediate adhesion and disadhesion. Our data (Heidemann *et al.,* 1990) would argue there is also a clutch mechanism at the rear, where the escalator fades, because we observed that the pulling force exerted by filopodia is intermittant, yet the obstacle remained attached throughout. That is, filopodial contractions were characterized by frequent changes in obstacle velocity and direction. In the context of a uniformly acting motor responsible for escalator movement and filopodial contraction, this suggests a rear clutch that slips in and out to effectively engage, then disengage, from the motor, producing the intermittant force. This intriguing model will require additional investigation, but seems to unify several robust motile phenomena associated with growth cone advance.

One additional motile phenomenon of growth cones that is less easy to reconcile with other observations is the movement of membrane antigens in an anterograde direction. Sheetz *et al.* (1990) used gold particles coated with antibodies to growth cone surface antigens, including the N-CAM cell adhesion molecule (Rutisheuser and Jessell 1988). These surface markers translocated from the central region of the growth cones of mouse brain neurons to collect at the periphery. Like other growth cone-associated movements, this one was poisoned by cytochalasin and so presumably is actin dependent. This reviewer, at least, is thoroughly puzzled as to how these membrane antigens move outward, apparently against the retrograde escalator. Their concentration at the leading edge may, arguably, have some relation to the structural specializations at the tips of growth cones.

## B. Relationship of Growth Cone Motility to Axonal Elongation

However the plethora of motile activities at the growth cone actually underlie its forward motion, there is uniform agreement that axonal elongation is intimately connected to growth cone advance. That is, the structures, processes, and reactions of hypertrophy of the axon—the addition of neurite mass—is connected in some direct way with the movement of the growth cone.

One direct link between growth cone motility and axonal elongation is clearly the tension exerted by the pulling growth cone. Axonal elongation occurs visibly over the course of seconds and minutes both from experimentally applied tension (Bray, 1984; Zheng *et al.,* 1991) and from tension produced by the growth cone (Lamoureux *et al.,* 1989). When axonal elongation is experimentally stimulated by "towing" with a needle, axons can elongate to produce ultrastructurally normal processes at physiological and far-above physiological rates for many hours (Bray, 1984; Zheng *et al.,* 1991, 1993). Indeed, in response to applications of tension, the process of axonal elongation bears a surprisingly robust similarity to the elongation of simple Newtonian fluid mechanical elements ("dashpots," e.g., the pistons found on screen doors to prevent slamming). That is, the rate of elongation of cultured chick neurons and PC12 cells is a simple linear function of the applied force under a variety of culture conditions and at all elongation rates (Fig. 3) (Dennerll *et al.,* 1989; Zheng *et al.,* 1991; Lamoureux *et al.,* 1992). The principal mechanical difference between axonal elongation in culture and the elongation behavior of dashpots is that a threshold tension (usually about 100–200 $\mu$dyn for chick sensory neurons) is required to begin axonal elongation, while ideal dashpots are sensitive to all tensions. The linearity of the relationship is robust; applied tension vs elongation rate data from 39 of 47 towed chick sensory axons showed a correlation coefficient of 0.9 or greater to the equation with the general form: growth rate = sensitivity × (applied tension − threshold tension). However, the quantitative aspects of this relationship (i.e., the threshold values and the slopes of the lines) vary among neurons as shown in Fig. 3. In this regard, the sensitivity of axonal elongation to tension can be quite high: in several neurons each additional microdyne of applied tension stimulated a greater than 5-$\mu$m/hr increase in elongation rate (Zheng *et al.,* 1991; Lamoureux *et al.,* 1992). This surprisingly simple and robust relationship between tension and elongation rate must be a major factor in determining how growth cone behavior actually causes additional lengths of axon to form, although understanding of mechanical input into biochemical systems is in its infancy.

Another important clue to the mechanisms connecting motility to mass addition is the temporal relationship between formation of additional

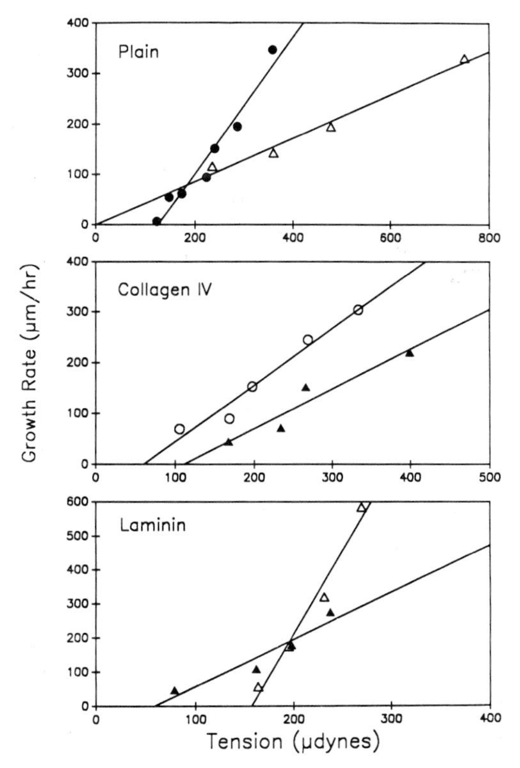


FIG. 3 Axonal elongation rates as a function of applied tension. Chick sensory neurons were cultured on untreated tissue culture plate plastic (plain), or the same dishes treated with collagen IV or laminin, substances known to stimulate neurite outgrowth. Each line reflects the data of a single neurite that was towed by a calibrated glass needle attached at the growth cone. For each line, a neurite was subjected to a step-function protocol of applied tensions: a constant force was applied for 30–60 min, the tension was incremented 25–200 $\mu$dyn for another 30–60 min, and so on. Each neurite was subjected to four to seven such steps and the axonal growth rate (elongation rate corrected for elastic stretching) at each tension was measured from the videotape record of the experiment. As shown here, the elongation rate is a robust linear function of applied tension in all axons tested (more than 50 to date) under all culture conditions at both physiological (growth cone-mediated) rates ($<125$ $\mu$m/hr) and rates greatly exceeding the physiological norm. However, axons vary with respect to the threshold tension required for growth ($x$ intercept) and the slope of the relationship (sensitivity to tension).

lengths of axon cylinder and the cytoplasmic movements of the growth cone and within it. Work from several groups now supports a unitary description of the cytological events underyling growth cone advance and assembly of the neurite shaft. Three phases of growth cone advance, first characterized by Goldberg and Burmeister (1986) as protrusion, engorgement, and consolidation, provide a good basis for the "duty cycle" of the growth cone. Although well-distinguished phases are often obscured by the wide variety of form and promiscuous motility of growth cones, lengthy observations of axonal elongation of several types of cultured neurons suggest this model as a reasonable basis for a general concept of the cytology of growth cone-mediated elongation (Fig. 4).

The protrusion phase is associated with the elongation of filopodia into the region ahead of the growth cone. Although the size and dynamic nature of filopodia make them rather difficult to study, there is widespread agreement that one important mechanism for filopodial formation is the assembly of an actin bundle pushing out a region of membrane. Okabe and Hirokawa (1991) showed that actin monomers assemble preferentially at the distal ends of filopodia. Forscher *et al.* (1992) demonstrated an experimentally induced formation of filopodium-like structures ("inductopodia") on the surface of the highly lamellipodial growth cone of cultured *Aplysia* neurons. Vinyl beads treated with a polycationic polymer were sprinkled onto the exposed surface of the growth cone, which caused the assembly of an actin bundle at the site of bead–membrane contact. This, in turn, produced a long, thin extrusion very similar in form to filopodia. Filopodia continuously extend and retract (Bray and Chapman, 1985; Heidemann *et al.,* 1990), and their behavior is poorly correlated with growth cone advance (Goldberg and Burmeister, 1986; Aletta and Green, 1988). To participate productively in growth cone advance, it appears that filopodia must become firmly adherent to an immovable surface, as suggested above. That is, it appears that only those filopodia able to exert tension play an active role in the next phase of growth cone advance, protrusion.

In engorgement, cytoplasm from the broad, central palmate region of the growth cone moves forward into the newly explored region of protrusion. Both the microtubule cytoskeleton and the fluid cytoplasm advance in this phase. Goldberg and Burmeister (1986) inferred the advance of microtubules by the movement of the membranous vesicles riding the microtubules in *Aplysia* neurons. The advance of microtubules has been directly observed in real time in cultured *Xenopus* neurons (Tanaka and Kirschner, 1995; Tanaka *et al.,* 1995). These workers imaged the real-time dynamics of fluorescently labeled microtubules in living growth cones using a sophisticated labeling and observation technique (Tanaka and Kirschner, 1991). They watched growth cone behavior and the underlying microtubule dynamics as growth cones encountered a substrate boundary at which the growth cone turned. Regions of the growth cone that subsequently became neurite shaft manifested a prior advance of microtubules. Experiments in which vinblastine was used to poison microtubule dynamics indicate that this invasion normally depends on both microtubule transport and microtubule assembly (Tanaka *et al.,* 1995).

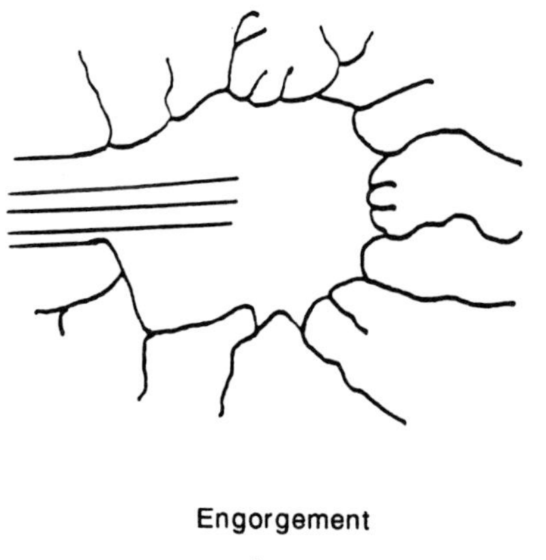


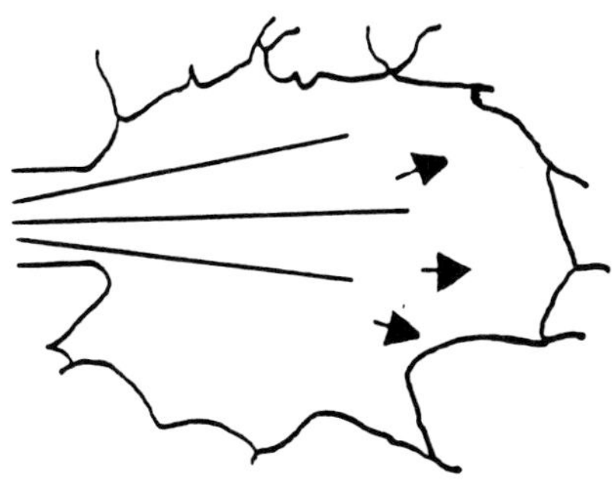


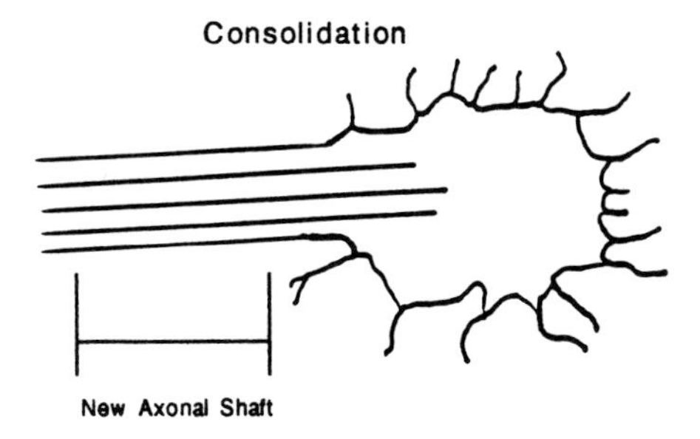


FIG. 4 A cycle of growth cone behavior associated with the addition of a new neurite shaft. As described originally by Goldberg and Burmeister (1986), growth cone advance leading to new lengths of axon can be divided into three phases. During the protrusion phase, the growth extends filopodia into the region ahead of the growth cone. During engorgement, the cytoplasm at the base of the growth cone moves forward in a concerted manner as seen by the advance of microtubules (both assembly and translocation are involved according to current evidence) and by a fluid flow of cytoplasm into the region of filopodial protrusion. Consolidation occurs when the extended regions of distal microtubules form a bundle, and cytoplasm appears to collapse around the bundle to form a new length of axonal shaft. This three-part sequence is something of an idealization in view of the complexity and variability of growth cone behavior, but essentially similar events have been observed in favorable cases in a variety of cultured neurons, both vertebrate and invertebrate.

The engorgement phase also includes a more easily seen flow of apparently fluid cytoplasm. This has been observed in a variety of neurons as the saltational advance of cytoplasm as a coherent front along or into the recently formed filopodial "tracks," rather like webbing suddenly growing between fingers (Bray and Chapman, 1985; Goldberg and Burmeister, 1986; Aletta and Greene, 1988; O'Connor *et al.,* 1990; Smith, 1994a; Tanaka and Kirschner, 1995). The appearance of this cytoplasmic engorgement as a fluid flow caused Goldberg and Burmeister (1986) and Aletta and Greene (1988) to propose that the growth cone is pushing forward because typical fluids are, indeed, moved by a pushing force. (For example, the higher hydrostatic pressure within an overturned glass drives the advancing spill.) However, cytoplasm is a highly unusual material showing both solid/elastic properties and fluid/flow properties (Elson, 1988; Kerst *et al.,* 1990). Heidemann *et al.* (1991) has proposed that the unusual material properties of cytoplasm allow it to be pulled forward by tension exerted along the filopodial actin bundle, which force shears the initially gellike cytoplasm to a fluid, which flows forward. That is, the cytoplasm of the growth cone undergoes a thixotropic transition from a gel to a sol (Kerst *et al.,* 1990), a fluid behavior long thought to be involved in animal cell motility (Seifriz, 1942; Taylor and Condeelis, 1979; Bray and White, 1988). This proposal connects the direct evidence for growth cone pulling via contractile filopodia with the cytoplasmic flow behavior within the growth cone widely observed during axonal advance and axon initiation. Like filopodial elongation, however, the protrusion of cytoplasmic veils to form lamellipodia is relatively easily reversed, and lamellipodia also engage in a fair amount of apparently unproductive backing and filling (Bray and Chapman, 1985; Aletta and Greene, 1988; Heidemann *et al.,* 1990; Tanaka and Kirschner, 1995; Tanaka *et al.,* 1995).

In the final step in the duty cycle of growth cone advance, the central region of the growth cone rounds into an additional length of cylindrical neurite shaft. Goldberg and Burmeister (1986) termed this consolidation, and current data suggest that this step is the *sine qua non* of growth cone-mediated axonal elongation. First, this is the step in which the characteristic, uniform-caliber cytoplasmic shaft actually becomes longer. Second, this step seems rather less reversible than either elongation or protrusion. The earlier two steps, protrusion and engorgement, appear to be part and parcel of the frenetic actin-based motility of the growth cone, with its promiscuous movements of filopodia and lamellipoda. That is, instances of protrusion and engorgement can occur, then reverse just as rapidly. However, consolidation involves the assembly or reorganization of microtubules to form an axial bundle, which is accompanied by the apparent "collapse" of growth cone cytoplasm around the bundle, thus adding an additional length of neurite cylinder (Goldberg and Burmiester, 1986; Tanaka *et al.,* 1995). As the least reversible of the three steps, the bundling of microtubules appears to be the final event that settles the frenetic motion of the growth cone into a new unit of axon.

Although much painstaking work went into gleaning this three-part cycle from the complexity of growth cone motion, in one sense it is only a vocubulary to begin asking mechanistic questions, all of which are unsettled. For example, which cytoplasmic events are the result of tension, and which are part of the cause-and-effect sequence for the production of tension by the growth cone? What changes are directly coupled to the tension input to cause the microtubule polymerization and membrane addition reactions? Does the tension provide a direct free energy input to alter chemical reaction steady states (Buxbaum and Heidemann, 1988, 1992)? Or is there a long, winding road typical of chemical signal transduction that begins with a particular protein mechanosensor, such as stretch-activated ion channels? The interpretation offered above for the engorgement of cytoplasm would suggest that the filopodia produce tension at some point during the protrusion phase; can one phase be shown to be the primary tension-exerting phase? How is this tension produced, e.g., where are the presumptive myosin motors located? Their location(s) must account for a wide variety of different behaviors, and all proposals have been explicitly speculative (Mitchison and Kirschner, 1988; Smith, 1988; Heidemann *et al.,* 1991; Lin *et al.,* 1994). What, if anything, have filopodial motions and forces to do with complex lamellipodial changes? For example, ruffling lamellipodia of growth cones have been observed to "twirl" short glass fibers above the growth cone surface (fifteen 360° rotations in 11 min; Heidemann *et al.,* 1990). What is the relationship between the actin cytoskeleton and microtubules during the duty cycle? Most opinion seems to regard actin motility as beginning the cycle, with microtubules subsequently following (Smith, 1988; Heidemann *et al.,* 1991; Bentley and O'Connor, 1994; Lin *et al.,* 1994), but Tanaka *et al.* (1995) have argued that dynamic microtubules are specifically required to allow growth cones to move forward (i.e., potentiating the putative actin motility). In truth, there is no evidence whatsoever on the cause–effect sequence that underlies the duty cycle or the generation of tension. Are microtubules pulled forward by tension during engorgement (Smith, 1994b)? Are they actively moving forward by microtubule sliding (Joshi and Baas, 1993)? Or do microtubules advance by assembling forward (Tanaka *et al.,* 1995), possibly in response to tension (Buxbaum and Heidemann, 1988, 1992)? Any or all of the above are consistent with the available data. Indeed, the promiscuity of growth cone behavior *vis-à-vis* the production of new axonal shaft may reflect a haphazard cause-and-effect relationship, e.g., addition of new shaft could depend on the stochastic occurrence of a sequence of independent cytoskeletal events. In a similar vein, we discussed earlier the possibility that the actual tension generation by the growth cone depends on something "catching," such as attachment of filopodial tips (Section III,A) to engage a continuously acting motor.

### C. Growth Cone Steering

Given a duty cycle of protrusion, engorgement, and consolidation for growth cone advance, it is certainly of interest to ask how these behaviors are changed in response to steering and navigational signals. Indeed, observations of growth cones at substrate borders in culture, or at guideposts *in situ,* provide what meager information we possess about the connections between the intrinsic and extrinsic aspects (see Section I) of the migration of neurons. Somewhat disappointingly, then, "all of the above" is the current answer to the question of how growth cones change their behavior to steer. That is, changes in any of the three behaviors seem capable of supporting steering "decisions."

One obvious way in which a growth cone might steer would be to control the protrusion phase of behavior so that filopodia were elaborated primarily in the direction of steering. Myers and Bastiani (1993) have observed this type of behavior in one steering decision of an identified neuron of grasshopper (Q1) that undergoes a highly stereotyped migration through the embryo. The axon of Q1 initially grows toward the head but then makes a sharp turn toward the midline at 32% of development. At the time of this decision, about 80% of the filopodia are oriented in the direction of the turn. This is not the result of selective stabilization of properly oriented filopodia or of any differences in the rate of extension or retraction relative to other filopodia. Rather, it appears that the bias of filopodia in the direction of steering arises from the selective protrusion of filopodia in the preferred orientation.

The growth cone of Q1 neurons also steers by filopodial dilation in response to contact with an appropriate guidepost cell, as discussed earlier in the context of the Ti1 neuron of grasshopper limb (O'Connor *et al.,* 1990; Myers and Bastiani, 1993). Insofar as filopodial dilation involves a concerted movement of cytoplasm into the filopodium, this type of steering event appears to be more closely related to a change in engorgement behavior than protrusion. However, any steering event that directly involves filopodia seems likely to reflect a close coupling between the tension-producing and the sensory role of filopodia, which clearly plays a major role in correct navigation by the growth cone. When growth cones are deprived of filopodia, they loose their way *in situ* (Bentley and Toroian-Raymond, 1986; Bentley and O'Connor, 1992; Chien *et al.,* 1993). Growth cones can collapse at the first filopodial contact with certain cues (Kapfhammer and Raper, 1987), and isolated filopodia can respond to environmental signals by filopodial contraction (Davenport *et al.,* 1993). Nevertheless, the properties of the navigational signal and how these stimulate the structural changes of filopodial protrusion and/or dilation are unknown but appear to be highly specific: the same identified neurons of grasshopper that steer by filopodial dilation at particular guidepost cells steer by different mechanisms elsewhere along their pathway (O'Connor *et al.,* 1990; Meyers and Bastiani, 1993).

In addition to filopodial dilation, other examples of engorgement behavior have also been shown to play an important role in steering decisions. Cultured *Aplysia* growth cones growing on a polylysine-treated surface advance slowly, but speed up when stimulated with hemolymph bound to the substratum (Burmeister *et al.,* 1991). A similar observation was made for growth of rat sympathetic neurons on laminin (Rivas *et al.,* 1992). In both studies, the major differences in growth cone behavior was the acceleration of engorgement following protrusions. Similarly, the Ti1 pioneer growth cone that makes some turns by filopodial dilation also turns at a different navigational signal by recruiting cytoplasm from one region of a well-spread growth cone to preferentially engorge another region of the same growth cone (O'Connor *et al.,* 1990). This newly engorged region now becomes the new direction for subsequent growth cone advance. Tanaka and Kirschner (1995) also observe selective engorgement as a process underlying instances of turning by *Xenopus* growth cones at a culture substrate boundary.

Consolidation, too, has been observed to underlie growth cone steering. In some instances of *Xenopus* growth cones exploring a substrate boundary, Tanaka and Kirschner (1995) observed microtubules within the growth cone bundling in a particular direction prior to a clear turn. Consolidation of new neurite shaft around this microtubule bundle subsequently turns the neurite shaft in a particular direction.

Thus, observations of growth cone behavior in response to environmental boundaries and to navigational signals suggest that steering can arise from changes in any of the basic phases of growth cone advance and that a given neuronal type or an individual identified neuron depends on more than one type of behavioral change to steer.

Older data supported an attractive proposal that growth cone steering was directly linked to intrinsic cytoskeletal mechanics via differential adhesion of growth cones to different regions (Bray, 1982, 1991; Letourneau, 1983; Bastiani and Goodman, 1984). In some cases, for example, growth cones preferentially steer onto culture surfaces that are more adhesive when presented with substratum choices, (Letourneau, 1975; Hammarback and Letourneau, 1986). It was postulated that growth cones preferentially steer onto more adhesive surfaces because growth cones or filopodia exerting more tension than the local adhesion limit pull free and are lost, leaving only those attached to more adhesive surfaces. Other studies, however, have questioned the generality of this differential adhesion mechanism. Several studies have shown that adhesion to culture substrata is poorly correlated with guidance preferences (Gundersen, 1987; Calof and Lander, 1991; Lemmon *et al.,* 1992), axonal elongation rate (Buettner and Pittman, 1991; Lemmon *et al.,* 1992), or fasciculation (Lemmon *et al.,* 1992). Most recently, Zheng *et al.* (1994) measured the force of adhesion of growth cones to various culture substrata, using calibrated glass needles. The principal advantage of this method is that it provides an estimate of adhesion forces in absolute units, which, in turn, allows a comparison with the previous measurements of tensions associated with resting neurites (Zheng *et al.,* 1991; Lamoureux *et al.,* 1992) and with growth cone advance (Lamoureux *et al.,* 1989; Heidemann *et al.,* 1991) on the various substrata. Strikingly, the forces required for growth cone attachment were significantly greater than the forces normally exerted by the neurite on its moorings. Thus, growth cones are not functioning near their adhesive limit and it seems unlikely that weaker adhesions would give way to stronger adhesions in the "tug of war" postulated by the differential adhesion mechanism. Further, few significant differences in the intensity of adhesion (force per unit area) of growth cones to different surfaces were observed, i.e., when corrected for area, growth cones adhered about equally to all surfaces (Zheng *et al.,* 1994). These data suggest no mechanical effect of navigational signals on growth cone steering and that "adhesion molecules" may be an inappropriate term for laminin and other surface active neurotropic substances. Steering in response to navigational cues may be transduced entirely by second messenger-type mechanisms (Doherty and Walsh, 1994).

## V. Proximate Mechanisms Underlying Axonal Assembly

The short answer to questions concerning proximate mechanisms underlying axonal assembly is, "We don't know." Presumably, the most fundamental level for thinking about axonal elongation is that new axonal shaft requires additional cytoskeletal elements and additional plasma membrane. Unfortunately, there is little agreement in the literature as to the site, mechanism, or regulation of these basic cytoplasmic processes underlying hypertrophy of the axon. Most awkwardly, many of the controversies are due to what appear to be fundamentally different experimental results, not merely different interpretations of data. For example, nine similar immunocytological experiments from 1987 to 1995 produced disparate results, yet all appear on their face to be a direct test of whether the microtubule cytoskeleton is moving en masse down the axon.

As noted in Section II, the axon does not synthesize its own protein or membrane elements, and thus addition of both cytoskeletal elements and membrane to the axon is dependent on axonal transport mechanisms. Membrane transport is closely linked with "fast axonal transport," while "slow axonal transport" is intimately linked to questions concerning axonal cytoskeleton addition, This natural dichotomy is adopted for the discussion here, and membrane and cytoskeletal additions are discussed in relation to these transport mechanisms. Like the coverage of microtubule-associated proteins, the immense number and variety of studies of axonal transport necessitate an unusually brief synopsis of this important topic. Once again, a large number of reviews and monographs on axonal transport is available for readers seeking a more detailed treatment (Grafstein and Forman, 1980; Weiss, 1982; Iqbal, 1986; Smith and Bisby, 1987; Brady, 1991; Bray and Mills, 1991; Nixon, 1991; Vallee and Bloom, 1991; Ochs and Brimijoin, 1993).

## A. Fast Axonal Transport and Membrane Addition

Our understanding of fast axonal transport has enjoyed major advances in the last decade. Cytological and biochemical methods have dissected the mechanism of fast axonal transport to a level of detail and confidence that is a welcome exception to the confusion and controversy noted above. Fast axonal transport is the saltatory movement of membranous organelles and vesicles in both directions within the axon. This transport mechanism supplies the axon with new plasma membrane and its associated proteins (e.g., voltage-dependent $Na^+$ and $K^+$ channels; Wonderlin and French, 1991), as well as mitochondria, synaptic vesicles, and enzymes associated with neurotransmission. When measured by kinetic methods *in situ,* proteins labeled at the cell body move down the axon at a net rate of about 200–400 mm/day (Lasek, 1968; Ochs *et al.,* 1969). Most studies during the 1960s and 1970s indicated a close link between fast axonal transport and microtubules (Kreutzberg, 1969; Fernandez *et al.,* 1971; Karlsson *et al.,* 1971). And speculative models proposed at that time for fast axonal transport are fundamentally similar to our current understanding, ATP-driven motors move vesicles along microtubule tracks (Schmitt, 1968; Ochs, 1972; Schwartz *et al.,* 1976).

The work that confirmed the railroad-like mechanism of fast transport and provided the details of the motor proteins arose from an improved method of microscopy and the finding that fast axonal transport continued in extruded axoplasm from squid giant axons (Allen *et al.,* 1982; Brady *et al.,* 1982). Bidirectional movement of axonal organelles had been observed microscopically in several of the earliest studies of individual axons (Matsumoto, 1920; Hughes, 1953; Nakai, 1956), but the resolution was too poor to confirm any direct links of membrane transport with microtubles. The late R. D. Allen introduced a substantial increase in the resolution of light microscopy, 10-fold better than the usual Rayleigh criterion for resolution, by video processing and enhancement of a differential interference contrast (Nomarski) optical signal (Allen *et al.,* 1981). This new method of microscopy allowed visualization of vesicles moving along single microtubules in axoplasm extruded from squid giant axon, like toothpaste from a tube (Brady *et al.,* 1982). Subsequent gentle fractionation of the squid axoplasm showed that not only did vesicles move along microtubules, but axonal microtubules would glide along glass slides treated with soluble fractions from axoplasm (Allen *et al.,* 1985; Schnapp *et al.,* 1985). Observation of vesicle movement and microtubule sliding then became the "enzymatic assay" used to purify the active factors required for motility. The purification of the active factors from squid axoplasm was aided by the finding that a nonhydrolyzable analog of ATP, AMP-PNP, causes this type of motility to stop, essentially freezing the motile complex in place on the microtubule, but could be released by ATP (Lasek and Brady, 1985). This selective binding allowed the active factors to be separated from crude soluble axoplasmic fractions by binding to exogenous, MAP-free microtubules and then pelleting the microtubules and releasing bound peptides with ATP (Brady, 1985; Vale *et al.,* 1985). The AMP-PNP-sensitive factor was later shown to be a previously unknown motor protein that was given the name kinesin.

Kinesin was shown to be a tetrameric protein, two light chains and two heavy chains, with a molecular mass of approximately 380 kDa. This molecule bears some convergent similarities to myosin, in that the molecule is elongated, with a head region that binds both ATP and the appropriate cytoskeletal filament and a tail region that binds to membrane in the case of kinesin (Bloom *et al.,* 1988; Kutznetzov *et al.,* 1988; Hirokawa *et al.,* 1989; Scholey *et al.,* 1989). The kinesin head ratchets along the microtubule in 8-nm intervals, the same as the tubulin dimer repeat (Svoboda *et al.,* 1993). Antibodies to kinesin bind primarily to vesicular structures in the cytoplasm, not to microtubules (Pfister *et al.,* 1989), and the binding of kinesin to membrane organelles occurs at high affinity *in vitro* (Schnapp *et al.,* 1992). These data indicate that kinesin is stably associated with membranous elements, which then row along microtubules via moving kinesin oars. There is now direct evidence that kinesin plays a major role in fast axonal transport. Antibodies to kinesin heavy chain infused into squid giant axon markedly inhibit the number and velocity of transported organelles (Brady *et al.,* 1990). Also, suppression of kinesin heavy chain expression in hippocampal neurons caused an inhibition of both dendritic and axonal outgrowth, and two proteins normally localized to the distal tips of neurons were confined to the cell body (Ferreira *et al.,* 1992).

The discovery of kinesin in axons led to much wider study, and neuronal kinesins are members of a large and growing family of proteins found throughout the eukaryotic domain, some of whose members are involved in mitotic and meiotic chromosomal movements (Goldstein, 1993). Only the ATP and microtuble-binding domains of the heavy chain of kinesin are conserved generally throughout the kinesin-related proteins. The divergence of other protein regions is thought to underlie the specificity of transport on microtubules. That is, different transport organelles associate with specific motor proteins, providing specificity of targeting (Okada *et al.,* 1995) Most, but not all, of the kinesin proteins are specific for transport toward the plus end of the microtubule, including the neuronal kinesin. Thus, kinesin is associated principally with anterograde transport, i.e., toward the distal tip of the axon.

Retrograde fast axonal transport, toward the cell body and the minus ends of microtubules, is mediated primarily by dynein, a motor protein long known to underlie the motility of cilia and flagella by minus end-directed motility. On the basis of the sensitivity of flagellar dynein to vanadate ions (Gibbons *et al.,* 1978), the finding that axonal transport of organelles in lobster axons was inhibited by low concentrations of vanadate first suggested a role for this protein (Forman, 1982; Forman *et al.,* 1983). Following isolation of kinesin, further analysis of axonal microtubule-associated proteins that are released by ATP treatment turned up a protein previously identified as a neuron-specific, high molecular weight MAP, MAP1C (Bloom *et al.,* 1984; Paschal *et al.,* 1987). This protein was then shown to support motility of both vesicles and microtubules on glass, similar to kinesin but in the opposite direction (Paschal *et al.,* 1987; Paschal and Vallee, 1987). Biochemical and ultrastructural analysis of MAP1C established its identity as a cytoplasmic dynein very similar to flagellar dynein (Shpetner *et al.,* 1988; Vallee *et al.,* 1988). The activity of dynein in fast axonal transport has not been as extensively studied as kinesin. The current model for dynein activity, however, is very similar to that of kinesin; dynein associates with vesicles (Hirokawa *et al.,* 1990), which are then able to move along microtubules.

Although the motors for fast axonal transport have been described, the brake, steering, accelerator, and loading/unloading mechanisms all remain mysterious. We know very little about the connections between the motors and the physiology of fast axonal transport. For example, the transport of the smallest observable vesicles (30–50 nm) moves primarily in the anterograde direction at a constant rate, while larger organelles undergo saltatory bidirectional motion (Allen *et al.,* 1982). Do the smallest vesicles interact only with kinesin and, if so, how? For the larger vesicles, what determines their velocity and direction of motion? *In vitro* work indicates that microtubules show saltatory bidirectional movements when gliding on microscope slides treated with both dynein and kinesin (Vale *et al.,* 1992). Thus, saltatory movement is characteristic of structures interacting with both motors (one might imagine having both motors could cause freezing in place). Current investigations into these questions focus on the "usual suspects." Both accessory proteins (Schroer and Sheetz, 1991; Toyoshima *et al.,* 1992) and phosphorylation of kinesin (Lee and Hollenbeck, 1995) have been shown to have effects on motor–membrane interactions, but these studies are at a preliminary stage.

Also poorly understood are the site and mechanism of plasma membrane addition, following transport down the axon. The membrane surface of growing rat sympathetic nerve has been estimated to expand at a rate of about 1 $\mu m^2$/min (Pfenninger and Maylie-Pfenninger, 1981). Most opinion supports the view that new membrane is added primarily at the growing tip. Beads used to mark the surface were found to remain stationary, suggesting that new membrane added at the tip and not along the axonal shaft (Bray, 1970). When lectins are used to label membrane glycoconjugates, new lectin receptors appear first at the periphery of growth cones (Feldman *et al.,* 1981; Pfenninger and Maylie-Pfenninger, 1981). More recently, Craig *et al.* (1995) have shown that newly synthesized protein following herpes simplex viral infection of hippocampal neurons first appeared at the distal tip of the growth cone. In contrast, Griffin *et al.* (1981) found that radiolabeled glycoproteins inserted all along the length of motor axons *in situ* (Griffin *et al.,* 1981). An influential study supports the latter view, showing that bulk lipid insertion occurs all along the axonal process (Popov *et al.,* 1993). Fluorescent lipid analogs were locally incorporated into growing *Xenopus* neurites in culture. The labeled regions were observed to move anterogradely, at a rate proportional to their proximity to the growth cone. The data were consistent with addition of membrane all along the neurite shaft and at the cell body. That is, any new lipid that adds proximal to the labeled region will "push out" the labeled lipids, causing their anterograde movement. The more distal the region of label, the greater the mass of unlabeled lipid that is added proximally; thus the label migration rate correlates with distal position. Of course, such disparate results are difficult to interpret. It may be that there is more than one membrane insertion mechanism: a mechanism for delivery at the tip, e.g., for synaptic vesicles, and another for general membrane addition and subtraction, as discussed below.

Neither the addition nor delivery of axonal membrane appears to be limiting for elongation. As noted earlier, cultured sensory neurons can be induced to elongate at four or five times the fastest physiological rate (i.e., the fastest rate supported by growth cone advance) for several hours by towing with a needle at constant force (Zheng *et al.,* 1991; Lamoureux *et al.,* 1992). Ultrastructural studies of the axon shaft and growth cone indicate a large pool of membranous organelles, much of it resembling endoplasmic reticulum-like elements, within the growth cone and the axonal shaft (Cheng and Reese, 1987; Dailey and Bridgman, 1991). At the growth cone, some of these vesicles have been shown to fuse with the plasma membrane, as shown by the uptake and "shadowing" of extracellular fluid markers in whole-mount, rapidly frozen growth cones (Dailey and Bridgman, 1993). In addition, the growth cone has long been known to support an active process of endocytic uptake (Birks *et al.,* 1972; Wessells *et al.,* 1974; Bunge, 1977). Thus, at the growth cone, and possibly elsewhere along the axon, the neuron must regulate the somewhat frenetic addition and substraction of plasma membrane.

Neither the mechanism for fusion and recycling per se, nor its regulation during growth, is clear. For the potentially special case of synaptic vesicle fusion and recycling, persuasive evidence indicates that this machinery is closely related to the general membrane trafficking pathway for secretion and triggered exocytosis (Bennett and Scheller, 1994; Rothman, 1994; Sudhof, 1995). In view of the polar delivery of some viral proteins by the same pathway used for vesicle sorting (see Section III,B) this could account for the findings of Craig *et al.,* (1995), in which viral proteins are delivered preferentially to the tip. Whether this protein machinery also underlies growth-associated membrane traffic is unclear. A potential model system for studying general membrane addition and subtraction in neurons has been reported by Morris and co-workers (Reuzeau *et al.,* 1995; Wan *et al.,* 1995). Cultured neurons of a freshwater snail were found to be unexpectedly resistant to osmotic shock, surviving in distilled water for an average of more than 1 hr. Significant membrane expansion occurs during the period of osmotic swelling as shown by increases in membrane capacitance. When neurons that had been exposed to hypoosomotic insult were given sudden osmolarity increases, large internal membrane vesicles rapidly formed (<1 min) within the cytoplasm by invagination of regions of the plasma membrane. Similar membrane responses to osmotic insult can also be observed in cultured mammalian neurons (C. Morris, personal communication). The authors postulate that vesicle formation is the result of "overloading" of the normal membrane addition/subtraction machinery in the face of rapid changes in membrane area. Interestingly, the authors also speculate that the membrane response is directly dependent on mechanical tension, consistent with tension effects on axonal elongation rates. It seems implausible that the neuron would possess a membrane addition/subtraction mechanism primarily to accommodate rapid osmotic changes of volume and area. It therefore seems likely that the same mechanism will also prove to be responsible for rapid growth-mediated changes in volume and area. The work from the Morris laboratory appears to be the first easily manipulated experimental system for the analysis of this mechanism.

### B. Slow Axonal Transport and the Cytoskeleton during Elongation

The analyses of fast axonal transport and its microtubule motors have been and are an area of major advance in our understanding of the cytoplasmic mechanisms underlying axonal outgrowth. The basic questions concerning slow axonal transport, however, have resisted clear solutions for more than 20 years despite a great deal of experimental activity. There is little or no agreement on the mechanism of this process; there exists long-standing controversy about the form of the cytoskeletal material being slowly transported; and this, in turn, has greatly confused our view of the cytoskeleton, particularly the microtubule cytoskeleton, during axonal growth. This is clearly an awkward state of affairs in view of the crucial importance of the axonal microtubule bundle to axonal elongation. There is general agreement only that cytoskeletal proteins ultimately come from the cell body, and that slow transport moves cytoskeletal material. Phrased in its most polar form, the slow transport controversy reflects two different models of cytoskeletal dynamics and slow transport. The first model is called the "structural hypothesis" and argues that the cytoskeletal elements of the axon are assembled primarily in the cell body and slow transport reflects the steady movement of assembled microtubules, as well as assembled intermediate filaments, and actin filaments to a lesser degree from the cell body down to the tip of the axon as a cytoskeletal complex (Lasek, 1982, 1986; Lasek *et al.*, 1992). The other model, which has no single recognized name, argues that slow transport reflects the movement of essentially soluble components (e.g., tubulin dimers or oligomeric complexes) down the axon (Ochs, 1982) and that axonal microtubules assemble primarily at the growth cone and then remain stationary (Bamburg, 1987). Although the cytoplasmic events postulated by the two models are not mutually exclusive (both assembly and transport could be occurring simultaneously throughout the neuron, involving many different forms of cytoskeletal elements), the two models make significantly different predictions about the organization and maintenance of the axonal cytoskeletal array. As a result, these two models and their differing subphenomena have served as a touchstone for much of the thinking about the spatial organization of the axonal cytoskeleton for the last 20 years.

Although the notion of a slow movement of cytoplasmic elements from the cell body down the axon is quite old (see Grafstein and Forman, 1980, for early history), the current ideas derive from metabolic labeling studies beginning in the mid-1960s. Several laboratories performed similar experiments in which proteins in the neuronal cell body were essentially pulse-labeled by injection of radiolabeled amino acids into a region of an animal containing a concentration of neuronal cell bodies, either a ganglion or the neural retina. The appropriate nerve fiber was subsequently analyzed along its length at various times for labeled protein (Droz and LeBlond, 1963; Karlson and Sjostrand, 1968; Lasek, 1968; McEwen and Grafstein, 1968; Ferndandez and Davison, 1969; Ochs and Johnson, 1969). In addition to a fast transport component discussed earlier, these experiments showed a slow peak of radioactivity moving down the axon at a rate of about 1 mm/day that accounted for 70–80% of the radioactivity leaving the cell body. Subsequent work showed more than one rate of slow protein movement and slow transport is often divided into a slow component (SCa) traveling at 0.3–1 mm/day and a faster component (SCb) moving at 2–4 mm/day (Willard *et al.,* 1974; Hoffman and Lasek, 1975; Lasek and Hoffman, 1976).

The key findings that led to the structural hypothesis were made by Hoffman and Lasek (1975). They reported that 75% of the slowly migrating protein reflected tubulin and neurofilament polypeptides, both moving at the SCa rate. Strikingly, the radioactivity moved down the nerve fibers as a coherent peak with little spreading and surprisingly little diminution of peak height over period of 3–4 months. Insofar as tubulin was known to be a soluble protein that could reversibly polymerize and depolymerize, these data implied that slow axonal transport was occurring as a stable complex of microtubules and neurofilaments moving down the axon as a coherent unit. That is, the unusually persistent peak of radioactivity represented a pulse-labeled region of a slowly moving, interconnected, population of microtubules and neurofilaments. Following the proposal of this interpretation (Lasek and Hoffman, 1976), a variety of reports provided data that were consistent with the structural hypothesis (Lasek, 1982, 1986). Two types of results appear to have been particularly influential. One was the analysis of a single short segment of axon (1–2 mm) following metabolic labeling that showed a well-defined peak of radioactivity moving through the segment and that the radioactive material was largely polymeric as assessed by differential extraction of the radioactivity (Brady *et al.,* 1984; Tashiro *et al.,* 1984; Filliatreau *et al.,* 1988). Studies of an organic neurotoxin, $\beta,\beta'$-iminodipropionitrile (IDPN), suggested that this substance inhibited the transport of intermediate filaments, causing their accumulation, and decoupled neurofilaments from tubulin, which continued to migrate (Griffin *et al.,* 1978, 1983; Papasozomenos *et al.,* 1981). With the discovery of microtubule-motor proteins in the axon, and the finding that actin moved primarily with the faster SCb component (Black and Lasek, 1980), the structural hypothesis was refined by proposing that slow axonal transport reflected the sliding of polymer and polymer networks relative to one another (Lasek, 1986), thus giving more degrees of freedom for cytoskeletal movement than in the original model.

The structural hypothesis became both highly influential and highly controversial (e.g., Alverez and Torres, 1985; Bamburg, 1987). The basic observation that metabolic labeling of proteins in the neuronal soma produced a sharp, unusually persistent peak of radioactivity was not universally confirmed. Several laboratories reported that the pulse-labeled peak of radioactivity broadened considerably as it moved down the axon, and also left a considerable shoulder of radioactive counts in the wake of the peak (e.g., Stromska and Ochs, 1977, 1981; Cancalon, 1979; Filliatreau and DiGiamberardino, 1982). Indeed, it remains difficult to reconcile the sharp persistent peak of radioactive protein with data on microtubule dynamics in the axon, as is discussed shortly. Also, further analysis of the movement of cytoskeletal material was not as consistent with stable cytoskeletal complexes as initially envisioned, e.g., tubulin was observed to move at both slow transport rates in some neurons, and MAPs did not always migrate with tubulin polypeptides (Nixon *et al.,* 1990).

However, much of the current interest in slow axonal transport has less to do with the state of transported elements (polymer/soluble unit) than with one's view of the mechanisms that underlie the organization of the axonal cytoskeleton. The structural hypothesis argues that the state of the cytoskeleton at a given time is due primarily to assembly events in the cell body and then to subsequent reorganization of extant filaments via transport throughout the axon (Lasek, 1982, 1986). An alternative view is that the axonal cytoskeleton, particularly the microtubules, are the result of polymerization/depolymerization remodeling of the cytoskeleton locally within the axon, and particularly at the growth cone during elongation (Bamburg, 1987). Indeed, in contrast to the emphasis of the structural hypothesis on microtubule assembly in the cell body, local application of antimicrotubule drugs to the cell body of cultured neurons had little effect on axonal elongation while drug application to the growth cone caused a rapid cessation of growth (Bamburg *et al.,* 1986). Extensive "local control" of axonal elongation was also shown in neurites severed from the cell body, in which the distal neurite fragment collapsed but then reorganized and reassembled microtubules to produce new neurite outgrowth without a cell body (Baas *et al.,* 1987). In a similar vein, the cytoskeletal architecture of mature neurons—in which many distal branches are supplied from a single "stem" branch arising from the cell body—predicts that axonal cytoskeletal filaments whose existence and position is due primarily to assembly in the cell body and subsequent transport must either be parceled out among the branches from the stem or change their velocity of transport as the moving column changes diameter, as in the cardiovascular system. But there is agreement that slow axonal transport occurs at a widely reproducible, steady rate, if nothing else. Also, the total number of microtubules in the terminal branches of single motor nerve fiber far surpasses the number of microtubules in the "stem" branch of the neurons (Zenker and Hohberg, 1973), and a similar situation can be demonstrated in cultured neurons (S.

Heidemann, unpublished observations). More recently, the number and length of microtubules in the minor processes of hippocampal neurons were compared with the microtubules of axons developing from the minor processes (Yu and Baas, 1994). Axons had about 10-fold the number of microtubules, the result of new recruitment by transport, or by new assembly nucleated by fragmented microtubules from the minor process.

Not surprisingly, in view of the history of dichotomous squabbles in biology, current opinion favors compromise mechanisms on all points. A report by Nixon and Logvinekno (1986) has been particularly influential in arguing the view that the cytoskeleton consists of both moving and stationary phases. These investigators analyzed a small region of nerve fiber for the disappearance of labeled neurofilament peptides after the passage of the SCa peak. They found that these labeled peptides were lost very slowly, suggesting an essentially stationary phase of neurofilaments. Subsequently, the notion of stationary and moving phases of cytoskeleton was extended to microtubules (Komiya and Tashiro, 1988; Tashiro and Komiya, 1989, 1991; Watson *et al.,* 1990). Similarly, the proposal of Joshi and Baas (1993; see Section III,A) explains the organization of axonal microtubules on the combined action of transport of existing filaments to serve as nuclei for any desired degree of local remodeling via assembly/transport. Indeed, as we noted in the earlier summary of the model, this mechanism is attractive precisely because one can plausibly imagine wide variation in the contributions from transport and assembly under different conditions. The degree of cytoskeletal filament transport required for this latter model has been observed directly. Fluorescent microtubules were observed to translocate in real time within *Xenopus* growth cones (Reinsch *et al.,* 1991; Tanaka and Kirschner, 1995; Tanaka *et al.,* 1995). Also, experiments in which microtubule drugs were used to poison assembly but permit the relocation of existing microtubules suggest that at least small numbers of microtubules can move from the cell body into the incipient axonal shaft (Baas and Ahmad, 1993; Zheng *et al.,* 1993; Smith, 1994b; Ahmad and Baas, 1995).

Despite this emerging consensus, it is probable that the controversy concerning slow axonal transport, its basic nature, and its function in cytoskeletal organization is likely to continue unabated into the forseeable future. For one thing, the current compromises do not address fundamental questions. All of the proposals of a motive force underlying slow axonal transport are rather speculative, whether involving motor proteins (Vallee and Bloom, 1991) like that responsible for fast transport but including chromotographic-like exchange of soluble transported proteins with the stationary cytoskeletal filament (Ochs, 1982), or, heretically, local protein synthesis in the axon and diffusion (Alverez and Torres, 1985). Investigations into this fundamental issue will keep controversy high.

Also, there remains serious interest in testing a basic idea of the structural hypothesis; bulk flow of the microtubule cytoskeleton supporting axonal elongation. As mentioned earlier, there have been many reports of basically similar immunocytological experiments aimed at detecting net translocation of the axonal microtubule bundles, which have produced a confusing variety of results. The basic idea is to inject into neurons tubulin that has been tagged, either with a conventional fluorophore (Keith, 1987; Lim *et al.,* 1989, 1990; Okabe and Hirokawa, 1990; Takeda *et al.,* 1995) or a photoactivated fluorophore (Reinsch *et al.,* 1991; Okabe and Hirokawa, 1992; Sabry *et al.,* 1995). This tubulin incorporates into axonal microtubules, and a region of the axonal microtubule bundle is marked by bleaching or activation. All groups agree that the axonal microtubules turnover their tubulin in the microtubule of the lattice with a half-life on the scale of 1 hr. This time scale for axonal tubulin dynamics is difficult to reconcile with the exceptional microtubule stability envisioned by the structural hypothesis interpretation of metabolic labeling studies. Instead, the fluorescence turnover data may indicate that the moving peak of polymer-associated radioactivity reflects a "chromotographic band" of labeled tubulin migrating down the axon, assembling and disassembling with microtubules as it goes (Stewart *et al.,* 1982). Nevertheless, three of eight communications report that marked bands of microtubules are observed to move toward the growth cone in elongating axons (Keith, 1987; Reinsch *et al.,* 1991; Okabe and Hirokawa, 1992), while six of eight report stationary microtubule bands in both cultured neurons (Lim *et al.,* 1989, 1990; Okabe and Hirokawa, 1990, 1992) and *in situ* (Sabry *et al.,* 1995; Takeda *et al.,* 1995). The number of reports seems inconsistent because Okabe and Hirokawa (1992) reproduced two previous, disparate results and found that microtubule bands were stationary in slowly growing axons but moved in rapidly growing axons. Interestingly, the rate of the moving bands in rapidly growing *Xenopus* axons was proportional to its distal position (Reinsch *et al.,* 1991; Okabe and Hirokawa, 1992). As with the incorporation of fluorescent lipid markers into axonal membrane discussed in the previous section (Popov *et al.,* 1993), this changing rate of the moving band suggests that the distal movement is due to proximal addition of tubulin in the bundle (Hirokawa and Okabe, 1992). It is likely that more fluorescence-labeling experiments of this sort will be reported in future.

Finally, in view of the agreement that some transport of filaments does occur in axons, it is of interest to determine the relative roles of microtubule transport (i.e., rearrangements of existing cytoskeleton), and new microtubule assembly in supporting axonal outgrowth. At this point, the reader should not be surprised to learn that reports using an apparently similar method produce conflicting results. The basic method is to use microtubule drugs [generally vinblastine, based on the elegant work of Jordan *et al.*

(1990, 1992)] such that existing microtubules are not depolymerized, but no new assembly occurs. This should permit functions dependent on reorganization of existing microtubules to occur, but disallow those events dependent on microtubule dynamics. One group (Baas and Ahmad, 1993; Ahmad and Baas, 1995) reports that cultured rat sympathetic neurons remain capable of extensive elongation in the presence of nanomolar concentrations of vinblastine. Also, the microtubules within such axons were uniformly polar as in control axon, the plus end distal to the cell body, indicating that microtubule transport is also uniformly polar into axons. However, three other groups have found that poisoning microtubule assembly without depolymerizing existing microtubules fails to support axonal elongation to any extent (Zheng *et al.,* 1993; Smith, 1994b; Tanaka *et al.,* 1995). In the most direct experiment, Tanaka *et al.* (1995) poisoned microtubule assembly with nanomolar vinblastine and imaged individual microtubules for both assembly and transport in cultured *Xenopus* neurons, while measuring growth cone motility and axonal elongation of the same cell. The poisoning of microtubule assembly was directly observable in these experiments as a suppression of the dynamic instability of microtubule assembly/disassembly in the growth cone, although movements of existing microtubules within the growth cone were observed to continue. Under these conditions, the growth cone actively explored its local surroundings, like a dog pulling on a short leash, but no axonal elongation was observed. Both Zheng *et al.* (1993) and Smith (1994b) found that poisoning the microtubule assembly without depolymerizing existing microtubules allowed some axonal initiation to occur, but that further elongation was essentially absent. These discrepant data are certain to be pursued further.

It is likely that the basic scenario of the structural hypothesis—the movement of "cytoskeletal elements relentlessly from the cell body to the axon tip" (Lasek *et al.,* 1992)—is not a usefully accurate description of the state of the axonal cytoskeleton, particularly the microtubule bundle. Given the many plot reversals in the story thus far, however, it would not be surprising were this view to change in future. Either way, it is difficult to escape the conclusion that the "tidy story" of the structural hypothesis has been extraordinarily useful in focusing attention on a wide variety of crucial issues concerning axonal cytoskeletal organization. Investigations were and will continue to be stimulated by the structural hypothesis into the cytoskeletal forms of transport, the interplay of transport and new cytoskeletal assembly in axonal elongation, the sites of new assembly, and the relationships among different kinetics of transport. Although it is disappointing that no single mechanism has arisen that satisfyingly accounts for all the data, this is surely due to the complexity of the phenomena, not to obfuscation by this or that proposal.

## VI. Concluding Remarks

Indeed, it is the complexity of the task of specifying neurogenetic shape that is ultimately so striking. As has been seen, even fundamental aspects of the cytoplasmic mechanisms of neurogenesis remain poorly understood. Although the growth cone has been observed and experimented on for decades, it remains unclear how it actually moves forward, and only with much painstaking work was a general duty cycle of behavior distilled from the complexity of its motion. The controversy over slow axonal transport should imbue scientists with a sense of humility. Of course, these answers may just await a critical finding or collaboration, as the visualization of fast axonal transport in isolated squid axoplasm held the key to an avalanche of new knowledge about microtubule motors. But there remain two broad areas that have hardly been explored. The necessary connection between extrinsic navigational signals and intrinsic steering mechanisms has already been mentioned. The existence of this level of integration is well recognized and is being vigorously pursued by new imaging techniques and better biological systems, such as the grasshopper limb system, in which access to natural navigational signals still permits high-resolution study of the cytoplasm. Although I'm biased by my own work on mechanical aspects of neurite outgrowth, I see another broad area of integration that will be needed for a satisfying understanding of the cytoplasmic mechanism of neurite outgrowth: how mechanical force drives the chemical assembly reactions of axonal elongation, as it apparently does. I find it astonishing that the "storm and fury" of cytoplasmic events—three ubiquitous and partly interconnected cytoskeletal systems moving, assembling, and disassembling with membrane coming and going, possibly everywhere—should be able to respond to tension with the overall simplicity of the elongation of a Newtonian fluid-mechanical element. Not only is our basic understanding of the connections between mechanical forces and biochemistry rudimentary, but also the cytoplasmic integration reflected by axonal, or bone, or muscle remodeling in response to force goes largely unremarked. The imaginary audience addressed at the start of this article, the beginning research worker, has no shortage of interesting questions to ask, or of important contributions to make.

## Acknowledgments

I thank my many colleagues for kindly sending me reprints and preprints of their work. Even the work that I did not cite (one has to stop somewhere!) was invaluable for adjusting my perspective and allowing me to see this field in its rich complexity. I am particularly grateful to Harish Joshi and Peter Baas for reading and commenting on the manuscript.

## References


Ahmad, F. J., and Baas, P. W. (1995). Microtubules released from the neuronal centrosome are transported into the axon. *J. Cell Sci.* **108,** 2761–2769.

Ahmad, F. J., Joshi, H. C., Centonze, V. E., and Baas, P. W. (1994). Inhibition of microtubule nucleation at the neuronal centrosome compromises axon growth. *Neuron* **12,** 271–280.

Aletta, J. M., and Greene, L. A. (1988). Growth cone configuration and advance: A time-lapse study using video-enchanced differential interference contrast microscopy. *J. Neurosci.* **8,** 1425–1435.

Allen, R. D., Allen, N. S., and Travis, J. L. (1981). Video-enhanced contrast, differential interference contrast (AVEC-DIC) microscopy: A new method capable of analyzing microtubule related movement in the reticulopodial network of *Allogromia laticollaris. Cell Motil.* **1,** 291–302.

Allen, R. D., Metuzalis, J., Tasaki, I., Brady, S. T., and Gilbert, S. P. (1982). Fast axonal transport in squid giant axons. *Science* **218,** 1127–1129.

Allen, R. D., Weiss, D. G., Hayden, J. H., Brown, D. T., Fujiwake, H., and Simpson, M. (1985). Gliding movement of and bidirectional transport along single native microtubules from squid axoplasm: Evidence for an active role of microtubules in cytoplasmic transport. *J. Cell Biol.* **100,** 1736–1752.

Alvarez, J., and Torres, J. C. (1985). Slow axoplasmic transport: A fiction? *J. Theor. Biol.* **112,** 627–651.

Argiro, V., Bunge, M. B., and Johnson, M. I. (1984). Correlation between growth form and movement and their dependence on neuronal age. *J. Neurosci.* **4,** 3051–3062.

Baas, P. W., and Ahmad, F. J. (1993). The transport properties of axonal microtubules establish their polarity orientation. *J. Cell Biol.* **120,** 1427–1437.

Baas, P. W., Black, M. M., and Banker, G. A. (1989). Changes in microtubule polarity orientation during the development of hippocampal neurons in culture. *J. Cell Biol.* **109,** 3085–3094.

Baas, P. W., Deitch, J. S., Black, M. M., and Banker, G. A. (1988). Polarity orientation of microtubules in hippocampal neurons: Uniformity in the axon and nonuniformity in the dendrite. *Proc. Natl. Acad. Sci. U.S.A.* **85,** 8335–8339.

Baas, P. W., and Heidemann, S. R. (1986). Microtubule reassembly from nucleating fragments during regrowth of amputated neurites. *J. Cell Biol.* **103,** 917–927.

Baas, P. W., and Joshi, H. C. (1992). Gamma-tubulin distribution in the neuron: Implications for the origins of neuritic microtubules. *J. Cell Biol.* **119,** 171–178.

Baas, P. W., Pienkonski, T. P., Cimbalnik, K. A., Toyama, K., Bakalis, S., Ahmad, F. J., and Kosik, K. S. (1994). Tau confers drug-stability but not cold stability to microtubules in living cells. *J. Cell Sci.* **107,** 135–143.

Baas, P. W., Pienkonski, T. P., and Kosik, K. S. (1991). Processes induced by tau expression in Sf9 cells have an axon-like microtubule organization. *J. Cell Biol.* **115,** 1333–1344.

Baas, P. W., Sinclair, G. I., and Heidemann, S. R. (1987). Role of microtubules in the cytoplasmic compartmentation of neurons. *Brain Res.* **420,** 73–81.

Bamburg, J. R. (1987). The axonal cytoskeleton: Stationary or moving matrix? *Trends Neurosci.* **11,** 248–249.

Bamburg, J. R., Bray, D., and Chapman, K. (1986). Assembly of microtubules at the tips of growing axons. *Nature (London)* **321,** 788–800.

Banker, G. A., and Cowan, W. M. (1979). Further observations on hippocampal neurons in dispersed cell culture. *J. Comp. Neurol.* **187,** 469–494.

Bartlett, W. P., and Banker, G. A. (1984). An electron microscopic study of the development of axons and dendrites by hippocampal neurons in culture. I. Cells which develop without intercellular contacts. *J. Neurosci.* **4,** 1944–1953.

Bassell, G. J., Singer, R. H., and Kosik, K. S. (1994). Association of poly(A) mRNA with microtubules in cultured neurons. *Neuron* **12,** 571–582.

Bastiani, M. J., and Goodman, C. S. (1984). The first growth cones in the central nervous system of the grasshopper embryo. *In* "Cellular and Molecular Biology of Neuronal Development" (I. B. Black, ed.), pp. 63–84. Plenum, New York.

Bennett, M. K., and Scheller, R. H. (1994). A molecular description of synaptic vesicle membrane trafficking. *Annu. Rev. Biochem.* **63,** 63–100.

Bentley, D., and O'Connor, T. P. (1992). Guidance and steering of peripheral pioneer growth cones in grasshopper embryos. *In* "The Nerve Growth Cone" (P. C. Letourneau, S. B. Kater, and E. R. Macagno, eds.), pp. 265–282. Raven, New York.

Bentley, D., and O'Connor, T. P. (1994). Cytoskeletal events in growth cone steering. *Curr. Opin. Neurobiol.* **4,** 43–48.

Bentley, D., and Toroian-Raymond, A. (1986). Disoriented pathfinding by pioneer neuron growth cones deprived of filopodia by cytochalasin treatment. *Nature* (*London*) **323,** 712–715.

Bernhardt, R., Huber, G., and Matus, A. (1985). Differences in the developmental patterns of 3 microtubule-associated proteins in the rat cerebellum. *J. Neurosci.* **5,** 977–991.

Binder, L. I., Frankfurter, A., Kim, H., Caceres, A., Payne, M. R., and Rebhun, L. I. (1984). Heterogeneity of microtubule-associated protein 2 during rat brain development. *Proc. Natl. Acad. Sci. U.S.A.* **81,** 5613–5617.

Birks, R. I., Mackey, M. C., and Weldon, P. R. (1972). Organelle formation from pinocytotic elements in neurites of cultured sympathetic ganglia. *J. Neurocytol.* **1,** 311–340.

Black, M. M., Aletta, J. M., and Greene, L. A. (1986). Regulation of microtubule composition and stability during nerve growth factor-promoted neurite outgrowth. *J. Cell Biol.* **103,** 545–557.

Black, M. M., and Baas, P. W. (1989). The basis of polarity in neurons. *Trends Neurosci.* **12,** 211–214.

Black, M. M., and Greene, L. A. (1982). Changes in colchicine susceptibility of microtubules associated with neurite outgrowth: Studies with nerve growth factor-responsive PC12 pheochromocytoma cells. *J. Cell Biol.* **95,** 379–386.

Black, M. M., and Lasek, R. J. (1980). Slow components of axonal transport: Two cytoskeletal networks. *J. Cell Biol.* **86,** 616–623.

Bloom, G. S., Schoenfeld, T. A., and Vallee, R. B. (1984). Widespread distribution of the major polypeptide component of MAP1 (microtubule-associated protein 1) in the nervous system. *J. Cell Biol.* **98,** 320–330.

Bloom, G. S., Wagner, M. C., Pfister, K. K., and Brady, S. T. (1988). Native structure and physical properties of bovine brain kinesin and identification of the ATP-binding subunit polypeptide. *Biochemistry* **27,** 3409–3416.

Bomsel, M., Parton, R., Kuznetsov, S. A., Schroer, T. A., and Gruenberg, J. (1990). Microtubule- and motor-dependent fusion *in vitro* between apical and basolateral endocytic vesicles from MDCK cells. *Cell* **62,** 719–731.

Bovolenta, P., and Mason, C. (1987). Growth cone morphology varies with position in the developing mouse visual pathway from retina to first targets. *J. Neurosci.* **7**(5), 1447–1460.

Brady, S. T. (1985). A novel brain ATPase with properties expected for the fast axonal transport motor. *Nature* (*London*) **317,** 73–75.

Brady, S. T. (1991). Molecular motors in the nervous system. *Neuron* **7,** 521–533.

Brady, S. T., Lasek, R. J., and Allen, R. D. (1982). Fast axonal transport in extruded axoplasm from squid giant axon. *Science* **218,** 1129–1131.

Brady, S. T., Pfister, K. K., and Bloom, G. S. (1990). A monoclonal antibody to the heavy chain of kinesin inhibits anterograde and retrograde axonal transport in isolated squid axoplasm. *Proc. Natl. Acad. Sci. U.S.A.* **87,** 1061–1065.

Brady, S. T., Tytell, M., and Lasek, R. J. (1984). Axonal transport and axonal tubulin: Biochemical evidence for cold-stability. *J. Cell Biol.* **99,** 1716–1724.

Bramblett, G. T., Goedert, M., Jakes, R., Merrick, S. E., Trojanowski, J. Q., and Lee, V. M.-Y. (1993). Abnormal tau phosphorylation at Ser396 in Alzheimer's disease recapitulates development and contributes to reduced microtubule binding. *Neuron* **10,** 1089–1099.

Bray, D. (1970). Surface movements during the growth of single explanted neurons. *Proc. Natl. Acad. Sci. U.S.A.* **65,** 905–910.

Bray, D. (1982). Filopodial contraction and growth cone guidance. *In* "Cell Behavior" (R. Bellair, A. Curtis, and G. Dunn, eds.), pp. 299–317. Cambridge University Press, Cambridge.

Bray, D. (1984). Axonal growth in response to experimentally applied tension. *Dev. Biol.* **102,** 379–389.

Bray, D. (1991). Cytoskeletal basis of nerve axon growth. *In* "The Nerve Growth Cone" (P. C. Letourneau, S. B. Kater, and E. R. Macagno, eds.), pp. 7–17. Raven, New York.

Bray, D., and Bunge, M. B. (1981). Serial analysis of microtubules in cultured rat sensory axons. *J. Neurocytol.* **10,** 589–605.

Bray, D., and Chapman, K. (1985). Analysis of microspike movements on the neuronal growth cone. *J. Neurosci.* **5,** 3204–3213.

Bray, D., Heath, J., and Moss, D. (1986). The membrane-associated cortex of animal cells: Its structure and mechanical properties. *J. Cell Sci. Suppl.* **4,** 71–88.

Bray, J. J., and Mills, R. G. (1991). Transport complexes associated with slow axonal flow. *Neurochem. Res.* **16,** 645–649.

Bray, D., and Vasiliev, J. (1989). Networks from mutants. *Nature (London)* **338,** 203–204.

Bray, D., and White, J. G. (1988). Cortical flow in animal cells. *Science* **239,** 883–888.

Bridgman, P. C., and Dailey, M. E. (1989). The organization of myosin and actin in rapid frozen nerve growth cones. *J. Cell Biol.* **108,** 95–109.

Bruckenstein, D. A., Lein, P. J., Higgins, D., and Fremeau, R. T., Jr. (1990). Distinct spatial localization of specific mRNAs in cultured sympathetic neurons. *Neuron* **5,** 809–819.

Buettner, H. M., and Pittman, R. N. (1991). Quantitative effects of laminin concentration on neurite outgrowth *in vitro. Dev. Biol.* **145,** 266–276.

Bunge, M. B. (1977). Initial endocytosis of peroxidase or ferritin by growth cones of cultured nerve cells. *J. Neurocytol.* **6,** 407–439.

Burgin, K. E., Ludin, B., Ferralli, J., and Matus, A. (1994). Binding of microtubules in transfected cells does not involve an autonomous dimerization site on the MAP2 molecule. *Mol. Biol. Cell* **5,** 511–517.

Burgoyne, R. D. (1991). "The Neuronal Cytoskeleton." Wiley-Liss, New York.

Burgoyne, R. D., and Cumming, R. (1984). Ontogeny of microtubule-associated protein 2 in rat cerebellum: Differential expression of the doublet polypeptides. *Neuroscience* **11,** 156–167.

Burmeister, D. W., Rivas, R. J., and Goldberg, D. J. (1991). Substrate-bound factors stimulate engorgement of growth cone lamellipodia during neurite elongation. *Cell Motil. Cytoskeleton* **19,** 255–268.

Burton, P. R. (1988). Dendrites of mitral cell neurons contain microtubules of opposite polarity. *Brain Res.* **473,** 107–115.

Butner, K. A., and Kirschner, M. W. (1991). Tau protein binds to microtubules through a flexible array of distributed weak sites. *J. Cell Biol.* **115,** 717–730.

Buxbaum, R. E. (1995). Biological levels. *Nature (London)* **373,** 567–568.

Buxbaum, R. E., and Heidemann, S. R. (1988). A thermodynamic model for force integration and microtubule assembly during axonal elongation. *J. Theor. Biol.* **134,** 379–390.

Buxbaum, R. E., and Heidemann, S. R. (1992). An absolute rate theory model for tension control of axonal elongation. *J. Theor. Biol.* **155,** 409–426.

Caceres, A., Binder, L. I., Payne, M. R., Bender, P., Rebhun, L., and Steward, O. (1984). Differential subcellular localization of tubulin and the microtubule-associated protein MAP2

in brain tissue as revealed by immunohistochemistry with monoclonal hybridoma antibodies. *J. Neurosci.* **4,** 394–410.
Caceres, A., and Kosik, K. (1990). Inhibition of neurite polarity by tau antisense oligonucleotides in primary cerebellar neurones. *Nature* (*London*) **343,** 461–463.
Caceres, A., Mautino, J., and Kosik, K. S. (1992). Suppression of MAP2 in cultured cerebellar macroneurons inhibits minor neurite formation. *Neuron* **9,** 607–618.
Caceres, A., Potrebic, S., and Kosik, K. S. (1991). The effect of tau antisense oligonucleotides on neurite formation of cultured cerebellar macroneurons. *J. Neurosci.* **11,** 1515–1523.
Calof, A. L., and Chikaraishi, D. M. (1989). Analysis of neurogenesis in a mammalian neuroepithelium: Proliferation and differentiation of an olfactory neuron precursor *in vitro. Neuron* **3,** 115–127.
Calof, A. L., and Lander, A. D. (1991). Relationship between neuronal migration and cell–substratum adhesion: Laminin and merosin promote olfactory neuronal migration but are anti-adhesive. *J. Cell Biol.* **115,** 779–794.
Cancalon, P. (1979). Influence of temperature on the velocity and on the isotope profile of slowly transported labeled proteins. *J. Neurochem.* **32,** 997–1007.
Chen, J., Kanai, Y., Cowan, N. J., and Hirokawa, N. (1992). Projection domains of MAP2 and tau determine spacings between microtubules in dendrites and axons. *Nature* (*London*) **360,** 674–677.
Cheng, T. P. O., and Reese, T. S. (1987). Recycling of plasmalemma in chick tectal growth cones. *J. Neurosci.* **7,** 1752–1759.
Chien, C.-B., Rosenthal, D. E., Harris, W. A., and Holt, C. E. (1993). Navigational errors made by growth cones without filopodia in the embryonic *Xenopus* brain. *Neuron* **11,** 237–251.
Cleveland, D. W., Hwo, S.-Y., and Kirschner, M. W. (1977). Physical and chemical properties of purified tau factor and the role of tau in microtubule assembly. *J. Mol. Biol.* **116,** 227–248.
Collins, F. (1978). Induction of neurite outgrowth by a conditioned-medium factor bound to the culture substratum. *Proc. Natl. Acad. Sci. U.S.A.* **75,** 5210–5213.
Cote, F., Collard, J., and Julien, J. (1993). Progressive neuronopathy in transgenic mice expressing the human neurofilament heavy gene: A mouse model of amyotrophic lateral sclerosis. *Cell* **73,** 35–46.
Couchie, D., Faivre-Bauman, A., Puymirat, J., Guilleminot, J., Tixier-Vidal, A., and Nunez, J. (1986). Expression of microtubule-associated proteins during the early stages of neurite extension by brain neurons cultured in a defined medium. *J. Neurochem.* **47,** 1255–1261.
Craig, A. M., and Banker, G. (1994). Neuronal polarity. *Annu. Rev. Neurosci.* **17,** 267–310.
Craig, A. M., Blackstone, C. D., Huganir, R. L., and Banker, G. (1993). The distribution of glutamate receptors in cultured rat hippocampal neurons: Postsynaptic clustering of AMPA-selective subunits. *Neuron* **10,** 1055–1068.
Craig, A. M., Wyborski, R. J., and Banker, G. (1995). Preferential addition of newly synthesized membrane protein at axonal growth cones. *Nature* (*London*) **375,** 592–594.
Dahl, D., and Bignami, A. (1986). Neurofilament phosphorylation in development—a sign of axonal maturation? *Exp. Cell Res.* **162,** 220–230.
Dailey, M. E., and Bridgman, P. C. (1991). Structure and organization of membrane organelles along distal microtubule segments in growth cones. *J. Neurosci. Res.* **30,** 242–258.
Dailey, M. E., and Bridgman, P. C. (1993). Vacuole dynamics in growth cones: Correlated EM and video observations. *J. Neurosci.* **13,** 3375–3393.
Davenport, R. W., Dou, P., Rehder, V., and Kater, S. B. (1993). A sensory role for neuronal growth cone filopodia. *Nature* (*London*) **361,** 721–723.
Davis, L., Burger, B., Banker, G., and Steward, O. (1990). Dendritic transport: Quantitative analysis of the time course of somatodendritic transport of recently synthesized RNA. *J. Neurosci.* **10,** 3056–3068.
DeCamilli, P., Miller, P. E., Navone, F., Theurkauf, W. F., and Vallee, R. B. (1984). Distribution of microtubule-associated protein 2 in the nervous system of the rat studied by immunofluorescence. *Neuroscience* **11,** 817–846.

Deitch, J. S., and Banker, G. A. (1993). An electron microscopic analysis of hippocampal neurons developing in culture: Early stages in the emergence of polarity. *J. Neurosci.* **13,** 4301–4315.

Dennerll, T. J., Lamoureux, P., Buxbaum, R. E., and Heidemann, S. R. (1989). The cytomechanics of axonal elongation and retraction. *J. Cell Biol.* **109,** 3073–3083.

Dinsmore, J., and Solomon, F. (1991). Inhibition of MAP-2 expression affects both morphological and cell division phenotypes of neuronal differentiation. *Cell* **64,** 817–826.

Doherty, P., and Walsh, F. S. (1994). Signal transduction events underlying neurite outgrowth stimulated by cell adhesion molecules. *Curr. Opin. Neurobiol.* **4,** 49–55.

Donahue, S. P., Wood, J. G., and English, A. W. (1988). On the role of the 200-kDa neurofilament protein at the developing neuromuscular junction. *Dev. Biol.* **130,** 154–166.

Dotti, C. G., and Banker, G. A. (1987). Experimentally induced alternation in the polarity of developing neurons. *Nature (London)* **330,** 254–256.

Dotti, C. G., Parton, R. G., and Simons, K. (1991). Polarized sorting of glypiated proteins in hippocampal neurons. *Nature (London)* **349,** 158–161.

Dotti, C. G., and Simons, K. (1990). Polarized sorting of viral glycoproteins to the axon and dendrites of hippocampal neurons in culture. *Cell* **62,** 63–72.

Dotti, C. G., Sullivan, C. A., and Banker, G. A. (1988). The establishment of polarity by hippocampal neurons in culture. *J. Neurosci.* **8,** 1454–1468.

Droz, B., and Leblond, C. P. (1963). Axonal migration of proteins in the central nervous system and peripheral nerves as shown by radioautography. *J. Comp. Neurol.* **121,** 325–346.

Drubin, D. G., Feinstein, S. C., Shooter, E. M., and Kirschner, M. W. (1985). Nerve growth factor-induced neurite outgrowth in PC 12 cells involves the coordinate induction of microtubule assembly and assembly-promoting factors. *J. Cell Biol.* **101,** 1799–1807.

Drubin, D. G., and Kirschner, M. W. (1986). Tau protein function in living cells. *J. Cell Biol.* **103,** 2739–2746.

Dye, R. B., Fink, S. P., and Williams, R. C., Jr. (1993). Taxol-induced flexibility of microtubules and its reversal by MAP-2 and tau. *J. Biol. Chem.* **268,** 6847–6850.

Eilers, U., Klumperman, J., and Hauri, H.-P. (1989). Nocodazole, a microtubule-active drug, interferes with apical protein delivery in cultured intestinal epithelial cells (Caco-2). *J. Cell Biol.* **108,** 13–22.

Elson, E. L. (1988). Cellular mechanics as an indicator of cytoskeletal structure and function. *Annu. Rev. Biophys. Biophys. Chem.* **17,** 397–430.

Eyer, J., and Peterson, A. (1994). Neurofilament-deficient axons and perikaryal aggregates in viable transgenic mice expressing a neurofilament-$\beta$-galactosidase fusion protein. *Neuron* **12,** 389–405.

Fath, K. R., and Lasek, R. J. (1988). Two classes of actin microfilaments are associated with the inner cytoskeleton of axons. *J. Cell Biol.* **107,** 613–621.

Feldman, E. L., Axelrod, D., Schwartz, M., Heacock, A. M., and Agranoff, B. W. (1981). Studies on the localization of newly added membrane in growing neurites. *J. Neurobiol.* **12,** 591–598.

Fernandez, H. L., Burton, P. R., and Samson, F. E. (1971). Axoplasmic transport in the crayfish nerve cord: The role of fibrillar constituents of neurons. *J. Cell Biol.* **51,** 176–192.

Fernandez, H. L., and Davison, P. F. (1969). Axoplasmic transport in the crayfish nerve cord. *Proc. Natl. Acad. Sci. U.S.A.* **64,** 512.

Ferralli, J., Doll, T., and Matus, A. (1994). Sequence analysis of MAP2 function in living cells. *J. Cell Sci.* **107,** 3115–3125.

Ferreira, A., Niclas, J., Vale, R. D., Banker, G., and Kosik, K. S. (1992). Suppression of kinesin expression in cultured hippocampal neurons using antisense oligonucleotides. *J. Cell Biol.* **117,** 595–606.

Filliatreau, G., Denoulet, P., de Nechaud, B., and Di Giamberardino, L. (1988). Stable and metastable cytoskeletal polymers carried by slow axonal transport. *J. Neurosci.* **8,** 2227–2233.

Filliatreau, G., and Di Giamberardino, L. (1982). Quantitative analysis of axonal transport of cytoskeletal proteins in chicken oculomotor nerve. *J. Neurochem.* **39,** 1033.

Forman, D. S. (1982). Vanadate inhibits saltatory organelle movement in a permeabilized cell model. *Exp. Cell Res.* **141,** 139–147.

Forman, D. S., Brown, K. J., and Livengood, D. R. (1983). Fast axonal transport in permeabilized lobster giant axons is inhibited by vanadate. *J. Neurosci.* **3,** 1279–1288.

Forscher, P., Lin, C. H., and Thompson, C. (1992). Novel form of growth cone motility involving site-directed actin filament assembly. *Nature* (*London*) **357,** 515–518.

Forscher, P., and Smith, S. J. (1988). Actions of cytochalasins on the organizaton of actin filaments and microtubules in a neuronal growth cone. *J. Cell Biol.* **107,** 1505–1516.

Francon, J., Fellous, A., Lennon, A. M., and Nunez, J. (1978). Requirement for "factors" for tubulin assembly during brain development. *Eur. J. Biochem.* **85,** 43–53.

Francon, J., Lennon, A. M., Fellous, A., Mareck, A., Pierre, M., and Nunez, J. (1982). Heterogeneity of microtubule-associated proteins and brain development. *Eur. J. Biochem.* **129,** 465–471.

Gibbons, I. R., Cosson, M. P., Evans, J. A., Gibbons, B. H., Houck, B., *et al.* (1978). Potent inhibition of dynein adenosinetriphosphatase and of the motility of cilia and sperm flagella by vanadate. *Proc. Natl. Acad. Sci. U.S.A.* **75,** 2220–2224.

Gilbert, T. G., Le Bivic, A., Quaroni, A., and Rodriguez-Boulan, E. (1991). Microtubular organization and its involvement in the biogenetic pathways of plasma membrane proteins in Caco-2 intestinal epithelial cells. *J. Cell Biol.* **113,** 275–288.

Gittes, F., Mickey, B., Nettleton, J., and Howard, J. (1993). Flexural rigidity of microtubules and actin filaments measured from thermal fluctuations in shape. *J. Cell Biol.* **120,** 923–934.

Glicksman, M. A., and Willard, M. (1985). Differential expression of the three neurofilament polypeptides. *Ann. N.Y. Acad. Sci.* **455,** 479–491.

Goedert, M., Spillantini, M. G., Cairns, N. J., and Crowther, R. A. (1992). Tau proteins of Alzheimer paired helical filaments: Abnormal phosphorylation of all six brain isoforms. *Neuron* **8,** 159–168.

Goedert, M., Spillantini, M. G., Jakes, R., Rutherford, D., and Crowther, R. A. (1989). Multiple isoforms of human microtubule-associated protein tau: Sequences and localization in neurofibrillary tangles of Alzheimer's disease. *Neuron* **3,** 519–526.

Goldberg, D. J., and Burmeister, D. W. (1986). Stages in axon formation: Observation of growth of *Aplysia* axons in culture using video-enhanced contrast-differential interference contrast microscopy. *J. Cell Biol.* **103,** 1921–1931.

Goldstein, L. S. B. (1993). With apologies to Scheherazade: Tails of 1001 kinesin motors. *Annu. Rev. Genet.* **27,** 319–351.

Goslin, K., and Banker, G. (1990). Rapid changes in the distribution of GAP-43 correlate with the expression of neuronal polarity during normal development and under experimental conditions. *J. Cell Biol* **110,** 1319–1331.

Grafstein, B., and Forman, D. S. (1980). Intracellular transport in neurons. *Physiol. Rev.* **60,** 1167–1283.

Griffin, J. W., Fahnestock, K. E., Price, D. L., and Hoffman, P. N. (1983). Microtubule-neurofilament segregation produced by $\beta,\beta'$-iminodipropionitrile: Evidence for the association of fast axonal transport with microtubules. *J. Neurosci.* **3,** 557–566.

Griffin, J. W., Hoffman, P. N., Clark, A. W., Carroll, P. T., and Price, D. L. (1978). Slow axonal transport of neurofilament proteins: Impairment by $\beta,\beta'$-iminodipropionitrile administration. *Science* **202,** 633–635.

Griffin, J. W., Price, D. L., Drachman, D. B., and Morris, J. (1981). Incorporation of axonally transported glycoproteins into axolemma during nerve regeneration. *J. Cell Biol.* **88,** 205–214.

Grundke-Iqbal, I., Iqbal, K., Tung, Y. C., Quinlan, M., Wisniewski, H. M., and Binder, L. I. (1986). Abnormal phosphorylation of the microtubule-associated protein tau in Alzheimer cytoskeletal pathology. *Proc. Natl. Acad. Sci. U.S.A.* **83,** 4913–4917.

Gundersen, R. W. (1987). Response of sensory neurites and growth cones to patterned substrata of laminin and fibronectin *in vitro. Dev. Biol.* **121,** 423–431.

Gundersen, R. W. (1988). Interference reflection microscopic study of dorsal root growth cones on different substrates: Assessment of growth cone–substrate contacts. *J. Neurosci. Res.* **21,** 298–306.

Hammarback, J. A., and Letourneau, P. C. (1986). Neurite extension across regions of low cell–substratum adhesivity: Implications for the guidepost hypothesis of axonal pathfinding. *Dev. Biol.* **117,** 655–662.

Harada, A., Oguchi, K., Okabe, S., Kuno, J., Terada, S., Ohshima, T., Sato-Yoshitake, R., Takei, Y., Noda, T., and Hirokawa, N. (1994). Altered microtubule organization in small-calibre axons of mice lacking tau protein. *Nature (London)* **369,** 488–491.

Harris, W. A., Holt, C. E., and Bonhoeffer, F. (1987). Retinal axons with and without their somata, growing to and arborizing in the tectum of *Xenopus* embryos: A time lapse video study of single fibers *in vivo. Development* **101,** 123–133.

Harrison, R. G. (1910). The outgrowth of the nerve fiber as a mode of cytoplasmic movement. *J. Exp. Zool.* **9,** 787–846.

Heidemann, S. R., and Buxbaum, R. E. (1994). Mechanical tension as a regulator of axonal development. *Neurotox.* **15,** 95–108.

Heidemann, S. R., Lamoureux, P., and Buxbaum, R. E. (1990). Growth cone behavior and production of traction force. *J. Cell Biol.* **111,** 1949–1957.

Heidemann, S. R., Lamoureux, P., and Buxbaum, R. E. (1991). On the cytomechanics and fluid dynamics of growth cone motility. *J. Cell Sci. Suppl.* **15,** 35–44.

Heidemann, S. R., Landers, J. M., and Hamborg, M. A. (1981). Polarity orientation of axonal microtubules. *J. Cell Biol.* **91,** 661–665.

Hirokawa, N. (1982). Cross linker system between neurofilaments, microtubules and membranous organelles revealed by the quick freeze, deep etch method. *J. Cell Biol.* **94,** 129–142.

Hirokawa, N. (1993). The neuronal cytoskeleton: Its role in neuronal morphogenesis and organelle transport. *In* "Neuronal Cytoskeleton: Morphogenesis, Transport and Synaptic Transmission" (N. Hirokawa, ed.). CRC Press, Boca Raton, Florida.

Hirokawa, N., Hisanaga, S., and Shiomura, Y. (1988a). MAP2 is a component of crossbridges between microtubules and neurofilaments *in vivo* and *in vitro.* Quick-freeze, deep etch immunoelectron microscopy and reconstitution studies. *J. Neurosci.* **8,** 2769–2779.

Hirokawa, N., and Okabe, S. (1992). Microtubules on the move? *Curr. Biol.* **2,** 193–195.

Hirokawa, N., Pfister, K. K., Yorifugi, H., Wagner, M. C., Brady, S. T., and Bloom, G. S. (1989). Submolecular domains of bovine brain kinesin identified by electron microscopy and monoclonal antibody decoration. *Cell* **56,** 867–878.

Hirokawa, N., Schiomura, Y., and Okabe, S. (1988b). Tau proteins: The molecular structure and mode of binding on microtubules. *J. Cell Biol.* **107,** 1449–1461.

Hirokawa, N., Yoshida, Y., and Sato-Yoshitake, R. (1990). Brain dynein localizes on both anterogradely and retrogradely transported membranous organelles. *J. Cell Biol.* **111,** 1027–1037.

Hoffman, P. N., Cleveland, D. W., Griffin, J. W., Landes, P. W., Cowan, N. J., and Price, D. L. (1987). Neurofilament gene expression: A major determinant of axonal caliber. *J. Cell Biol.* **84,** 3472–3476.

Hoffman, P. N., Griffin, J. W., and Price, D. W. (1984). Control of axonal caliber by neurofilament transport. *J. Cell Biol.* **99,** 705–714.

Hoffman, P. N., and Lasek, R. J. (1975). The slow component of axonal transport. Identification of major structural polypeptides of the axon and their generality among mammalian neurons. *J. Cell Biol.* **66,** 351–366.

Hollenbeck, P. J., and Bray, D. (1987). Rapidly transported organelles containing membrane and cytoskeletal components: Their relation to axonal growth. *J. Cell Biol.* **105,** 2827–2835.

Horio, T., and Hotani, Y. (1986). Visualization of the dynamic instability of individual microtubules by dark-field microscopy. *Nature (London)* **321,** 605–607.

Hughes, A. (1953). The growth of embryonic neurites. A study on cultures of chick neural tissues. *J. Anat.* **87,** 150–162.

Hugon, J. S., Bennett, G., Pothier, P., and Ngoma, Z. (1987). Loss of microtubules and alteration of glycoprotein migration in organ cultures of mouse intestine exposed to nocodazole or colchicine. *Cell Tissue Res.* **248,** 653–662.

Hyams, J. S., and Lloyd, C. W. (1994). "Microtubules." Wiley-Liss, New York.

Iqbal, Z. (1986). "Axoplasmic Transport." CRC Press, Boca Raton, Florida.

Jacobson, M. (1991). "Developmental Neurobiology." Plenum, New York.

Jareb, M., Esch, T., Craig, A. M., and Banker, G. (1993). The development of polarity by hippocampal neurons in culture. *J. Cell Biochem.* **17B,** 267.

Jordan, M. A., Thrower, D., and Wilson, L. (1992). Effect of vinblastine, podophyllotoxin and nocodazole on mitotic spindles: Implications for the role of microtubule dynamics in mitosis. *J. Cell Sci.* **102,** 402–416.

Jordan, M. A., and Wilson, L. (1990). Kinetic analysis of tubulin exchange at microtubules ends at low vinblastine concentrations. *Biochemistry* **29,** 2730–2739.

Joshi, H. C. (1994). Microtubule organizing centers and γ-tubulin. *Curr. Opin. Cell Biol.* **6,** 55–62.

Joshi, H. C., and Baas, P. W. (1993). A new perspective on microtubules and axon growth. *J. Cell Biol.* **121,** 1191–1196.

Jung, L. J., and Scheller, R. H. (1991). Peptide processing and targeting in the neuronal secretory pathway. *Science* **251,** 1330–1335.

Kanai, Y., Takemura, R., Okhima, H., Mori, H., Ihara, Y., Yanagisawa, M., Masaki, T., and Hirokawa, N. (1989). Expression of multiple tau isoforms and microtubule bundle formation in fibroblasts transfected with a single tau cDNA. *J. Cell Biol.* **109,** 1173–1184.

Kapfhammer, J. P., and Raper, J. A. (1987). Collapse of growth cone structure on contact with specific neurites in culture. *J. Neurosci.* **7,** 201–212.

Karlsson, J. O., and Sjostrand, J. (1971). Synthesis, migration and turnover of protein in retinal ganglion cells. *J. Neurochem.* **18,** 749–767.

Karlsson, J. O., and Sjostrand, J. (1968). Transport of labeled proteins in the optic nerve and tract of the rabbit. *Brain Res.* **11,** 431.

Keith, C. M. (1987). Slow transport of tubulin in the neurites of differentiated PC12 cells. *Science* **235,** 337–339.

Kerst, A., Chmielewski, C., Livesay, C., Buxbaum, R. E., and Heidemann, S. R. (1990). Liquid crystal domains and thixotropy of F-actin suspensions. *Proc. Natl. Acad. Sci. U.S.A.* **87,** 4241–4245.

Killisch, I., Dotti, C. G., Laurie, D. J., *et al.* (1991). Expression patterns of GABA-A receptor subtypes in developing hippocampal neurons. *Neuron* **7,** 927–936.

Kleiman, R., Banker, G., and Steward, O. (1990). Differential subcellular localization of particular mRNAs in hippocampal neurons in culture. *Neuron* **5,** 821–830.

Kleitman, N., and Johnson, M. I. (1989). Rapid growth cone translocation on laminin is supported by lamellipodial not filopodial structures. *Cell Motil. Cytoskeleton* **13,** 288–300.

Knops, J., Kosik, K. S., Lee, G., Pardee, J. D., Cohen-Gould, L., and McConlogue, L. (1991). Overexpression of tau in a non-neuronal cell induces long cellular processes. *J. Cell Biol.* **114,** 725–733.

Koenig, E., Kinsman, S., Repasky, E., and Sultz, L. (1985). Rapid mobility of motile varicosities and inclusions containing spectrin, actin and calmodulin in regenerating axons. *J. Neurosci.* **5,** 715–729.

Komiya, Y., and Tashiro, T. (1988). Effects of taxol on slow and fast axonal transport. *Cell Motil. Cytoskeleton* **11,** 151–156.

Kosik, K. S., and Finch, E. A. (1987). MAP2 and tau segregate into axonal and dendritic domains after the elaboration of morphologically distinct neurites: An immunocytochemical study of cultured rat cerebrum. *J. Neurosci.* **7,** 3142–3153.

Kreutzberg, G. (1969). Neuronal dynamics and axonal flow. IV. Blockage of intraaxonal transport by colchicine. *Proc. Natl. Acad. Sci. U.S.A.* **62,** 722–728.

Kuczmarski, E. R., and Rosenbaum, J. L. (1979). Studies on the organization and localization of actin and myosin in neurons. *J. Cell Biol.* **80,** 356–371.

Kutznetsov, S. A., Vaisberg, E. A., Shanina, N. A., Magretova, N. A., Chernyak, N. M., and Gelfand, V. I. (1988). The quaternary structure of bovine brain kinesin. *EMBO J.* **7,** 353–356.

Lafont, F., Burkhardt, J. K., and Simons, K. (1994). Involvement of microtubule motors in basolateral and apical transport in kidney cells. *Nature* (*London*) **372,** 801–803.

Lamoureux, P., Buxbaum, R. E., and Heidemann, S. R. (1989). Direct evidence that growth cones pull. *Nature* (*London*) **340,** 159–162.

Lamoureux, P., Zheng, J., Buxbaum, R. E., and Heidemann, S. R. (1992). A cytomechanical investigation of neurite growth on different culture surfaces. *J. Cell Biol.* **118,** 655–661.

Landis, S. C. (1983). Neuronal growth cones. *Annu. Rev. Physiol.* **45,** 567–580.

Lasek, R. (1968). Axoplasmic transport in cat dorsal root ganglion cells: As studied with (3H)-1-leucine. *Brain Res.* **7,** 360–377.

Lasek, R. J. (1982). Translocation of the cytoskeleton in neurons and axonal growth. *Phil. Trans. R. Soc. Lond. B* **299,** 313–327.

Lasek, R. J. (1986). Polymer sliding in axons. *J. Cell Sci. Suppl.* **5,** 161–179.

Lasek, R. J., and Brady, S. T. (1985). AMP-PNP facilitates attachment of transported vesicles to microtubules in axoplasm. *Nature* (*London*) **316,** 645–647.

Lasek, R. J., and Hoffman, P. N. (1976). The neuronal cytoskeleton, axonal transport and axonal growth. *Cell Motil. Cold Spring Harbor Conf. Cell Prolif.* **3,** 1021–1049.

Lasek, R. J., Paggi, P., and Katz, M. J. (1992). Slow axonal transport mechanisms move neurofilaments relentlessly in mouse optic axons. *J. Cell Biol.* **117,** 607–616.

LeClerc, N., Kosik, K. S., Cowan, N., Pienkowski, T. P., and Baas, P. W. (1993). Process formation in S19 cells induced by the expression of a MAP2C-like construct. *Proc. Natl. Acad. Sci. U.S.A.* **70,** 6223–6227.

Lee, K.-D., and Hollenbeck, P. J. (1995). Phosphorylation of kinesin *in vivo* correlates with organelle association and neurite outgrowth. *J. Biol. Chem.* **270,** 5600–5605.

Lee, V. M.-Y., Balin, B. J., Otvos, L., and Trojanowski, J. Q. (1991). A68—a major subunit of paired helical filaments and derivatized forms of normal tau. *Science* **251,** 675–678.

Lefcort, F., and Bentley, D. (1989). Organization of cytoskeletal elements and organelles preceding growth cone emergence from an identified neuron *in situ. J. Cell Biol.* **108,** 1737–1749.

Lemmon, V., Burden, S. M., Payne, H. R., Elmslie, G. J., and Hlavin, M. J. (1992). Neurite growth on different substrates: Permissive versus instructive influences and the role of adhesive strength. *J. Neurosci.* **12**(3), 818–826.

Letourneau, P. C. (1979). Cell–substratum adhesion of neurite growth cones, and its role in neurite elongation. *Exp. Cell Res.* **124,** 127–138.

Letourneau, P. C. (1983). Axonal growth and guidance. *Trends Neurosci.* **6,** 451–456.

Lewis, A. K., and Bridgman, P. C. (1992). Nerve growth cone lamellipodia contain two populations of actin filaments that differ in organization and polarity. *J. Cell Biol.* **119,** 1219–1243.

Lewis, S. A., Ivanov, I. E., Lee, G. H., and Cowan, N. J. (1989). Organization of microtubules in dendrites and axons is determined by a short hydrophobic zipper in microtubule-associated and tau. *Nature* (*London*) **342,** 498–505.

Lim, S. S., Edson, K. J., Letourneau, P. C., and Borisy, G. G. (1990). A test of microtubule translocation during neurite elongation. *J. Cell Biol.* **111,** 123–130.

Lim, S. S., Sammak, P. J., and Borisy, G. G. (1989). Progressive and spatially differentiated stability of microtubules in developing neuronal cells. *J. Cell Biol.* **109,** 253–263.

Lin, C.-H., and Forscher, P. (1995a). Cytoskeletal remodeling during growth cone–target interactions. *J. Cell Biol.* **121,** 1369–1383.

Lin, C.-H., and Forscher, P. (1995b). Growth cone advance is inversely proportional to retrograde F-actin flow. *Neuron* **14,** 763–771.

Lin, C.-H., Thompson, C. A., and Forscher, P. (1994). Cytoskeletal reorganization underlying growth cone guidance. *Curr. Opin. Neurobiol.* **4,** 640–647.

Lyser, K. M. (1964). Early differentiation of motor neuroblasts in the chick embryo as studied by electron microscopy. *Dev. Biol.* **10,** 433–466.

Lyser, K. M. (1968). Early differentiation of motor neuroblasts in the chick embryo as studied by electron microscopy. II. Microtubules and neurofilaments. *Dev. Biol.* **17,** 117–142.

Mandelkow, E., and Mandelkow, E.-M. (1995). Microtubules and microtubule-associated proteins. *Curr. Opin. Cell Biol.* **7,** 72–81.

Mareck, A., Fellous, A., Francon, J., and Nunez, J. (1980). Changes in composition and activity of microtubule-associated proteins during brain development. *Nature* (*London*) **284,** 353–355.

Marotta, C. A. (1983). "Neurofilaments." University of Minnesota Press, Minneapolis, Minnesota.

Marsh, L., and Letourneau, P. C. (1984). Growth of neurites without filopodial or lamellipodial activity in the presence of cytochalasin B. *J. Cell Biol.* **99,** 2041–2047.

Matsumoto, T. (1920). The granules, vacuoles and mitochondria in the sympathetic nerve fibres cultivated in vitro. *Johns Hopkins Hosp. Bull.* **31,** 91–93.

Matter, K., Bucher, K., and Hauri, H.-P. (1990). Microtubule perturbation retards both the direct and the indirect apical pathway but does not affect sorting of plasma membrane proteins in intestinal epithelial cells (Caco-2). *EMBO J.* **9,** 3163–3170.

Mattson, M. P., Guthrie, P. B., Hughes, B. C., and Kater, S. B. (1989). Roles for mitotic history in the generation and degeneration of hippocampal neuroarchitecture. *J. Neurosci.* **9,** 1223–1232.

Matus, A. (1994). Stiff microtubules and neuronal morphology. *Trends Neurosci.* **17,** 19–22.

McEwen, B. S., and Grafstein, B. (1968). Fast and slow components in axonal transport of protein. *J. Cell Biol.* **38,** 494–508.

McIntosh, J. R., and Porter, M. E. (1989). Enzymes for microtubule-dependent motility. *J. Biol. Chem.* **264,** 6001–6004.

Mitchison, T., and Kirschner, M. (1984). Dynamic instability of microtubule growth. *Nature* (*London*) **312,** 237–242.

Mitchison, T., and Kirschner, M. (1988). Cytoskeletal dynamics and nerve growth. *Neuron* **1,** 761–772.

Mizushima-Sugano, J., Maeda, T., and Miki-Nomura, T. (1983). Flexural rigidity of singlet microtubules estimated from statistical analysis of their contour lengths and end-to-end distances. *Biochim. Biophys. Acta* **755,** 257–262.

Morris, J. R., and Lasek, R. J. (1982). Stable polymers of the axonal cytoskeleton: The axoplasmic ghost. *J. Cell Biol.* **92,** 192–198.

Muller, R., Kindler, S., and Garner, C. C. (1994). The MAP1 family. *In* "Microtubules" (J. S. Hyams and C. W. Lloyd, eds.), pp. 141–154. Wiley-Liss, New York.

Murphy, D. B., and Borisy, G. G. (1975). Association of high-molecular weight proteins with microtubules and their role in microtubule assembly *in vitro. Proc. Natl. Acad. Sci. U.S.A.* **72,** 2696–2700.

Myers, P. Z., and Bastiani, M. J. (1993). Growth cone dynamics during the migration of an identified commissural growth cone. *J. Neurosci.* **13,** 127–143.

Nakai, J. (1956). Dissociated dorsal root ganglia in tissue culture. *Am. J. Anat.* **99,** 81–129.

Nixon, R. A. (1991). Axonal transport of cytoskeletal proteins. *In* "The Neuronal Cytoskeleton" pp. 283–307. Wiley-Liss, New York.

Nixon, R. A., Fischer, I., and Lewis, S. E. (1990). Synthesis, axonal transport, turnover of the high molecular weight microtubule-associated protein MAP1A in mouse retinal ganglion cells: Tubulin and MAP1A display distinct transport kinetics. *J. Cell Biol.* **110,** 437–448.

Nixon, R. A., and Logvinenko, K. B. (1986). Multiple fates of newly synthesized neurofilament proteins: Evidence for a stationary neurofilament network distributed nonuniformly along axons of retinal ganglion cell neurons. *J. Cell Biol.* **102,** 647–659.

Nixon, R. A., Paskevich, P. A., Sihag, R. K., and Thayer, C. Y. (1994). Phosphorylation on carboxyl terminus domains of neurofilament proteins in retinal ganglion cell neurons *in vivo:* Influences on regional neurofilament accumulation, interneurofilament spacing, and axon caliber. *J. Cell Biol.* **126,** 1031–1046.

O'Connor, T. P., and Bentley, D. (1993). Accumulation of actin in subsets of pioneer growth cone filopodia in response to neural and epithelial guidance cues *in situ. J. Cell Biol.* **123,** 935–948.

O'Connor, T. P., Duerr, J. S., and Bentley, D. (1990). Pioneer growth cone steering decisions mediated by a single filopodial contact *in situ. J. Neurosci.* **10,** 3935–3949.

Ochs, S. (1972). Fast transport of materials in mammalian nerve fibers. *Science* **176,** 252–260.

Ochs, S. (1982). On the mechanism of axoplasmic transport. *In* "Axoplasmic Transport" (D. G. Weiss, ed.), pp. 342–350. Springer-Verlag, Berlin.

Ochs, S., and Brimijoin, W. S. (1993). Axonal transport. *In* "Peripheral Neuropathy" (P. J. Dyck, P. K. Thomas, J. W. Griffin, P. A. Low, and J. F. Poduslo, eds.), pp. 331–360. W. B. Saunders, Philadelphia, Pennsylvania.

Ochs, S., and Johnson, J. (1969). Fast and slow phases of axoplasmic flow in ventral root nerve fibers. *J. Neurochem.* **16,** 845–853.

Ochs, S., Sabri, M. I., and Johnson, J. (1969). Fast transport system of materials in mammalian nerve fibers. *Science* **163,** 686–687.

Ohara, O., Gahara, Y., Miyake, T., Teraoka, H., and Kitamura, T. (1993). Neurofilament deficiency in quail caused by nonsense mutation in neurofilament-L gene. *J. Cell Biol.* **121,** 387–395.

Okabe, S., and Hirokawa, N. (1990). Turnover of fluorescently labelled tubulin and actin in the axon. *Nature* (*London*) **343,** 479–482.

Okabe, S., and Hirokawa, N. (1991). Actin dynamics in growth cones. *J. Neurosci.* **11**(7), 1918–1929.

Okabe, S., and Hirokawa, N. (1992). Differential behavior of photoactivated microtubules in growing axons of mouse and frog neurons. *J. Cell Biol.* **117,** 105–120.

Okada, Y., Yamazaki, H., Sekine-Aizawa, Y., and Hirokawa, N. (1995). The neuron-specific kinesin superfamily protein KIF1A is a unique monomeric motor for anterograde axonal transport of synaptic vesicle precursors. *Cell* **81,** 769–780.

Palay, S. L., and Chan-Palay, V. (1974). "Cerebellar Cortex." Springer-Verlag, New York.

Papandrikopoulou, A., Doll, T. Tucker, R. P., Garner, C. G., and Matus, A. (1989). Embryonic MAP2 lacks the crosslinking sidearm sequences and dendritic targeting signals of adult MAP2. *Nature* (*London*) **340,** 650–652.

Papsozomenos, S., Autulio-Giambetti, L., and Gambetti, P. (1981). Reorganization of axoplasmic organelles following $\beta,\beta'$-iminodipropionitrile administration. *J. Cell Biol.* **91,** 866–871.

Paschal, B. M., Shpetner, H. S., and Vallee, R. B. (1987). MAP1C is a microtubule-activated ATPase that translocates microtubules *in vitro* and has dynein-like properties. *J. Cell Biol.* **105,** 1273–1282.

Paschal, B. M., and Vallee, R. B. (1987). Retrograde transport by the microtubule associated protein MAP1C. *Nature* (*London*) **330,** 181–183.

Pepperkok, R., Bre, M. H., Davoust, J., and Kreis, T. E. (1990). Microtubules are stabilized in confluent epithelial cells but not in fibroblasts. *J. Cell Biol.* **111,** 3003–3012.

Peters, A., Palay, S. L., and de Webster, H. (1976). "The Fine Structure of the Nervous System." W. B. Saunders, Philadelphia, Pennsylvania.

Pfenninger, K. H. (1986). Of nerve growth cones, leukocytes and memory: Second messenger systems and growth-regulated proteins. *Trends Neurosci.* **9,** 562–565.

Pfenninger, K. H., and Maylie-Pfenninger, M. F. (1981). Lectin labeling of sprouting neurons. I. Regional distribution of surface glycoconjugates. *J. Cell Biol.* **89,** 536–546.

Pfister, K. K., Wagner, M. C., Stenoien, D. L., Brady, S. T., and Bloom, G. S. (1989). Monoclonal antibodies to kinesin heavy and light chains stain vesicle-like structures, but not microtubules in cultured cells. *J. Cell Biol.* **108,** 1453–1463.

Phillips, L. L., Autelio-Gambetti, L., and Lasek, R. J. (1983). Bodians silver method reveals molecular variation on the evolution of neurofilament proteins. *Brain Res.* **278,** 219–233.

Popov, S. A., Brown, A., and Poo, M.-M. (1993). Forward plasma membrane flow in growing nerve processes. *Science* **259,** 244–246.

Purves, D., and Lichtman, J. W. (1985). "Principles of Neural Development." Sinauer Associates, Sunderland, Massachusetts.

Reiderer, B., and Matus, A. (1985). Differential expression of distinct microtubule-associated protein during brain development. *Proc. Natl. Acad. Sci. U.S.A.* **82,** 6006–6009.

Reinsch, S. S., Mitchison, T. J., and Kirschner, M. (1991). Microtubule polymer assembly and transport during axonal elongation. *J. Cell Biol.* **115,** 365–379.

Reuzeau, C., Mills, L. R., Harris, J. A., and Morris, C. E. (1995). Discrete and reversible vacuole-like dilations induced by osmomechanical perturbation of neurons. *J. Membr. Biol.* **145,** 33–47.

Rindler, M. J., Ivanov, I. E., and Sabatini, D. D. (1987). Microtubule-acting drugs lead to the nonpolarized delivery of the influenza hemagglutinin to the cell surface of polarized Madin-Darby canine kidney cells. *J. Cell Biol.* **104,** 231–241.

Rivas, R. J., Burmeister, D. W., and Goldberg, D. J. (1992). Rapid effects of laminin on the growth cone. *Neuron* **8,** 107–115.

Rodriguez-Boulan, E., and Nelson, W. J. (1989). Morphogenesis of the polarized epithelial cell phenotype. *Science* **245,** 718–725.

Rodriguez-Boulan, E., and Pendergast, M. (1980). Polarized distribution of viral envelope proteins in the plasma membrane of infected epithelial cells. *Cell* **20,** 45–54.

Rodriguez-Boulan, E., and Sabatini, D. D. (1978). Asymmetric budding of viruses in epithelial monolayers: A model system for study of epithelial polarity. *Proc. Natl. Acad. Sci. U.S.A.* **75,** 5071–5075.

Rothman, J. E. (1994). Mechanisms of intracellular protein transport. *Nature* (*London*) **372,** 55–63.

Rutishauser, U., and Jessell, T. M. (1988). Cell adhesion molecules in vertebrate neural development. *Physiol. Rev.* **68,** 819–857.

Sabry, J., O'Connor, T. P., and Kirschner, M. W. (1995). Axonal transport of tubulin in Ti1 pioneer neurons *in situ. Neuron* **14,** 1247–1256.

Sakaguchi, T., Okada, M., Kitamura, T., and Kawasaki, K. (1993). Reduced diameter and conduction velocity of myelinated fibers in the sciatic nerve of a neurofilament-deficient mutant quail. *Neurosci. Lett.* **153,** 65–68.

Sammak, P. J., Gorbsky, G. J., and Borisy, G. G. (1987). Microtubule dynamics *in vivo:* A test of mechanisms of turnover. *J. Cell Biol.* **104,** 395–405.

Saxton, W. M., Stemple, D. L., Leslie, R. J., Salmon, E. D., Zavortink, M., and McIntosh, J. R. (1984). Tubulin dynamics in cultured mammalian cells. *J. Cell Biol.* **99,** 2175–2186.

Schmitt, F. O. (1968). Fibrous proteins–neuronal organelles. *Proc. Natl. Acad. Sci. U.S.A.* **60,** 1092–1101.

Schnapp, B. J., and Reese, T. S. (1982). Cytoplasmic structure in rapid frozen axons. *J. Cell Biol.* **94,** 667–679.

Schnapp, B. J., Reese, T. S., and Bechtold, R. (1992). Kinesin is bound with high affinity to squid axon organelles that move to the plus-end of microtubules. *J. Cell Biol.* **119,** 389–399.

Schnapp, B. J., Vale, R. D., Sheetz, M. P., and Reese, T. S. (1985). Single microtubules from squid axoplasm support bidirectional movement of organelles. *Cell* **40,** 455–462.

Schoenfeld, T. A., and Obar, R. A. (1994). Diverse distribution and function of fibrous microtubule-associated proteins in the nervous system. *Int. Rev. Cytol.* **151,** 67–137.

Scholey, J. M., Heuser, J., Yang, J. T., and Goldstein, L. S. B. (1989). Identification of globular mechanochemical heads of kinesin. *Nature (London)* **338,** 355–357.

Schroer, T. A., and Sheetz, M. P. (1991). Two activators of microtubule-based vesicle transport. *J. Cell Biol.* **115,** 1309–1318.

Schwartz, J. H., Goldman, J. E., Aubron, R., and Goldberg, D. J. (1976). Axonal transport of vesicles carrying serotonin in the metacerebral ganglion of *Aplysia californica. Cold Spring Harbor Symp. Quant. Biol.* **40,** 83–92.

Seifritz, W. (1942). "The Structure of Protoplasm." Iowa State College Press, Ames, Iowa.

Selkoe, D. J. (1994). Normal and abnormal biology of the beta-amyloid precursor protein. *Annu. Rev. Neurosci.* **17,** 489–517.

Sharp, G. A., Weber, K., and Osborn, M. (1982). Centriole number and process formation in established neuroblastoma cells and primary dorsal root ganglion neurones. *Eur. J. Cell Biol.* **29,** 97–103.

Shaw, G. (1991). Neurofilament proteins. *In* "The Neuronal Cytoskeleton" (R. D. Burgoyne, ed.), pp. 185–214. Wiley-Liss, New York.

Sheetz, M. P., Baumrind, N. L., Wayne, D. B., and Pearlman, A. L. (1990). Concentration of membrane antigens by forward transport and trapping in neuronal growth cones. *Cell* **61,** 231–241.

Shiomura, Y., and Hirokawa, H. (1987). Colocalization of MAP1 and MAP2 on the neuronal microtubule *in situ* revealed with double-labeling immunelectron microscopy. *J. Cell Biol.* **104,** 1575–1578.

Shpetner, H. S., Paschal, B. M., and Vallee, R. B. (1988). Characterization of the microtubule-activated ATPase of brain cytoplasmic Dynein (MAP-1C). *J. Cell Biol.* **107,** 1001–1009.

Skoufias, D. A., and Scholey, J. M. (1993). Cytoplasmic microtubule-based motor proteins. *Curr. Opin. Cell Biol.* **5,** 95–104.

Sloboda, R. D., Dentler, W. L., and Rosenbaum, J. L. (1976). Microtubule-associated proteins and the stimulation of tubulin assembly *in vitro. Biochemistry* **15,** 4497–4505.

Smith, C. (1994a). The initiation of neurite outgrowth by sympathetic neurons grown *in vitro* does not depend on assembly of microtubules. *J. Cell Biol.* **127,** 1407–1418.

Smith, C. (1994b). Cytoskeletal movements and substrate interactions during initiation of neurite outgrowth by sympathetic neurons *in vitro. J. Neurosci.* **14,** 384–398.

Smith, R. S., and Bisby, M. A (1987). "Axonal Transport." Alan R. Liss, New York.

Smith, S. J. (1988). Neuronal cytomechanics: The actin based motility of growth cones. *Science* **242,** 708–715.

Solomon, F. (1979). Detailed neurite morphologies of sister neuroblastoma cells are related. *Cell* **16,** 165–169.

Solomon, F. (1980). Neuroblastoma cells recapitulate their detailed neurite morphologies after reversible microtubule disassembly. *Cell* **21,** 333–338.

Speidel, C. C. (1933). Studies of living nerves. II. Activities of amoeboid growth cones, sheath cells and myelin segments as revealed by prolonged observation of individual nerve fibers in frog tadpoles. *Am. J. Anat.* **52,** 1–75.

Stevens, J. K., Trogadis, J., and Jacobs, J. R. (1988). Development and control of axial neurite form: A serial electron microscopic analysis. *In* "Intrinsic Determinants of Neuronal Form and Function" (R. J. Lasek and M. M. Black, eds.), pp. 115–146. Alan R. Liss, New York.

Steward, O., and Banker, G. A. (1992). Getting the message from the gene to the synapse: Sorting and intracellular transport of RNA in neurons. *Trends Neurosci.* **5,** 180–186.

Steward, O., and Fass, B. (1983). Polyribosomes associated with dendrite spines in the denervated dentate gyrus: Evidence for local regulation of protein synthesis during reinnveration. *Prog. Brain Res.* **58,** 131–136.

Stewart, G. H., Horwitz, B., and Gross, G. W. (1982). A chromatographic model of axoplasmic transport. *In* "Axoplasmic Transport" (D. G. Weiss, ed.), pp. 414–422. Springer-Verlag, Berlin.

Strittmatter, W. J., and Roses, A. D. (1995). Apolipoprotein E and Alzheimer disease. *Proc. Natl. Acad. Sci. U.S.A.* **92,** 4725–4727.

Strittmatter, W. J., Saunders, A. M., Goedert, M., Weisgraber, K. H., Dong, L.-M., Jakes, R., Huang, D. Y., Pericak-Vance, M., Schmechel, D., and Roses, A. D. (1994). Isoform-specific interactions of apolipoprotein E with microtubule-associated protein tau: Implications for Alzheimer disease. *Proc. Natl. Acad. Sci. U.S.A.* **91,** 11183–11186.

Stromska, D. P., and Ochs, S. (1977). Slow outflow patterns of labeled proteins in cat and rat sensory nerves. *Fed. Proc.* **36,** 485.

Stromska, D. P., and Ochs, S. (1981). Patterns of slow transport in sensory nerves. *J. Neurobiol.* **12,** 441–453.

Sudhof, T. C. (1995). The synaptic vesicle cycle: A cascade of protein–protein interactions. *Nature* (*London*) **375,** 645–653.

Svoboda, K., Schmidt, C. F., Schnapp, B. J., and Block, S. M. (1993). Direct observation of kinesin stepping by optical tapping interferometry. *Nature* (*London*) **365,** 721–727.

Takeda, S., Funakoshi, T., and Hirokawa, N. (1995). Tubulin dynamics in neuronal axons of living zebrafish embryos. *Neuron* **14,** 1257–1264.

Tanaka, E., Ho, T., and Kirschner, M. W. (1995). The role of microtubule dynamics in growth cone motility and axonal growth. *J. Cell Biol.* **128,** 139–155.

Tanaka, E. M., and Kirschner, M. W. (1991). Microtubule behavior in the growth cones of living neurons during axon elongation. *J. Cell Biol.* **115,** 345–363.

Tanaka, E., and Kirschner, M. W. (1995). The role of microtubules in growth cone turning at substrate boundaries. *J. Cell Biol.* **128,** 127–137.

Tashiro, T., and Komiya, Y. (1989). Stable and dynamic forms of cytoskeletal proteins in slow axonal transport. *J. Neurosci.* **9,** 760–768.

Tashiro, T., and Komiya, Y. (1991). Changes in organization and axonal transport of cytoskeletal proteins during regeneration. *J. Neurochem.* **56,** 1557–1663.

Tashiro, T., Kurokawa, M., and Komiya, Y. (1984). Two populations of axonally transported tubulin, differentiated by the interaction with neurofilaments. *J. Neurochem.* **43,** 1220–1225.

Taylor, D. L., and Condeelis, J. S. (1979). Cytoplasmic structure and contractility in amoeboid cells. *Int. Rev. Cytol.* **56,** 57–144.

Tennyson, U. M. (1965). Electron microscopic study of the developing neuroblast of the dorsal root ganglion of the rabbit embryo. *J. Comp. Neurol.* **124,** 267–318.

Tiedge, H., Fremeau, R. T., Jr., Weinstock, P. H., Arancio, O., and Brosius, J. (1991). Dendritic location of neural BC1 RNA. *Proc. Natl. Acad. Sci. U.S.A.* **88,** 2093–2097.

Topp, K. S., Bisla, K., Saks, N. D., and La Vail, J. H. (1995). Centripetal transport of herpes simplex virus in human retinal pigment epithelial cells *in vitro. Neuroscience* (in press).

Tosney, K. W., and Landmesser, L. T. (1985). Growth cone morphology and trajectory in the lumbosacral region of the chick embryo. *J. Neurosci.* **5**(9), 2345–2358.

Tosney, K. W., and Wessells, N. K. (1983). Neuronal motility: The ultrastructure of veils and microspikes correlates with their motile activities. *J. Cell Sci.* **61,** 389–411.

Toyoshima, I., Yu, H., Steuer, E. R., and Sheetz, M. P. (1992). Kinectin, a major kinesin-binding protein on ER. *J. Cell Biol.* **118,** 1121–1131.

Tsui, H. C. T., Lankford, K. L., and Klein, W. L. (1985). Differentiation of neuronal growth cones: Specialization of filopodial tips for adhesive interactions. *Proc. Natl. Acad. Sci. U.S.A.* **82,** 8256–8260.

Tsukita, S., and Ishikawa, H. (1981). The cytoskeleton in myelinated axons: A serial section study. *Biomed. Res.* **2,** 424–437.

Tucker, R. P. (1990). The roles of microtubule-associated proteins in brain morphogenesis: A review. *Brain Res. Rev.* **15,** 101–120.

Tucker, R. P., Binder, L. I., and Matus, A. I. (1988a). Neuronal microtubule-associated proteins in the embryonic avian spinal cord. *J. Comp. Neurol.* **271,** 44–55.

Tucker, R. P., Binder, L. I., Viereck, C., Hemmings, B. A., and Matus, A. I. (1988b). The sequential appearance of low- and high-molecular weight forms of MAP2 in the developing cerebellum. *J. Neurosci.* **8,** 4503–4512.

Umeyama, T., Okabe, S., and Kanai, Y. (1993). Dynamics of microtubules bundled by microtubule associated protein 2C (MAP2C). *J. Cell Biol.* **120,** 451–465.

Vale, R. D. (1990). Microtubule-based motor proteins. *Curr. Opin. Cell Biol.* **2,** 15–22.

Vale, R. D., Malik, F., and Brown, D. (1992). Directional instability of microtubule transport in the presence of kinesin and dynein, two opposite polarity motor proteins. *J. Cell Biol.* **119,** 1589–1596.

Vale, R. D., Reese, T. S., and Sheetz, M. S. (1985). Identification of a novel force-generating protein, kinesin, involved in microtubule-based motility. *Cell* **42,** 39–50.

Vallee, R. B. (1980). Structure and phosphorylation of microtubule-associated protein 2 (MAP2). *Proc. Natl. Acad. Sci. U.S.A.* **77,** 3206–3210.

Vallee, R. (1993). Molecular analysis of the microtubule motor dynein. *Proc. Natl. Acad. Sci. U.S.A.* **90,** 8769–8772.

Vallee, R. B., and Bloom, G. W. (1991). Mechanisms of fast and slow axonal transport. *Annu. Rev. Neurosci.* **14,** 59–92.

Vallee, R. B., Wall, J. S., Paschal, B. M., and Shpetner, H. S. (1988). Microtubule associated protein 1C from brain is a two-headed cytosolic dynein. *Nature* (*London*) **332,** 561–563.

Viereck, C., Tucker, R. P., Binder, L. L., and Matus, A. (1988). Phylogenetic conservation of brain microtubule-associated proteins MAP2 and tau. *Neuroscience* **26,** 893–904.

Walker, R., and Sheetz, M. P. (1993). Cytoplasmic microtubule-associated motors. *Annu. Rev. Biochem.* **62,** 429–451.

Wan, X., Harris, J. A., and Morris, C. E. (1995). Responses of neurons to extreme osmomechanical stress. *J. Membr. Biol.* **145,** 21–31.

Watson, D. F., Hoffman, P. N., and Griffin, J. W. (1990). The cold stability of microtubules increases during axonal maturation. *J. Neurosci.* **10,** 3344–3352.

Weingarten, M. D., Lockwood, A. H., Hwo, S.-Y., and Kirschner, M. W. (1975). A protein factor essential for microtubule assembly. *Proc. Natl. Acad. Sci. U.S.A.* **72,** 1858–1862.

Weiss, D. G. (1982). "Axoplasmic Transport." Springer-Verlag, Berlin.

Weisshaar, B., Doll, T., and Matus, A. (1992). Reorganization of the microtubular cytoskeleton by embryonic microtubule-associated protein 2 (MAP2c). *Development* **116,** 1151–1161.

Wessells, N. K. (1982). Axon elongation: A special case of cell locomotion. *In* "Cell Behavior" (R. Bellairs, A. Curtis, and G. Dunn, eds.), pp. 225–246. Cambridge University Press, Cambridge, UK.

Wessells, N. K., Luduena, M. A., Letourneau, P. C., Wrenn, J. T., and Spooner, B. S. (1974). Thorotrast uptake and transit in embryonic glia, heart fibroblasts and neurons *in vitro. Tissue Cell* **6,** 757–776.

Wiche, G. Oberkanins, C., and Himmler, A. (1992). Molecular structure and function of microtubule-associated proteins. *Int. Rev. Cytol.* **124,** 217–273.

Willard, M., Cowan, W. M., and Vagelos, P. R. (1974). The polypeptide composition of intro-axonally transported proteins: Evidence for four transport velocities. *Proc. Natl. Acad. Sci. U.S.A.* **71,** 2183–2187.

Wille, H., Drewes, G., Biermat, J., Mandelkow, E. M., and Mandelkow, E. (1992). Alzheimer-like paired helical filaments and antiparallel dimers formed from microtubule-associated protein tau *in vitro. J. Cell Biol.* **118,** 573–584.

Wonderlin, W. F., and French, R. J. (1991). Ion channels in transit: Voltage-gated Na and K channels in axoplasmic organelles of the squid *Loligo pealei. Proc. Natl. Acad. Sci. U.S.A.* **88,** 4391–4395.

Wu, D.-Y., and Goldberg, D. J. (1993). Regulated tyrosine phosphorylation at the tips of growth cone filopodia. *J. Cell Biol.* **123,** 653–664.

Xu, Z., Cork, L., Griffin, J., and Cleveland, D. (1993). Increased expression of neurofilament NF-L produces morphological alterations that resemble the pathology of human motor neuron disease. *Cell* **73,** 23–33.

Yamada, K. M., Spooner, B. S., and Wessels, N. K. (1970). Axon growth: Role of microfilaments and microtubules. *Proc. Natl. Acad. Sci. U.S.A.* **66,** 1206–1212.

Yisraeli, J. K., Sokol, S., and Melton, D. A. (1990). A two-step model for the localizatoin of maternal mRNA in *Xenopus* oocytes: Involvement of microtubules and microfilaments in the translocation and anchoring of Vg1 mRNA. *Development* **108,** 289–298.

Young, J. Z. (1944). Contraction, turgor and the cytoskeleton of nerve fibres. *Nature* (*London*) **153,** 333–335.

Yu, W., and Baas, P. W. (1994). Changes in microtubule number and length during axon differentiation. *J. Neurosci.* **14,** 2818–2829.

Yu, W., Centonze, V. E., Ahmad, F. J., and Baas, P. W. (1993). Microtubule nucleation and release from the neuronal centrosome. *J. Cell Biol.* **122,** 349–359.

Zenker, W., and Hohberg, E. (1973). An alpha-nerve fibre: Number of neurotubules in the stem fibre and in the terminal branches. *J. Neurocytol.* **2,** 143–148.

Zheng, J., Buxbaum, R. E., and Heidemann, S. R. (1993). Investigation of microtubule assembly and reorganization accompanying tension-induced neurite initiation. *J. Cell Sci.* **104,** 1239–1250.

Zheng, J., Buxbaum, R. E., and Heidemann, S. R. (1994). Measurements of growth cone adhesion to culture surfaces by micromanipulation. *J. Cell Biol.* **127,** 2049–2060.

Zheng, J., Lamoureux, P., Santiago, V., Dennerll, T., Buxbaum, R. E., and Heidemann, S. R. (1991). Tensile regulation of axonal elongation and initiation. *J. Neurosci.* **11,** 1117–1125.

# INDEX

## A



## B



## C

## D

## E

## L



## M



## N

## R



## S

## U



## V



## W



## Y



## Z

ISBN 0-12-364569-7